Pitman Research Notes in Mathematics Series

Submission of proposals for consideration

Suggestions for publication, in the form of outlines and representative samples, are invited by the Editorial Board for assessment. Intending authors should approach one of the main editors or another member of the Editorial Board, citing the relevant AMS subject classifications. Alternatively, outlines may be sent directly to the publisher's offices. Refereeing is by members of the board and other mathematical authorities in the topic concerned, throughout the world.

Preparation of accepted manuscripts

On acceptance of a proposal, the publisher will supply full instructions for the preparation of manuscripts in a form suitable for direct photo-lithographic reproduction. Specially printed grid sheets can be provided and a contribution is offered by the publisher towards the cost of typing. Word processor output, subject to the publisher's approval, is also acceptable.

Illustrations should be prepared by the authors, ready for direct reproduction without further improvement. The use of hand-drawn symbols should be avoided wherever possible, in order to maintain maximum clarity of the text.

The publisher will be pleased to give any guidance necessary during the preparation of a typescript, and will be happy to answer any queries.

Important note

In order to avoid later retyping, intending authors are strongly urged not to begin final preparation of a typescript before receiving the publisher's guidelines. In this way it is hoped to preserve the uniform appearance of the series.

Longman Scientific & Technical
Longman House
Burnt Mill
Harlow, Essex, CM20 2JE
UK
(Telephone (0279) 426721)

Titles in this series. A full list is available on request from the publisher.

51 Subnormal operators
J B Conway
52 Wave propagation in viscoelastic media
F Mainardi
53 Nonlinear partial differential equations and their applications: Collège de France Seminar. Volume I
H Brezis and J L Lions
54 Geometry of Coxeter groups
H Hiller
55 Cusps of Gauss mappings
T Banchoff, T Gaffney and C McCrory
56 An approach to algebraic K-theory
A J Berrick
57 Convex analysis and optimization
J-P Aubin and R B Vintner
58 Convex analysis with applications in the differentiation of convex functions
J R Giles
59 Weak and variational methods for moving boundary problems
C M Elliott and J R Ockendon
60 Nonlinear partial differential equations and their applications: Collège de France Seminar. Volume II
H Brezis and J L Lions
61 Singular Systems of differential equations II
S L Campbell
62 Rates of convergence in the central limit theorem
Peter Hall
63 Solution of differential equations by means of one-parameter groups
J M Hill
64 Hankel operators on Hilbert Space
S C Power
65 Schrödinger-type operators with continuous spectra
M S P Eastham and H Kalf
66 Recent applications of generalized inverses
S L Campbell
67 Riesz and Fredholm theory in Banach algebra
B A Barnes, G J Murphy, M R F Smyth and T T West
68 Evolution equations and their applications
K Kappel and W Schappacher
69 Generalized solutions of Hamilton–Jacobi equations
P L Lions
70 Nonlinear partial differential equations and their applications: Collège de France Seminar. Volume III
H Brezis and J L Lions
71 Spectral theory and wave operators for the Schrödinger equation
A M Berthier
72 Approximation of Hilbert space operators I
D A Herrero
73 Vector valued Nevanlinna theory
H J W Ziegler
74 Instability, nonexistence and weighted energy methods in fluid dynamics and related theories
B Straughan
75 Local bifurcation and symmetry
A Vanderbauwhede
76 Clifford analysis
F Brackx, R Delanghe and F Sommen
77 Nonlinear equivalence, reduction of PDEs to ODEs and fast convergent numerical methods
E E Rosinger
78 Free boundary problems, theory and applications. Volume I
A Fasano and M Primicerio
79 Free boundary problems, theory and applications. Volume II
A Fasano and M Primicerio
80 Symplectic geometry
A Crumeyrolle and J Grifone
81 An algorithmic analysis of a communication model with retransmission of flawed messages
D M Lucantoni
82 Geometric games and their applications
W H Ruckle
83 Additive groups of rings
S Feigelstock
84 Nonlinear partial differential equations and their applications: Collège de France Seminar. Volume IV
H Brezis and J L Lions
85 Multiplicative functionals on topological algebras
T Husain
86 Hamilton–Jacobi equations in Hilbert spaces
V Barbu and G Da Prato
87 Harmonic maps with symmetry, harmonic morphisms and deformations of metric
P Baird
88 Similarity solutions of nonlinear partial differential equations
L Dresner
89 Contributions to nonlinear partial differential equations
C Bardos, A Damlamian, J I Díaz and J Hernández
90 Banach and Hilbert spaces of vector-valued functions
J Burbea and P Masani
91 Control and observation of neutral systems
D Salamon
92 Banach bundles, Banach modules and automorphisms of C^*-algebras
M J Dupré and R M Gillette
93 Nonlinear partial differential equations and their applications: Collège de France Seminar. Volume V
H Brezis and J L Lions
94 Computer algebra in applied mathematics: an introduction to MACSYMA
R H Rand
95 Advances in nonlinear waves. Volume I
L Debnath
96 FC-groups
M J Tomkinson
97 Topics in relaxation and ellipsoidal methods
M Akgül
98 Analogue of the group algebra for topological semigroups
H Dzinotyiweyi
99 Stochastic functional differential equations
S E A Mohammed

100 Optimal control of variational inequalities
V Barbu
101 Partial differential equations and dynamical systems
W E Fitzgibbon III
102 Approximation of Hilbert space operators Volume II
C Apostol, L A Fialkow, D A Herrero and D Voiculescu
103 Nondiscrete induction and iterative processes
V Ptak and F-A Potra
104 Analytic functions – growth aspects
O P Juneja and G P Kapoor
105 Theory of Tikhonov regularization for Fredholm equations of the first kind
C W Groetsch
106 Nonlinear partial differential equations and free boundaries. Volume I
J I Díaz
107 Tight and taut immersions of manifolds
T E Cecil and P J Ryan
108 A layering method for viscous, incompressible L_p flows occupying R^n
A Douglis and E B Fabes
109 Nonlinear partial differential equations and their applications: Collège de France Seminar. Volume VI
H Brezis and J L Lions
110 Finite generalized quadrangles
S E Payne and J A Thas
111 Advances in nonlinear waves. Volume II
L Debnath
112 Topics in several complex variables
E Ramírez de Arellano and D Sundararaman
113 Differential equations, flow invariance and applications
N H Pavel
114 Geometrical combinatorics
F C Holroyd and R J Wilson
115 Generators of strongly continuous semigroups
J A van Casteren
116 Growth of algebras and Gelfand–Kirillov dimension
G R Krause and T H Lenagan
117 Theory of bases and cones
P K Kamthan and M Gupta
118 Linear groups and permutations
A R Camina and E A Whelan
119 General Wiener–Hopf factorization methods
F-O Speck
120 Free boundary problems: applications and theory. Volume III
A Bossavit, A Damlamian and M Fremond
121 Free boundary problems: applications and theory. Volume IV
A Bossavit, A Damlamian and M Fremond
122 Nonlinear partial differential equations and their applications: Collège de France Seminar. Volume VII
H Brezis and J L Lions
123 Geometric methods in operator algebras
H Araki and E G Effros
124 Infinite dimensional analysis–stochastic processes
S Albeverio
125 Ennio de Giorgi Colloquium
P Krée
126 Almost-periodic functions in abstract spaces
S Zaidman
127 Nonlinear variational problems
A Marino, L Modica, S Spagnolo and M Degliovanni
128 Second-order systems of partial differential equations in the plane
L K Hua, W Lin and C-Q Wu
129 Asymptotics of high-order ordinary differential equations
R B Paris and A D Wood
130 Stochastic differential equations
R Wu
131 Differential geometry
L A Cordero
132 Nonlinear differential equations
J K Hale and P Martinez-Amores
133 Approximation theory and applications
S P Singh
134 Near-rings and their links with groups
J D P Meldrum
135 Estimating eigenvalues with *a posteriori/a priori* inequalities
J R Kuttler and V G Sigillito
136 Regular semigroups as extensions
F J Pastijn and M Petrich
137 Representations of rank one Lie groups
D H Collingwood
138 Fractional calculus
G F Roach and A C McBride
139 Hamilton's principle in continuum mechanics
A Bedford
140 Numerical analysis
D F Griffiths and G A Watson
141 Semigroups, theory and applications. Volume I
H Brezis, M G Crandall and F Kappel
142 Distribution theorems of L-functions
D Joyner
143 Recent developments in structured continua
D De Kee and P Kaloni
144 Functional analysis and two-point differential operators
J Locker
145 Numerical methods for partial differential equations
S I Hariharan and T H Moulden
146 Completely bounded maps and dilations
V I Paulsen
147 Harmonic analysis on the Heisenberg nilpotent Lie group
W Schempp
148 Contributions to modern calculus of variations
L Cesari
149 Nonlinear parabolic equations: qualitative properties of solutions
L Boccardo and A Tesei
150 From local times to global geometry, control and physics
K D Elworthy

151 A stochastic maximum principle for optimal control of diffusions
U G Haussmann
152 Semigroups, theory and applications. Volume II
H Brezis, M G Crandall and F Kappel
153 A general theory of integration in function spaces
P Muldowney
154 Oakland Conference on partial differential equations and applied mathematics
L R Bragg and J W Dettman
155 Contributions to nonlinear partial differential equations. Volume II
J I Díaz and P L Lions
156 Semigroups of linear operators: an introduction
A C McBride
157 Ordinary and partial differential equations
B D Sleeman and R J Jarvis
158 Hyperbolic equations
F Colombini and M K V Murthy
159 Linear topologies on a ring: an overview
J S Golan
160 Dynamical systems and bifurcation theory
M I Camacho, M J Pacifico and F Takens
161 Branched coverings and algebraic functions
M Namba
162 Perturbation bounds for matrix eigenvalues
R Bhatia
163 Defect minimization in operator equations: theory and applications
R Reemtsen
164 Multidimensional Brownian excursions and potential theory
K Burdzy
165 Viscosity solutions and optimal control
R J Elliott
166 Nonlinear partial differential equations and their applications: Collège de France Seminar. Volume VIII
H Brezis and J L Lions
167 Theory and applications of inverse problems
H Haario
168 Energy stability and convection
G P Galdi and B Straughan
169 Additive groups of rings. Volume II
S Feigelstock
170 Numerical analysis 1987
D F Griffiths and G A Watson
171 Surveys of some recent results in operator theory. Volume I
J B Conway and B B Morrel
172 Amenable Banach algebras
J-P Pier
173 Pseudo-orbits of contact forms
A Bahri
174 Poisson algebras and Poisson manifolds
K H Bhaskara and K Viswanath
175 Maximum principles and eigenvalue problems in partial differential equations
P W Schaefer
176 Mathematical analysis of nonlinear, dynamic processes
K U Grusa
177 Cordes' two-parameter spectral representation theory
D F McGhee and R H Picard
178 Equivariant K-theory for proper actions
N C Phillips
179 Elliptic operators, topology and asymptotic methods
J Roe
180 Nonlinear evolution equations
J K Engelbrecht, V E Fridman and E N Pelinovski
181 Nonlinear partial differential equations and their applications: Collège de France Seminar. Volume IX
H Brezis and J L Lions
182 Critical points at infinity in some variational problems
A Bahri
183 Recent developments in hyperbolic equations
L Cattabriga, F Colombini, M K V Murthy and S Spagnolo
184 Optimization and identification of systems governed by evolution equations on Banach space
N U Ahmed
185 Free boundary problems: theory and applications. Volume I
K H Hoffmann and J Sprekels
186 Free boundary problems: theory and applications. Volume II
K H Hoffmann and J Sprekels
187 An introduction to intersection homology theory
F Kirwan
188 Derivatives, nuclei and dimensions on the frame of torsion theories
J S Golan and H Simmons
189 Theory of reproducing kernels and its applications
S Saitoh
190 Volterra integrodifferential equations in Banach spaces and applications
G Da Prato and M Iannelli
191 Nest algebras
K R Davidson
192 Surveys of some recent results in operator theory. Volume II
J B Conway and B B Morrel
193 Nonlinear variational problems. Volume II
A Marino and M K V Murthy
194 Stochastic processes with multidimensional parameter
M E Dozzi
195 Prestressed bodies
D Iesan
196 Hilbert space approach to some classical transforms
R H Picard
197 Stochastic calculus in application
J R Norris
198 Radical theory
B J Gardner
199 The C^*-algebras of a class of solvable Lie groups
X Wang
200 Stochastic analysis, path integration and dynamics
K D Elworthy and J C Zambrini

201 Riemannian geometry and holonomy groups
S Salamon
202 Strong asymptotics for extremal errors and polynomials associated with Erdős type weights
D S Lubinsky
203 Optimal control of diffusion processes
V S Borkar
204 Rings, modules and radicals
B J Gardner
205 Two-parameter eigenvalue problems in ordinary differential equations
M Faierman
206 Distributions and analytic functions
R D Carmichael and D Mitrovic
207 Semicontinuity, relaxation and integral representation in the calculus of variations
G Buttazzo
208 Recent advances in nonlinear elliptic and parabolic problems
P Bénilan, M Chipot, L Evans and M Pierre
209 Model completions, ring representations and the topology of the Pierce sheaf
A Carson
210 Retarded dynamical systems
G Stepan
211 Function spaces, differential operators and nonlinear analysis
L Paivarinta
212 Analytic function theory of one complex variable
C C Yang, Y Komatu and K Niino
213 Elements of stability of visco-elastic fluids
J Dunwoody
214 Jordan decomposition of generalized vector measures
K D Schmidt
215 A mathematical analysis of bending of plates with transverse shear deformation
C Constanda
216 Ordinary and partial differential equations. Volume II
B D Sleeman and R J Jarvis
217 Hilbert modules over function algebras
R G Douglas and V I Paulsen
218 Graph colourings
R Wilson and R Nelson
219 Hardy-type inequalities
A Kufner and B Opic
220 Nonlinear partial differential equations and their applications: Collège de France Seminar. Volume X
H Brezis and J L Lions
221 Workshop on dynamical systems
E Shiels and Z Coelho
222 Geometry and analysis in nonlinear dynamics
H W Broer and F Takens
223 Fluid dynamical aspects of combustion theory
M Onofri and A Tesei
224 Approximation of Hilbert space operators. Volume I. 2nd edition
D Herrero
225 Operator theory: proceedings of the 1988 GPOTS–Wabash conference
J B Conway and B B Morrel
226 Local cohomology and localization
J L Bueso Montero, B Torrecillas Jover and A Verschoren
227 Nonlinear waves and dissipative effects
D Fusco and A Jeffrey
228 Numerical analysis 1989
D F Griffiths and G A Watson
229 Recent developments in structured continua. Volume III
D De Kee and P Kaloni
230 Boolean methods in interpolation and approximation
F J Delvos and W Schempp
231 Further advances in twistor theory. Volume I
L J Mason and L P Hughston
232 Further advances in twistor theory. Volume II
L J Mason and L P Hughston
233 Geometry in the neighborhood of invariant manifolds of maps and flows and linearization
U Kirchgraber and K Palmer
234 Quantales and their applications
K I Rosenthal
235 Integral equations and inverse problems
V Petkov and R Lazarov
236 Pseudo-differential operators
S R Simanca
237 A functional analytic approach to statistical experiments
I M Bomze
238 Quantum mechanics, algebras and distributions
D Dubin and M Hennings
239 Hamilton flows and evolution semigroups
J Gzyl
240 Topics in controlled Markov chains
V S Borkar
241 Invariant manifold theory for hydrodynamic transition
S Sritharan
242 Lectures on the spectrum of $L^2(\Gamma\backslash G)$
F L Williams
243 Progress in variational methods in Hamiltonian systems and elliptic equations
M Girardi, M Matzeu and F Pacella
244 Optimization and nonlinear analysis
A Ioffe, M Marcus and S Reich
245 Inverse problems and imaging
G F Roach
246 Semigroup theory with applications to systems and control
N U Ahmed
247 Periodic-parabolic boundary value problems and positivity
P Hess
248 Distributions and pseudo-differential operators
S Zaidman
249 Progress in partial differential equations: the Metz surveys
M Chipot and J Saint Jean Paulin
250 Differential equations and control theory
V Barbu

251 Stability of stochastic differential equations with respect to semimartingales
X Mao
252 Fixed point theory and applications
J Baillon and M Théra
253 Nonlinear hyperbolic equations and field theory
M K V Murthy and S Spagnolo
254 Ordinary and partial differential equations. Volume III
B D Sleeman and R J Jarvis
255 Harmonic maps into homogeneous spaces
M Black
256 Boundary value and initial value problems in complex analysis: studies in complex analysis and its applications to PDEs 1
R Kühnau and W Tutschke
257 Geometric function theory and applications of complex analysis in mechanics: studies in complex analysis and its applications to PDEs 2
R Kühnau and W Tutschke
258 The development of statistics: recent contributions from China
X R Chen, K T Fang and C C Yang
259 Multiplication of distributions and applications to partial differential equations
M Oberguggenberger
260 Numerical analysis 1991
D F Griffiths and G A Watson
261 Schur's algorithm and several applications
M Bakonyi and T Constantinescu
262 Partial differential equations with complex analysis
H Begehr and A Jeffrey
263 Partial differential equations with real analysis
H Begehr and A Jeffrey
264 Solvability and bifurcations of nonlinear equations
P Drábek
265 Orientational averaging in mechanics of solids
A Lagzdins, V Tamuzs, G Teters and A Kregers
266 Progress in partial differential equations: elliptic and parabolic problems
C Bandle, J Bemelmans, M Chipot, M Grüter and J Saint Jean Paulin
267 Progress in partial differential equations: calculus of variations, applications
C Bandle, J Bemelmans, M Chipot, M Grüter and J Saint Jean Paulin
268 Stochastic partial differential equations and applications
G Da Prato and L Tubaro
269 Partial differential equations and related subjects
M Miranda
270 Operator algebras and topology
W B Arveson, A S Mishchenko, M Putinar, M A Rieffel and S Stratila
271 Operator algebras and operator theory
W B Arveson, A S Mishchenko, M Putinar, M A Rieffel and S Stratila
272 Ordinary and delay differential equations
J Wiener and J K Hale
273 Partial differential equations
J Wiener and J K Hale
274 Mathematical topics in fluid mechanics
J F Rodrigues and A Sequeira
275 Green functions for second order parabolic integro-differential problems
M G Garroni and J F Menaldi

J F Rodrigues and A Sequeira (Editors)
CMAF, University of Lisbon, Portugal

Mathematical topics in fluid mechanics

Proceedings of the summer course held in Lisbon, Portugal, September 9–13, 1991

CRC Press
Taylor & Francis Group
Boca Raton London New York

CRC Press is an imprint of the
Taylor & Francis Group, an informa business

CRC Press
Taylor & Francis Group
6000 Broken Sound Parkway NW, Suite 300
Boca Raton, FL 33487-2742

First issued in paperback 2019

ISSN 0269-3674

ISBN-13: 978-0-582-20954-1 (hbk)
ISBN-13: 978-0-367-40254-9 (pbk)

British Library Cataloguing in Publication Data

A catalogue record for this book is
available from the British Library

Library of Congress Cataloging-in-Publication Data

Mathematical topics in fluid mechanics : proceedings of the summer course held in Lisbon, Portugal, September 9–13, 1991 / J.F. Rodrigues, editor and A. Sequeira, editor.
p. cm. -- (Pitman research notes in mathematics series ; 274)
Includes bibliographical references.
1. Fluid mechanics--Mathematics--Congresses. I. Rodrigues, J. F. (José Francisco) II. Sequeira, A. (Adélia) III. Series.
TA357.M378 1992
620. . 1'06--dc20 92-31700
CIP

Visit the Taylor & Francis Web site at
http://www.taylorandfrancis.com

and the CRC Press Web site at
http://www.crcpress.com

Contents

Preface

These are the proceedings of the summer course on *Mathematical Topics in Fluid Mechanics*, held in Lisbon, Portugal, from September 9 to 13, 1991.

The mathematical theory of fluid mechanics is an important and active area of current research in the mathematical sciences. It was not our objective to give a compreensive account of the whole field, but rather to provide an advanced introduction and overview of some topics with the presentation of some current mathematical methods and recent research results on new problems arising in classical and non-classical fluid mechanics.

We are grateful to the invited lectures, D. Cioranescu, G.-H. Cottet, J.-C. Saut, V.A. Solonnikov, B. Straughan and A. Valli, for their excellent colaboration in this summer course and for having agreed to contribute to this book with their lecture notes, surveys and articles with new results.

We also acknowledge the short communications presented by some young researchers, corresponding to their seminars given during the last afternoon of the course. They are collected at the end of this volume.

The financial and organization support of the following institutions, which made the summer course possible, are gratefully acknowledged:

Centro de Matemática e Aplicações Fundamentais (CMAF)
Departamento de Matemática da Universidade de Lisboa
Fundação Calouste Gulbenkian
Instituto Nacional de Investigação Científica (INIC)
Junta Nacional de Investigação Científica e Tecnológica (JNICT)
Ambassade de France au Portugal.

Special thanks are due to our organization assistant, Dr. Carlos Albuquerque, to the meeting secretaries, M. Margarida Pereira and M. Odete Ramalho, and to the TEX operator at the CIISA, Carlos Perpétuo, who prepared the final version of this book.

Lisbon, June 1992

Adélia Sequeira
José Francisco Rodrigues

List of Participants

C. Albuquerque (Lisboa)
J.M. André (Lisboa)
P. Avilez Valente (Porto)
O. Ban (Paris/Orsay)
D. Cioranescu (Paris), *Lecturer*
L. Consiglieri (Lisboa)
G.-H. Cottet (Grenoble), *Lecturer*
C. Dias (Lisboa)
J.-P. Dias (Lisboa)
S. Fabre (Palaiseau)
J.S. Ferreira (Rio de Janeiro)
P. Ferreira (Lisboa)
G. Galiano Casas (Madrid)
W. Horn (Essen)
F. James (Palaiseau)
M. Levitin (Moscow)
J. Liang (Lisboa)
L. Loura (Lisboa)
B. Louro (Lisboa)
F. Marir (Paris)
J. Oliveira (Lisboa)
J.F. Padial Molina (Madrid)
H. Pina (Lisboa)
C. Pires (Lisboa)
M.F. Rocha (Palaiseau)
J.-F. Rodrigues (Lisboa), *Lecturer*
B. Rubino (Pisa)
L. Santos (Braga)
J.C. Saut (Paris/Orsay), *Lecturer*
K. Schilling (Bonn)
A. Sequeira (Lisboa), *Lecturer*
R. Severino (Braga)
H. Shahgholian (Stockholm)
Y. Skrynnikov (Moscow)
A. Skvortsov (Moscow)
A. Soares (Coimbra)
V.A. Solonnikov (St.Petersburg), *Lecturer*
B. Straughan (Glasgow), *Lecturer*
L. Tello del Castillo (Madrid)
L. Trabucho (Lisboa)
A. Valli (Trento), *Lecturer*
J. Vasconcelos (Viana do Castelo)

Lectures

Quelques Exemples de Fluides Newtoniens Generalisés

DOINA CIORANESCU

Introduction

Considérons un fluide incompressible occupant une région Ω de $\mathbb{R}^N$. Désignons par $\sigma = (\sigma_{ij})_{ij}$ le tenseur de contraintes et par $D = (D_{ij})_{ij}$ le tenseur des taux des déformations

$$D_{ij} = \frac{1}{2}\Big(\frac{\partial u_i}{\partial x_j} + \frac{\partial u_j}{\partial x_i}\Big)$$

où u désigne le champ des vitesses du fluide.

On appelle fluides newtoniens les milieux continus dont la loi de comportement est donnée par

$$\sigma = -pI + 2\mu D$$

où p (inconnue) est la pression et I est le tenseur identité. On peut ainsi caractériser les fluides newtoniens par le fait qu'il y a une relation linéaire entre le déviateur $\tau = \sigma + pI$ du tenseur des contraintes et le tenseur des taux des déformations. La dénomination de non newtonien est appliquée à tous les autres fluides.

Dans ce qui suit on présente deux types de fluides non newtoniens qui serons des cas particuliers de la classe des fluides obéissants à la loi de Reiner (cf. M. Reiner [9] et R. S. Rivlin [10]) qui est de la forme

$$\sigma = -pI + \varphi_1(D_{II}, D_{III})D + \varphi_2(D_{II}, D_{III})D^2 \tag{1.1}$$

où D_{II}, D_{III} sont les deuxième et troisième principaux invariants de D. Avec la convention de sommation des indices répétés (qui sera utilisée d'ailleurs dans toute la suite), la définition de ces invariants est

$$D_{II} = \frac{1}{2}D_{ij}D_{ij}$$
$$D_{III} = \frac{1}{3}D_{ij}D_{jk}D_{ki}.$$

On s'intéresse ici à la loi de Reiner restreinte aux fluides appelés newtoniens généralisés qui correspondent au cas où $\varphi_2 = 0$ et $\varphi_1 = \varphi_1(D_{II})$ est indépendant de D_{III} i. e.

$$\sigma = -pI + \varphi_1(D_{II})D. \tag{*}$$

Le but de ce qui suit est de donner des lois générales relatives à cette situation et d'étudier les problèmes mathématiques qui en découlent. Dans ces notes on présente deux types de définitions pour (*).

Dans le premier chapitre on définit σ comme un élément du sous différentiel d'une fonction ϕ et on donne ensuite des théorèmes d'existence et d'unicité pour les équations de mouvement de ces fluides particuliers lorsque ϕ vérifie certaines hypothèses. Ces hypothèses s'appliques en particulier pour les cas où ϕ est de forme polynômiale. Il faut remarquer qu'un très grand nombre de fluides cités dans la littérature (Wilkinson [12], Govier-Aziz [1]) rentre dans le cadre de cette loi générale. On peut en mentionner les fluides dilatants, les fluides pseudoplastiques ou le fluide de Bingham. Tous ces exemples serons traîtés au dernier paragraphe de ce chapitre. Appliqués à ces exemples, les théorèmes qui seront prouvés ici généralisent des résultats connus (voir H. Brezis [2], G. Duvaut - J. L. Lions [6], O. A. Ladyzhenskaya [7], J. L. Lions [8]), l'idée de ce chapitre étant de donner une théorie unifiée des problèmes mathématiques associés aux fluides non newtoniens definis par une loi de type sous différentiel. Une version abrégée des résultats de ce premier chapitre est contenue dans D. Cioranescu [4].

Le deuxième chapitre est consacré aux fluides où la fonction $\varphi_1 = \varphi_1(D_{II})$ est telle qu'en particulier, elle a une limite finie à l'infini. Parmi les fluides définis par une fonction de ce type, citons ceux de Carreau, de Eyring-Prandtl, de Williamson ou de Cross (leurs définitions sont rappelées ou debut du deuxième chapitre). Comme dans le premier chapitre, on étudie les problèmes variationnels attachés à ce cas et on donne des résultats d'existence et d'unicité de la solution. Ces résultats, ainsi que d'autres problèmes comme la stabilité ou le comportement asymptotique des solutions, sont contenus dans la thèse [3] de S. Boujena (voir aussi pour une autre approche P. Slobodesky [11]).

Les notes présentées ici font partie d'un cours assuré dans le cadre du DEA d'Analyse Numérique à l'Université Paris VI en 1989 et 1990, dont la rédaction est en préparation (voir D. Cioranescu - J. Saint Jean Paulin [5]).

CHAPITRE I – LOI DE TYPE SOUS-DIFFERENTIEL

1 – Loi Génerale de Comportement

Soit ϕ une fonction définie de $[L^2(\Omega)]^{N\times N} \rightarrow]-\infty,+\infty]$. Définissons alors le déviateur $\tau = \sigma + pI$ par

$$\tau \in \partial\phi(D(u)). \tag{1.2}$$

Dans (1.2), $\partial\phi(D(u))$ est le sous-différentiel de ϕ au point $D(u)$ de coordonnées $D_{ij}(u)$. Grâce aux propriétés du sous-différentiel, τ vérifie

$$\begin{cases} \tau \in [L^2(\Omega)]^{N\times N} \\ \phi(w) - \phi(D(u)) - (\tau, w - D(u)) \geq 0\ ,\ \forall w \in [L^2(\Omega)]^{N\times N}. \end{cases} \tag{1.3}$$

Pour simplifier l'écriture, on introduit la notation

$$\phi(D(u)) = \varphi(u).$$

Soit $v \in H_0^1(\Omega)$. Ecrivons la formule (1.3) au point $w = D(v)$. Avec cette notation on a

$$\varphi(v) - \varphi(u) - \int_\Omega \tau_{ij}(u) D_{ij}(v-u)dx \geq 0. \tag{1.4}$$

Rappelons l'équation du mouvement d'un fluide incompressible adhérant à la paroi $\partial\Omega$

$$(1.5)\qquad \begin{cases} \dfrac{\partial u_i}{\partial t} + u_j \dfrac{\partial u_i}{\partial x_j} = \dfrac{\partial \sigma_{ij}}{\partial x_j} + f_i \quad \text{dans } \Omega \\ \text{div } u = 0 \quad \text{dans } \Omega \\ u|_{\partial\Omega} = 0. \end{cases}$$

Multiplions cette équation par $(v - u)$ où v est une fonction test telle que

$$\begin{cases} \text{div } v = 0 \text{ dans } \Omega \\ v|_{\partial\Omega} = 0. \end{cases}$$

En intégrant par parties il vient

$$(1.6)\qquad \int_\Omega \sigma_{ij}(u) D_{ij}(v - u)\, dx = \int_\Omega (f_i - \frac{\partial u_i}{\partial t} - u_j \frac{\partial u_i}{\partial x_j})(v_i - u_i)\, dx.$$

En utilisant (1.4), cette relation devient

$$(1.7)\qquad \varphi(v) - \varphi(u) \geq \int_\Omega (f_i - \frac{\partial u_i}{\partial t} - u_j \frac{\partial u_i}{\partial x_j})(v_i - u_i)\, dx.$$

Evidemment, si φ n'est pas différentiable, u est solution d'une inéquation variationnelle. Par contre, si φ est différentiable, (1.7) se simplifie en une équation

$$(1.8)\qquad (\varphi'(u), D(v)) = \int_\Omega (f_i - \frac{\partial u_i}{\partial t} - u_j \frac{\partial u_i}{\partial x_j}) v_i\, dx.$$

Cas particulier. Supposons que

$$(1.9)\qquad \varphi(u) = \int_\Omega \overline{\varphi}(u) dx = \int_\Omega \overline{\phi}(D(u)) dx$$

avec $\overline{\phi}$ convexe s.c.i. propre et différentiable. D'après la définition de la dérivée fonctionnelle, l'équation (1.8) devient

$$(1.10)\qquad \int_\Omega \overline{\phi}'_{D_{II}(u)}(D(u)) \cdot D_{ij}(v) dx = \int_\Omega (f_i - \frac{\partial u_i}{\partial t} - u_j \frac{\partial u_i}{\partial x_j}) v_i\, dx$$

où naturellement $\overline{\phi}'_{D_{II}(u)}$ désigne la dérivée de ϕ par rapport à $D_{II}(u)$. L'identité (1.10) signifie que le tenseur des contraintes σ est défini par

$$\tau_{ij} = \overline{\phi}'_{D_{II}(u)}(D(u)) \cdot D_{ij}(u),$$

donc en revenant à la loi de comportement (1.1) on a

$$\varphi_1(D_{II}) = \overline{\phi}'_{D_{II}(u)}(D(u)).$$

Exemples.

1. Fluide newtonien. Un fluide newtonien est défini par la loi de comportement

$$\sigma_{ij} = -p\delta_{ij} + 2\mu\, D_{ij}$$

oú $\mu > 0$ est la viscosité du fluide et δ_{ij} est le symbole de Kronecker. On a donc une fonction $\varphi = \varphi_N$ vérifiant (1.9), elle est définie par

$$\varphi_N(u) = 2\mu \int_\Omega D_{II}(u)dx$$

Evidemment

$$\overline{\phi}'_{D_{II}}(u) = 2\mu.$$

Avec cette loi on a bien entendu, à résoudre les équations de Navier-Stokes.

2. Fluide de Bingham. Les fluides de Bingham sont caractérisés par le seuil de plasticité $g > 0$ tel que

$$\begin{cases} (\sigma_{II})^{1/2} < g \Longrightarrow D_{ij} = 0 \\ (\sigma_{II})^{1/2} \geq g \Longrightarrow \sigma_{ij} = -pI + 2\mu\, D_{ij} + \dfrac{g\, D_{ij}}{(D_{II})^{1/2}} \end{cases}$$

oú σ_{II} est le deuxième invariant du tenseur σ. D'après cette loi de comportement et avec les notations ci-dessus on a

$$\varphi_B(u) = 2\mu \int_\Omega D_{II}(u)dx + 2g \int_\Omega (D_{II}(u))^{\frac{1}{2}}dx = \varphi_N(u) + \psi(u).$$

On remarque que

$$\psi(u) = 2g \int_\Omega \overline{\psi}(D(u))dx$$

avec

$$\overline{\psi}(D(u)) = (D_{II}(u))^{\frac{1}{2}}$$

qui est une fonctionnelle non différentiable. On aura donc à résoudre une inéquation variationnelle que nous rappellerons au dernier paragraphe de ce chapitre.

3. Fluide pseudoplastique. On considère l'équation de Sisko

$$\sigma_{ij} = 2\mu D_{ij}(u) + \alpha(D_{II}(u))^{\frac{\gamma-1}{2}} D_{ij}(u)\,, \quad 0 < \gamma < 1.$$

Alors, la fonction φ correspondante s'écrit

$$\varphi_P(u) = \varphi_N(u) + \psi_1(u)$$

avec

$$\psi_1(u) = \frac{2\alpha}{\gamma+1} \int_\Omega (D_{II}(u))^{\frac{\gamma+1}{2}} dx.$$

4. Fluide dilatant. On a dans ce cas

$$\varphi_D(u) = \varphi_N(u) + \psi_2(u)$$

avec

$$\psi_2(u) = \frac{2\alpha}{\gamma+1}\int_\Omega (D_{II}(u))^{\frac{\gamma+1}{2}} dx\,, \quad \gamma > 1.$$

2 – Théorèmes d'Existence et d'Unicité

On se donne un ouvert Ω borné de $\mathbb{R}^N$ $(N \geq 2)$, de frontière $\partial\Omega$ régulière. Considérons dans Ω le mouvement d'un fluide incompressible adhérant à la paroi $\partial\Omega$. D'après le paragraphe précédent nous avons à étudier l'inéquation variationnelle (1.7). Si l'on introduit une fonction test w, régulière telle que div $w = 0$, on peut réécrire (1.7) sous la forme

$$(2.1) \qquad \begin{cases} (u', w-u) + b(u,u,w-u) + \varphi(w) - \varphi(u) - \varphi \geq (f, w-u) & \text{dans } \Omega \\ \text{div } u = 0 \quad \text{dans } \Omega \\ u|_{\partial\Omega} = 0 \\ u(0) = 0 \end{cases}$$

où

$$b(u,v,w) = \int_\Omega u_i \frac{\partial u_j}{\partial x_i} w_j dx.$$

Pour faciliter la présentation des résultats nous avons pris ici une condition initiale nulle. On peut montrer- à travers quelques complications techniques- qu'ils restent encore valables si on a une condition initiale u_0 non nulle.

2.1. Cadre fonctionnel

On considère les espaces

$\mathcal{V} = \{v | v \in \mathcal{D}(\Omega)^N, \text{div}\, v = 0\}$
$\mathcal{W}^p$ = adhérence de $\mathcal{V}$ dans $[W^{1,p}(\Omega)]^N = \{v | v \in [W^{1,p}(\Omega)]^N, \text{div}\, v = 0\}$
$\mathcal{V}_s$ = adhérence de $\mathcal{V}$ dans $[W^{s,2}(\Omega)]^N$
H = adhérence de $\mathcal{V}$ dans $[L^2(\Omega)]^N = \{v | v \in [L^2(\Omega]^N,\ \text{div}\, v = 0,\ v \cdot n|_{\partial\Omega} = 0\}$

où n est la normale extérieure à Ω. On remarque que

$$\mathcal{W}^2 = \mathcal{V}_1 = V$$

où V est l'espace classique dans l'étude des équations de Navier-Stokes.

Dans la suite $((\cdot,\cdot))$ $\|\cdot\|$ désignent le produit scalaire, respectivement la norme dans V, $(\cdot,\cdot)$, $|\cdot|$ ceux de H, tandis que les normes sur les autres espaces introduits ci-dessus, par exemple $\mathcal{W}^p$, seront notées par $\|\cdot\|_{\mathcal{W}^p}$.

En identifiant H à son dual on a les injections suivantes:

$$\mathcal{W}^p \subset \mathcal{W}^{p-1} \subset \ldots \subset \mathcal{W}^2 = V \subset H \subset V' \subset \ldots \subset (\mathcal{W}^{p-1})' \subset (\mathcal{W}^p)', \qquad \forall p \geq 3. \tag{2.2}$$

2.2. Hypothèses sur la fonctionnelle φ

Avec les notations du paragraphe 1 on a

$$\varphi(u) = \overline{\phi}(D(u)) = \phi(D_{II}(u)).$$

On suppose que $\varphi, \overline{\phi}, \phi$ sont des fonctionnelles convexes s.c.i. propres. On fait les hypothèses suivantes:

H.1. La fonction ϕ est de la forme

$$\phi = \nu_i \phi_i + g\psi$$

où $1 \leq i \leq k$, $\nu_i \geq 0$, $g \geq 0$, les fonctions ϕ_i et ψ sont convexes s.c.i. propres et ψ est non différentiable.

H.2. Les fonctions ϕ_i sont G-différentiables et telles qu'il existe $q_i \geq p_i > 0$ et $\alpha_i > 0$ avec

$$((\phi_i)'_{D_{II}(u)} D_{ij}(u), D_{ij}(u)) \geq \alpha_i \Big[\int_\Omega (D_{II}(u))^{\frac{p_i}{2}} dx \Big]^{\frac{q_i}{p_i}}, \; \forall i \in \{1, \ldots, k\}. \tag{2.3}$$

De plus $(\phi_i)'_{D_{II}(u)} D_{ij}(u) \in L^{p'_i}(\Omega), \dfrac{1}{p'_i} + \dfrac{1}{p_i} = 1$ et vérifie

$$\| (\phi_i)'_{D_{II}(u)} D_{ij}(u) \|_{L^{p'_i}(\Omega)} \leq C_i \Big[\int_\Omega (D_{II}(u))^{\frac{p_i}{2}} dx \Big]^{\frac{q_i}{2}}, \; \forall i \in \{1, \ldots, k\} \tag{2.4}$$

où C_i sont des constantes strictement positives.

H.3. Il existe une famille de fonctions ψ_ε convexes différentiables telles que l'on ait

$$\int_0^T \psi_\varepsilon(v(t))\, dt \to \int_0^T \psi(v(t))\, dt, \quad \forall v \in L^2(0, T; \mathcal{W}^2), \tag{2.5}$$

où on a utilisé les notations

$$\psi(v(t)) = \psi(D_{II}(v(t)))$$
$$\psi_\varepsilon(v(t)) = \psi_\varepsilon(D_{II}(v(t))).$$

On suppose de plus, que

$$\psi'_\varepsilon(0) = 0.$$

H.4. Pour toute suite $\{v_\varepsilon\}$ telle que

$$v_\varepsilon \rightharpoonup v \qquad \text{dans } L^2(0, T; \mathcal{W}^2) \text{ faible}$$

et telle que

$$|\int_0^T \psi_\varepsilon(v_\varepsilon(t))\,dt| \le C \qquad (C \text{ constante indépendante de } \varepsilon),$$

on a

$$\lim_{\varepsilon\to 0}\int_0^T \psi_\varepsilon(v_\varepsilon(t))\,dt \ge \int_0^T \psi(v(t))\,dt. \tag{2.6}$$

H.5. On a

$$\|(\psi_\varepsilon)'_{(D_{II}(u))}\, D_{ij}(u)\|_{L^{p_i'}(\Omega)} \le \overline{C}_i \Big[\int_\Omega (D_{II}(u))^{\frac{p_i}{2}}\,dx\Big]^{\frac{q_i}{2}},\ \forall i\in\{1,...,k\}$$

où $\overline{C}_i$ sont des constantes strictement positives.

2.3. Formulation variationnelle

Soit $s > 1 + \dfrac{N}{2}$. Il résulte des théorèmes d'inclusion de Sobolev que

$$D_i(v) = \{\frac{\partial v_1}{\partial x_i}, ..., \frac{\partial v_N}{\partial x_i}\} \in L^\infty(\Omega)\ , \quad \forall v \in \mathcal{V}_s$$

et donc

$$\mathcal{V}_s \subset \mathcal{W}^p \subset H \subset (\mathcal{W}^p)' \subset (\mathcal{V}_s)'\ , \quad \forall p \ge 2. \tag{2.7}$$

On peut munir $\mathcal{W}^p$ de la norme

$$|||v|||_p = [\int_\Omega (D_{II}(v))^{\frac{p}{2}} dx]^{\frac{1}{p}}$$

qui est équivalente à celle induite par $W^{1,p}(\Omega)$.

Introduisons enfin les notations

$$\begin{cases} \varphi_i(v) = \phi_i(D_{II}(v)) \\ (A_i(v), w) = (\varphi_i'(v), w). \end{cases} \tag{2.8}$$

D'après l'hypothèse H.2. (2.3), l'opérateur A_i est coercif sur $\mathcal{W}^{p_i}$ pour tout $i \in \{1, ..., k\}$, et on a

$$(A_i(v), v) \ge \alpha_i(|||v|||_{p_i})^{q_i}. \tag{2.9}$$

D'autre part, grâce à (2.4), A_i envoie $\mathcal{W}^{p_i}$ dans $(\mathcal{W}^{p_i})'$.

De plus, les A_i sont des opérateurs monotones et hémicontinus. En effet,

$$u \longmapsto A_i(u)$$

correspond au gradient de la fonctionelle

$$u \longmapsto \varphi_i(u)$$

avec φ_i convexe et G-différentiable.

On pose

$$A(u) = \sum_{i}^{k} \nu_i A_i(u) \tag{2.10}$$

et on définit

$$\mathcal{U}_{ad} = \{v | v \in \bigcap_{i=1}^{k} L^{q_i}(0,T;\mathcal{V}_s), v' \in \bigcap_{i=1}^{k} L^{q_i'}(0,T;H), v(0) = 0\}. \tag{2.11}$$

Dans (2.1), on peut prendre comme fonction test $w = v(t)$ pp, où $v \in \mathcal{U}_{ad}$. Alors (2.1) implique

$$\begin{aligned} \int_0^T \{(v', v-u) + (A(u), v-u) + b(u,u,v,-u) + g\psi(v) - g\psi(u)\}dt \\ \geq \int_0^T (f, v-u)dt, \quad \forall v \in \mathcal{U}_{ad}. \end{aligned} \tag{2.12}$$

On cherche donc u vérifiant (2.12) et

$$\begin{cases} \operatorname{div} \ u = 0 & \text{dans } \Omega \\ u|_{\partial\Omega} = 0 \\ u(0) = 0. \end{cases} \tag{2.13}$$

2.4. Enoncés des principaux théorèmes

Théorème 2.1 (Existence). *Soit $f \in \bigcap_{i=1}^{k} L^{q_i'}(0,T;(\mathcal{W}^{p_i})')$. Alors il existe u vérifiant (2.12) et telle que*

$$\begin{cases} u \in (\bigcap_{i=1}^{k} L^{q_i}(0,T;\mathcal{W}^{p_i})) \cap L^\infty(0,T;H) \\ u' \in (\bigcap_{i=1}^{k} L^{q_i'}(0,T;\mathcal{V}_s)\ , \quad \forall s > 1 + \dfrac{N}{2} \\ u(0) = 0 \end{cases}$$

avec

$$p_i \geq \frac{3N}{N+2}\ , \ q_i \geq \frac{2p_i}{p_i - 1}\ , \ i \in \{1,...,k\}. \tag{2.14}$$

Théorème 2.2 (Existence). *Le résultat énoncé dans le Théorème 2.1 est encore valable si*

$$p_i = q_i\ , \quad 2 \leq p_i \leq 3 \quad \text{et } N \text{ quelconque.} \tag{2.15}$$

Remarque 2.3. Si l'on applique le Théorème 2.1 dans le cas particulier $p_i = q_i$, la deuxième relation (2.14) devient

$$p_i \geq \frac{2p_i}{p_i - 1} \quad , \quad i \in \{1, ..., k\},$$

c'est-à-dire $p_i \geq 3$. Donc, d'après le Théorème 2.1, on a l'existence d'une solution si

$$p_i = q_i \geq \text{Max}\,(3, \frac{3N}{N+2}) = 3 \quad , \quad \forall N \; , \; i \in \{1, ..., k\}.$$

En appliquant le Théorème 2.1 si $p_i = q_i \geq 3$ et le Théorème 2.2 si $3 > p_i = q_i \geq 2$, on en déduit l'existence d'une solution pour N quelconque si

$$p_i = q_i \geq 2.$$

3 – Démonstration des Théorèmes d'Existence

3.1. Démonstration du Théorème 2.1

Cette démonstration s'effectue en cinq étapes qui sont

1. birégularisation du problème (2.12)
2. définition des solutions approchées u_m
3. estimations a priori
4. passage à la limite en m dans l'équation birégularisée
5. retour à l'inéquation variationnelle (2.12).

3.1.1. Birégularisation.

Tout d'abord on approche l'inéquation (2.12) par des équations grâce aux hypothèses H.3-H.5 en régularisant la fonctionnelle ψ non différentiable par les fonctionnelles ψ_ε différentiables. La deuxième régularisation vient du fait que l'on oblige la solution à être dans un espace $L^r(0,T;\mathcal{V}_s)$ et pas seulement dans des espaces du type $L^r(0,T;\mathcal{W}^p)$. Pour cela on utilise une méthode de viscosité artificielle. On introduit des opérateurs monotones hémicontinus $\widetilde{A_i}$ de $\mathcal{V}_s$ dans $\mathcal{V}'_s$ tels que

$$\text{(3.1)} \qquad \begin{cases} (\widetilde{A_i} w, w) \geq \beta(\|w\|_{\mathcal{V}_s})^{q_i} \; , \quad \beta > 0 \\ \sup\limits_{v \in \mathcal{V}_s} \dfrac{|(\widetilde{A_i}\, w, v)|}{\|v\|_{\mathcal{V}_s}} \leq C(\|w\|_{\mathcal{V}_s})^{q_i - 1}. \end{cases}$$

La première étape est la résolution de l'équation "birégularisée"

$$\text{(3.2)} \qquad (u'_{\varepsilon\delta}, v) + (A(u_{\varepsilon\delta}), v) + b(u_{\varepsilon\delta}, u_{\varepsilon\delta}, v) + g(\psi'_\varepsilon(u_{\varepsilon\delta}), v) + \\ + \delta \sum_{i=1}^{k} (\widetilde{A_i}(u_{\varepsilon\delta}), v) = (f, v) \, , \quad \forall v \in \mathcal{V}_s$$

où δ est positif et les ψ_ε sont données par les hypothèses H.3-H.5.

Dans la suite on utilise comme base spéciale les vecteurs propres w_j de l'injection compacte de $\mathcal{V}_s$ dans H définis par

$$(3.3) \qquad (w_j, v)_{\mathcal{V}_s} = \lambda_j (w_j, v) \quad , \quad \forall v \in \mathcal{V}_s \quad \text{(pas de sommation en } j\text{)}.$$

Les valeurs propres λ_j sont telles que

$$0 < \lambda_1 < \ldots < \lambda_j < \ldots \qquad \lambda_m \to +\infty \text{ pour } m \to +\infty.$$

3.1.2. Définition des solutions approchées u_m.

On utilise la méthode de Faedo-Galerkin: on cherche $u_m = u_{m\varepsilon\delta}$ sous la forme

$$u_m = \sum_{j=1}^{m} g_{jm}(t) w_j$$

satisfaisant à

$$(3.4) \quad \begin{cases} (u'_m, w_i) + (A(u_m), w_i) + b(u_m, u_m, w_i) + g(\psi'_\varepsilon(u_m), w_i) + \delta \sum_{i=1}^{k} (\widetilde{A_i}(u_m, w_i) = \\ \qquad = (f, w_i) \ , \quad 1 \leq i \leq m \\ u_m(0) = 0. \end{cases}$$

Ce système est un système de m équations différentielles ordinaires en g_{jm} et, d'après le théorème de Cauchy, il a une solution locale u_m sur un intervalle $[0, T_m]$. On va obtenir des estimations a priori qui montrent qu'en fait $T_m = T$.

3.1.3. Estimations a priori.

On multiplie (3.4) par g_{im} et on somme en i. En utilisant (2.9) et (3.1), on a

$$\frac{1}{2}\frac{\partial}{\partial t}(|u_m|^2) + \sum_{i=1}^{k} \alpha_i \nu_i (|||u_m|||_{p_i})^{q_i} + g(\psi'_\varepsilon(u_m), u_m) + \delta \sum_{i=1}^{k} \beta(\|u_m\|_{\mathcal{V}_s})^{q_i} \leq (f, u_m).$$

Cette inéquation donne, puisque grâce à H.3. $(\psi'_\varepsilon(u_m), u_m)$ est positif,

$$(3.5) \quad \begin{cases} u_m \in \text{ borné de } L^\infty(0,T;H) \cap (\bigcap_{i=1}^{k} L^{q_i}(0,T;\mathcal{W}^{p_i}) \\ \delta^{1/q_i} u_m \in \text{ borné de } L^{q_i}(0,T;\mathcal{V}_s), \quad i \in \{1, \ldots, k\} \end{cases}$$

indépendamment de ε, δ et m.

On a tout de suite par les propriétés de la forme b,

$$b(u_m, u_m, w_i) = (h_m, w_i)$$

avec $h_m \in$ borné de $L^\infty(0,T;\mathcal{V}_s')$. On a aussi, grâce à (2.4)

$$A_i(u_m) \in \text{ borné de } L^{q_i'}(0,T;(\mathcal{W}^{p_i})') \subset L^{q_i'}(0,T;\mathcal{V}_s'),$$

et également, d'après les hypothèses sur $\widetilde{A}_i$ et ψ_ε'

$$\begin{cases} \widetilde{A}_i(u_m) \in \text{ borné de } L^{q_i'}(0,T;\mathcal{V}') \\ \psi_\varepsilon'(u_m) \in \text{ borné de } L^{q_i'}(0,T;(\mathcal{W}^{p_i})') \subset L^{q_i'}(0,T;\mathcal{V}_s'), \quad i \in \{1,\dots,k\}, \end{cases}$$

d'où

$$u_m' \in \text{ borné de } \bigcap_{i=1}^{k} L^{q_i'}(0,T;\mathcal{V}_s'), \tag{3.6}$$

indépendamment de m, ε, δ.

3.1.4. Passage à la limite en m.

On extrait de la suite $\{u_m\}$ une sous-suite notée $\{u_n\}$ telle que

$$\begin{cases} u_n \rightharpoonup u_{\varepsilon\delta} & \text{dans } \bigcap_{i=1}^{k} L^{q_i}(0,T;\mathcal{W}^{p_i}) \text{ faible} \\ & \text{dans } L^\infty(0,T;H) \text{ faible} \star \text{ et donc} \\ & \text{dans } \bigcap_{i=1}^{k} L^{q_i}(0,T;H) \text{ fort car aussi} \\ u_n' \rightharpoonup u_{\varepsilon\delta}' & \text{dans } \bigcap_{i=1}^{k} L^{q_i'}(0,T;\mathcal{V}_s') \text{ faible;} \\ A_i(u_n) \rightharpoonup \chi_i^{\varepsilon\delta} & \text{dans } L^{q_i'}(0,T;(\mathcal{W}^{p_i})') \text{ faible} \\ \tilde{A}_i(u_n) \rightharpoonup \xi_i^{\varepsilon\delta} & \text{dans } L^{q_i'}(0,T;\mathcal{V}_s') \text{ faible} \\ \delta^{1/q_i} u_n \rightharpoonup \delta^{1/q_i} u_{\varepsilon\delta} & \text{dans } L^{q_i}(0,T;\mathcal{V}_s) \text{ faible} \\ \psi_\varepsilon'(u_n) \rightharpoonup \theta^{\varepsilon\delta} & \text{dans } \bigcap_{i=1}^{k} L^{q_i'}(0,T;(\mathcal{W}^{p_i})') \text{ faible.} \end{cases} \tag{3.7}$$

Comme pour les équations de Navier-Stokes, on a

$$b(u_n, u_n, w_j) \rightharpoonup b(u_{\varepsilon\delta}, u_{\varepsilon\delta}, w_j) \text{ dans } \mathcal{D}'(]0,T[)$$

par exemple, et donc $u_{\varepsilon\delta}$ vérifie

$$\begin{cases} (u_{\varepsilon\delta}', v) + \sum_{i=1}^{k} \nu_i(\chi_i^{\varepsilon\delta}, v) + b(u_{\varepsilon\delta}, u_{\varepsilon\delta}, v) + g(\theta^{\varepsilon\delta}, v) + \delta \sum_{i=1}^{k} (\xi_i^{\varepsilon\delta}, v) = \\ \qquad = (f, v), \quad \forall v \in \mathcal{V}_s \\ u_{\varepsilon\delta}(0) = 0. \end{cases} \tag{3.8}$$

Il faut maintenant montrer que

$$(3.9)\qquad \sum_{i=1}^{k}\nu_i\chi_i^{\varepsilon\delta}+g\theta^{\varepsilon\delta}+\delta\sum_{i=1}^{k}\xi_i^{\varepsilon\delta}=\sum_{i=1}^{k}\nu_iA_i(u_{\varepsilon\delta})+g\psi'_\varepsilon(u_{\varepsilon\delta})+\delta\sum_{i=1}^{k}\widetilde{A}_i(u_{\varepsilon\delta}).$$

Pour cela, on va utiliser un raisonnement de monotonie. Soit

$$\tilde v\in\bigcap_{i=1}^{k}L^{q_i}(0,T;\mathcal{V}_s)$$

avec

$$\tilde v'\in\bigcap_{i=1}^{k}L^{q'_i}(0,T;\mathcal{V}'_s),$$
$$\tilde v(0)=0.$$

On pose

$$X_n=\int_0^T\{(u'_n-\tilde v',u_n-\tilde v)+(A(u_n)-A(\tilde v),u_n-\tilde v)+g(\psi'_\varepsilon(u_n)-\psi'_\varepsilon(\tilde v),u_n-\tilde v)+$$
$$+\delta\sum_{i=1}^{k}(\widetilde{A}_i(u_n)-\widetilde{A}_i(v),u_n-\tilde v)\}dt.$$

Grâce à la monotonie des divers termes, on a

$$X_n\geq 0.$$

Comme δ est fixé, on a

$$u_n\in\bigcap_{i=1}^{k}L^{q_i}(0,T;\mathcal{V}_s)$$

et alors, d'après (3.8)

$$X_n=\int_0^T(f,u_n)dt-\int_0^T\{(u'_n,\tilde v)+(A(u_n),\tilde v)+(A(\tilde v),u_n-\tilde v)+$$
$$+g(\psi'_\varepsilon(u_n),\tilde v)+g(\psi'_\varepsilon(\tilde v),u_n-\tilde v)+\delta\sum_{i=1}^{k}[(\widetilde{A}_i(u_n),\tilde v)+(\widetilde{A}_i(\tilde v),u_n-\tilde v)]\}dt.$$

Alors, pour $n\to\infty$, en utilisant les convergences (3.7), on a

$$X_n\to X_{\varepsilon\delta},$$

où

$$X_{\varepsilon\delta}=\int_0^T(f,u_{\varepsilon\delta})dt-\int_0^T\{(u'_{\varepsilon\delta},\tilde v)+(\tilde v',u_{\varepsilon\delta}-\tilde v)+\sum_{i=1}^{k}\nu_i(\chi^{\varepsilon\delta},\tilde v)+(A(\tilde v),u_{\varepsilon\delta}-\tilde v)+$$
$$+g(\theta^{\varepsilon\delta},\tilde v)+g(\psi'_\varepsilon(\tilde v),u_{\varepsilon\delta}-\tilde v)+\delta\sum_{i=1}^{k}[(\xi_i^{\varepsilon\delta},\tilde v)+(\widetilde{A}_i(\tilde v),u_{\varepsilon\delta}-\tilde v)]\}dt.$$

De plus, $X_{\varepsilon\delta}$ est positif puisque X_n l'est. En utilisant maintenant l'équation (3.8) avec $v = u_{\varepsilon\delta}$, on a

$$X_{\varepsilon\delta} = \int_0^T \{(u'_{\varepsilon\delta} - \tilde{v}', u_{\varepsilon\delta} - \tilde{v}) + \sum_{i=1}^k \nu_i(\chi_i^{\varepsilon\delta} - A_i(\tilde{v}), u_{\varepsilon\delta} - \tilde{v})$$
$$+ g(\theta^{\varepsilon\delta} - \psi'_\varepsilon(\tilde{v}), u_{\varepsilon\delta} - \tilde{v}) + \delta \sum_{i=1}^k (\xi_i^{\varepsilon\delta} - \widetilde{A}_i(\tilde{v}), u_{\varepsilon\delta} - \tilde{v})]\}dt.$$

On prend dans cette inégalité (en suivant l'idée de Minty)

$$\tilde{v} = u_{\varepsilon\delta} - \lambda\tilde{w}$$

avec

$$\lambda > 0, \quad \tilde{w} \in \bigcap_{i=1}^k L^{q_i}(0,T;\mathcal{V}_s)\ ,\ \tilde{w}' \in \bigcap_{i=1}^k L^{q'_i}(0,T;\mathcal{V}'_s), \quad \tilde{w}(0) = 0.$$

Ainsi, en divisant par λ, on a

$$0 \le X_{\varepsilon\delta} = \lambda \int_0^T (\tilde{w}', \tilde{w})dt + \int_0^T \{\sum_{i=1}^k \nu_i(\chi_i^{\varepsilon\delta} - A_i(u_{\varepsilon\delta} - \lambda\tilde{w}), \tilde{w})$$
$$+ g(\theta^{\varepsilon\delta} - \psi'_\varepsilon(u_{\varepsilon\delta} - \lambda\tilde{w}), \tilde{w}) + \delta \sum_{i=1}^k (\xi_i^{\varepsilon\delta} - \widetilde{A}_i(u_{\varepsilon\delta} - \lambda\tilde{w}), \tilde{w})\}dt.$$

On fait ensuite $\lambda \to 0$. Grâce à l'hémicontinuité de chaque terme, on a

$$0 \le \int_0^T \{((\sum_{i=1}^k \nu_i\chi_i^{\varepsilon\delta} + g\,\theta^{\varepsilon\delta} + \delta \sum_{i=1}^k (\xi_i^{\varepsilon\delta}, \tilde{w}) -$$
$$- (\sum_{i=1}^k \nu_i A_i(u_{\varepsilon\delta}) + g\,\psi'_\varepsilon(u_{\varepsilon\delta}) + \delta \sum_{i=1}^k \widetilde{A}_i(u_{\varepsilon\delta}), \tilde{w})\}dt.$$

Comme ceci est vrai quel que soit $\tilde{w}$, on a (3.9).

Soit maintenant $v \in \mathcal{U}_{\text{ad}}$ dans (3.8). Tous les termes ont un sens et de plus, grâce aux estimations a priori, on peut intégrer (4.8) de 0 à T si l'on montre que

$$t \longmapsto b(u(t), u(t), w(t)) \in L^1(0,T) \quad \text{si} \quad u \in \bigcap_{i=1}^k L^{q_i}(0,T;\mathcal{W}^{p_i}) \cap L^\infty(0,T;H).$$

C'est le seul endroit de la démonstration du Théorème 2.1 où intervient l'hypothèse (2.14). Par un théorème d'injection de Sobolev, on a

$$W^{1,p_i}(\Omega) \subset L^{q_i}(\Omega),$$

avec

$$\frac{1}{\rho_i} = \frac{1}{p_i} - \frac{1}{n} \quad \text{si} \quad \frac{1}{p_i} - \frac{1}{n} > 0$$
$$\rho_i \quad \text{quelconque} \quad \text{si} \quad \frac{1}{p_i} - \frac{1}{n} \le 0.$$

Comme le cas $p_i \geq n$ est facile, nous supposons que $p_i < n$. Alors

$$L^{q_i}(0,T; \mathcal{W}^{p_i} \subset L^{q_i}(0,T; L^{\rho_i}(\Omega)).$$

(i) Si $q_i \leq \rho_i$, on a l'inclusion

$$L^{q_i}(0,T; L^{\rho_i}(\Omega)) \subset L^{q_i}(0,T; L^{q_i}(\Omega)),$$

et $b(u,u,v) \in L^1(0,T)$ si

$$\frac{1}{p_i} + \frac{2}{q_i} \leq 1,$$

ce qui implique qu'il faut avoir

$$\frac{2p_i}{p_i - 1} \leq q_i \leq \rho_i = \frac{Np_i}{N - p_i}$$

d'où

$$p_i \geq \frac{3N}{N+2}.$$

(ii) Par contre, si $q_i > \rho_i$, on a

$$L^{q_i}(0,T; L^{\rho_i}(\Omega)) \subset L^{\rho_i}(0,T; L^{\rho_i}(\Omega)),$$

et donc il faut avoir

$$\frac{1}{p_i} + \frac{2}{\rho_i} \leq 1$$

ce qui donne

$$p_i \geq \frac{3N}{N+2} \quad \text{et} \quad q_i \leq \rho_i = \frac{Np_i}{N - p_i}.$$

Donc $b(u,u,v) \in L^1(0,T)$ si p_i et q_i vérifient les hypothèses (2.14) du théorème. On a alors, en utilisant (3.8) et (3.9)

$$\int_0^T \{(u'_{\varepsilon\delta}, v) + \sum_{i=1}^k \nu_i(A_i(u_{\varepsilon\delta}), v) + b(u_{\varepsilon\delta}, u_{\varepsilon\delta}, v) + g(\psi'_\varepsilon(u_{\varepsilon\delta}), v)) + \tag{3.10}$$

$$+ \delta \sum_{i=1}^k (\widetilde{A}_i(u_{\varepsilon\delta}), v)\}dt = \int_0^T (f,v)dt, \quad \forall v \in \mathcal{U}_{\text{ad}}.$$

3.1.5. Passage à la limite en ε et δ.

Introduisons la fonction $Y_{\varepsilon\delta}$ définie par

$$Y_{\varepsilon\delta} = \int_0^T \{(v', v - u_{\varepsilon\delta}) + (A(u_{\varepsilon\delta}, v - u_{\varepsilon\delta}) + b(u_{\varepsilon\delta}, u_{\varepsilon\delta}, v - u_{\varepsilon\delta}) + g\,\psi_\varepsilon(v) -$$

$$- g\,\psi_\varepsilon(u_{\varepsilon\delta}) + \delta \sum_{i=1}^k (\widetilde{A}_i(u_{\varepsilon\delta}), v - u_{\varepsilon\delta}) - (f, v - u_{\varepsilon\delta})\}dt. \tag{3.11}$$

D'après (3.10), on peut mettre $Y_{\varepsilon\delta}$ sous la forme

$$Y_{\varepsilon\delta} = \int_0^T (v' - u'_{\varepsilon\delta}, v - u_{\varepsilon\delta})dt + g \int_0^T [\psi_\varepsilon(v) - \psi_\varepsilon(u_{\varepsilon\delta}) - (\psi'_\varepsilon(u_{\varepsilon\delta}), v - u_{\varepsilon\delta})]\, dt$$

qui vérifie, grâce à la coercivité de ψ_ε

$$Y_{\varepsilon\delta} \geq 0.$$

On déduit alors de (3.11)

$$\int_0^T \{(v', v - u_{\varepsilon\delta}) + (A(u_{\varepsilon\delta}), v) + b(u_{\varepsilon\delta}, u_{\varepsilon\delta}, v) + g\, \psi_\varepsilon(v) + \delta \sum_{i=1}^{k} (\widetilde{A}_i(u_{\varepsilon\delta}), v) - (f, v - u_{\varepsilon\delta})\} dt \geq$$

$$\geq \int_0^T (A(u_{\varepsilon\delta}), u_{\varepsilon\delta})dt + g \int_0^T \psi_\varepsilon(u_{\varepsilon\delta})dt, \tag{3.12}$$

puisque le terme $\delta \sum_{i=1}^{k} (\widetilde{A}_i(u_{\varepsilon\delta}), u_{\varepsilon\delta})$ est positif d'après (3.1).

Grâce aux estimations a priori, on peut extraire une sous-suite, encore notée $\{u_{\varepsilon\delta}\}$ telle que, pour ε et δ tendant vers zéro, l'on ait

$$\begin{aligned} u_{\varepsilon\delta} \rightharpoonup u \quad & \text{dans } \bigcap_{i=1}^{k} L^{q_i}(0, T; \mathcal{W}^{p_i}) \text{ faible} \\ & \text{dans } L^\infty(0, T; H) \text{ faible } \star \\ u'_{\varepsilon\delta} \rightharpoonup u' \quad & \text{dans } \bigcap_{i=1}^{k} L^{q'_i}(0, T; \mathcal{V}^{p_i}) \text{ faible} \end{aligned}$$

et on passe à la limite dans (3.12). On obtient

$$\begin{aligned} \int_0^T \{(v', v - u) + (A(u), v) + b(u, u, v) + g\, \psi(v) - (f, v - u)\}\, dt \geq \\ \geq \liminf \int_0^T (A(u_{\varepsilon\delta}, u_{\varepsilon\delta})\, dt + \liminf g \int_0^T \psi_\varepsilon(u_{\varepsilon\delta})dt \geq \\ \geq \int_0^T (A(u), u)\, dt + g \int_0^T \psi(u)dt. \end{aligned}$$

Le dernier passage à la limite s'effectue grâce à la monotonie de l'opérateur A et à l'hypothèse H.4. sur les fonctionnelles ψ_ε. Ainsi l'on a

$$\begin{aligned} \int_0^T \{(v', v - u) + (A(u), v) + b(u, u, v) + g\, \psi(v) - g\, \psi(u)\} dt \geq \\ \geq \int_0^T (f, v - u)dt, \ \forall v \in \mathcal{U}_{\text{ad}} \end{aligned} \tag{3.13}$$

ce qui achève la démonstration du Théorème 2.1.

3.2. Démonstration du théorème 2.2

On suppose que $p_i = q_i$. Regardons de nouveau le terme $b(u, u, v)$. On veut montrer qu'il est dans $L^1(0, T)$. C'est le seul endroit dans la démonstration du Théorème 2.1

où l'on avait utilisé les hypothèses (2.14) sur p_i et q_i. On a, en utilisant des résultats d'interpolation

$$u \in L^{q_i}(0,T;\mathcal{W}^{p_i} \cap L^\infty(0,T;H) \subset L^{q_i}(0,T;L^\gamma(\Omega)) \cap L^\infty(0,T;H) \subset L^\rho(0,T;L^\sigma(\Omega))$$

où

$$\frac{1}{\gamma} = \frac{1}{p_i} - \frac{1}{N}, \qquad \frac{1}{\rho} = \frac{\theta}{\infty} + \frac{1-\theta}{q_i} \qquad \text{et} \qquad \frac{1}{\sigma} = \frac{\theta}{2} + \frac{1-\theta}{\gamma}.$$

D'autre part, $v \in L^{p_i}(0,T;\mathcal{V}_s)$ avec $s > \frac{N}{2}+1$, donc $D_i v \in L^\infty(\Omega)$, ce qui entraîne que $v \in L^{p_i}(0,T;\mathcal{W}^\alpha)$ pour tout α. On peut alors évaluer la norme L^1 de $b(u,u,v)$

$$\int_0^T |b(u,u,v)|dt \leq \int_0^T (|u|^2_{L^\sigma(\Omega)} |\text{grad}\, v|_{[L^{\sigma'}(\Omega)]^N})dt$$

avec $\dfrac{1}{\sigma'} + \dfrac{2}{\sigma} = 1$. On utilise alors le fait que $v \in L^{p_i}(0,T;\mathcal{W}^\alpha)$ avec $\alpha = \sigma'$ et on applique la formule de Hölder donc

$$\int_0^T |b(u,u,v)|dt \leq (\int_0^T |u|^\rho_{L^\sigma(\Omega)}dt)^{\frac{2}{\rho}} (\int_0^T |v|^\beta_{\mathcal{W}^\alpha}dt)^{\frac{1}{\beta}}$$

avec $\dfrac{1}{\beta} + \dfrac{2}{\rho} = 1$, donc

$$\beta = \frac{\rho}{\rho - 2}.$$

On détermine β pour que l'on ait $\beta = p_i$ d'où

$$p_i = \frac{\rho}{\rho-2}, \text{ i.e. } \rho = \frac{2p_i}{p_i - 1}.$$

Finalement, on obtient

$$\frac{p_i - 1}{2p_i} = \frac{1-\theta}{p_i}$$

d'où

$$\theta = \frac{3 - p_i}{2}.$$

Comme $\theta \in (0,1)$, on doit avoir l'inégalité (2.15) pour obtenir le résultat souhaité.

4 – Unicité

De manière générale, on n'a que des résultats partiels d'unicité. Les difficultés pour démontrer des théorèmes d'unicité sont d'abord dues au fait que l'opérateur est en général non linéaire, d'où l'impossibilité d'appliquer des lemmes du type Gronwall. Cette difficulté disparaît si l'on suppose que parmi les opérateurs A_i se trouver le Laplacien $-\Delta$. D'autre part, u' et u ne sont pas dans des espaces en dualité. Une possibilité de contourner cette deuxième difficulté est de supposer $\psi = 0$; on est alors en présence d'une équation

$$(4.1) \qquad \begin{cases} u' + A(u) + \sum_{i=1}^{k} u_i D_i u = f - \nabla p \quad \text{dans } \Omega, \\ \operatorname{div} u = 0 \quad \text{dans } \Omega, \\ u = 0 \quad \text{sur } \partial\Omega, \\ u(0) = 0. \end{cases}$$

Les hypothèses (2.4) sur A_i entraînent

$$A(u) \in \bigcap_{i=1}^{k} L^{q_i}(0,T;(\mathcal{W}^{p_i})')$$

et d'après les théorèmes d'existence 2.1 et 2.2, le système (4.1) implique

$$u' \in \bigcap_{i=1}^{k} L^{q_i'}(0,T;(\mathcal{W}^{p_i})').$$

On peut donc prendre le produit de (4.1) par u et on a ainsi le théorème suivant:

Théorème 4.1 (Unicité). *On se place dans le cas où $\psi = 0$ et $p_i = q_1 = 2$ et on fait l'hypothèse (2.14) (resp. (2.15)). On suppose de plus que*

$$(4.2) \qquad p_i > \frac{N}{2}, \qquad q_i \geq \frac{2p_i}{2p_i - N}, \qquad \forall i \in \{2, ..., k\}.$$

Alors la solution donnée par le Théorème 2.1 (ou (2.2) est unique.

Démonstration: Soient u_1 et u_2 deux solutions de (4.1). On pose

$$u^* = u_1 - u_2.$$

Alors, on a

$$((u^*)', u^*) + (A(u_1) - A(u_2), u_1 - u_2) + b(u^*, u_1, u^*) = 0$$

d'où

$$(4.3) \qquad \frac{1}{2}\frac{\partial}{\partial t}|u^*|^2 + \alpha_1 \|u^*\|^2 + b(u^*, u_1, u^*) \leq 0.$$

Donc pour avoir unicité, il suffit que

$$(4.4) \qquad u \in L^S(0,T;[L^r(\Omega)]^N) \quad \text{avec} \quad \frac{2}{S} + \frac{N}{r} \leq 1, \ r > N.$$

Admettons provisoirement que (4.4) est réalisé et démontrons l'unicité. Comme dans le cas des équations de Navier-Stokes, on a la majoration

$$|b(u^*, u_1, u^*)| \leq C_1 \|u_1(t)\|_{L^r} |u^*(t)|^{\frac{2}{S}} \|u^*(t)\|^{1+\frac{N}{r}} \leq \alpha_1 \|u^*(t)\|^2 + C_2 M(t) |u^*(t)|^2$$

où

$$M(t) = (\|u_1(t)\|_{L^r})^S$$

et $M(t) \in L^1$ si $u_1 \in L^S(0,T;[L^r(\Omega)]^N)$. On utilise cette majoration de $|b(u^*, u_1, u^*)|$ dans (4.3) et on applique le lemme de Gronwall ce qui démontre l'unicité. Pour achever la démonstration, il reste à établir (4.4).

Comme auparavant, on sait que

$$u \in L^{q_i}(0,T;\mathcal{W}^{p_i}) \cap L^\infty(0,T;H) \subset L^{q_i}(0,T;[L^\gamma(\Omega)]^N) \cap L^\infty(0,T;H) \subset L^\rho(0,T;[L^\sigma(\Omega)]^N)$$

avec

$$\frac{1}{\gamma} = \frac{1}{p_i} - \frac{1}{N}, \quad \frac{1}{\rho} = \frac{\theta}{\infty} + \frac{1-\theta}{q_i} \quad \text{et} \quad \frac{1}{\sigma} = \frac{\theta}{2} + \frac{1-\theta}{\gamma}.$$

On aura donc (4.3) si l'on peut choisir θ tel que

$$\frac{2}{\rho} + \frac{N}{\sigma} \leq 1, \quad \sigma > N,$$

c'est-à-dire

$$(1-\theta)\left(\frac{2}{q_i} + \frac{N-p_i}{p_i} - \frac{N}{2}\right) \leq 1 - \frac{N}{2} \leq 0$$

et

$$(1-\theta)\left(\frac{N-p_i}{Np_i} - \frac{1}{2}\right) < \frac{1}{N} - \frac{1}{2} \leq 0.$$

Comme $0 < (1-\theta) \leq 1$, on doit avoir (en changeant les signes)

$$\frac{N}{2} - \frac{2}{q_i} - \frac{N-p_i}{p_i} \geq \frac{N}{2} - 1 > 0$$

et

$$\frac{1}{2} - \frac{N-p_i}{Np_i} > \frac{1}{2} - \frac{1}{N} > 0.$$

Avec p_i et q_i vérifiant (4.2), ces conditions sont remplies.

Dans le cas où $\psi \neq 0$, l'approche précédente ne peut plus être appliquée étant donné que l'on a une *inéquation* variationnelle.

5 – Exemples — Applications des Théorèmes

5.1. Fluides dilatants

Comme on l'a vu auparavant, le comportement d'un fluide dilatant est donné par

$$\sigma_{ij}^D = 2\mu\, D_{ij} + \alpha\, D_{II}^{(\gamma-1)/2} D_{ij} \quad \text{avec } \gamma > 1.$$

On a vérifié que ceci correspond au cas où

$$\varphi_D(u) = \varphi_1(u) + \varphi_2(u)$$

avec

$$\varphi_1(u) = 2\mu \int_\Omega D_{II}(u)dx$$
$$\varphi_2(u) = \frac{2}{\gamma+1} \int_\Omega (D_{II}(u))^{\frac{\gamma+1}{2}} dx$$

donc on a

$$p_1 = q_1 = 2 \quad \text{et} \quad p_2 = q_2 = \gamma + 1$$

et on a toujours existence d'une solution.

Dans la littérature, on trouve aussi comme définition de fluides dilatants

$$\sigma_{ij}^D = \alpha D_{II}^{(\gamma-1)/2} D_{ij}.$$

On doit alors avoir

$$\gamma + 1 \geq \frac{3N}{N+2} \quad \text{et} \quad \gamma \geq 2.$$

Si la deuxième inégalité est vérifiée, la première l'est aussi.

Donc si $\gamma \geq 2$, on a existence de la solution dans $L^{\gamma+1}(0,T;\mathcal{W}^{\gamma+1})$.

5.2. Exemple de Ladyzhenskaya

Une variante des équations de Navier-Stokes introduite par Ladyzhenskaya consiste à considérer le cas où la fonction φ est donnée par

$$\varphi(u) = \nu \int_\Omega D_{II}(u)dx + \frac{\alpha}{2}\left[\int_\Omega D_{II}(u)\, dx\right]^2$$

ce qui s'écrit

$$\varphi(u) = \varphi_1(u) + \varphi_2(u)$$

avec

$$\varphi_1(u) = \nu \int_\Omega D_{II}(u)dx\ , \quad p_1 = q_1 = 2$$
$$\varphi_2(u) = \frac{\alpha}{2}\left[\int_\Omega D_{II}(u)dx\right]^2\ , \quad p_2 = 2 \text{ et } q_2 = 4.$$

Le Théorème 2.1 donne donc l'existence dans $L^2(0,T;L^4(\Omega))$ pour $N \leq 4$. Pour l'unicité, on doit avoir, d'après le Théorème 4.1,

$$p_i > \frac{N}{2}\ , \ p_i \geq \frac{3n}{N+2}\ , \ q_i \geq \frac{2p_i}{2p_i - N}\ , \ \forall i \in \{1,2\}.$$

La première inégalité est vérifiée si $N < 4$ et la deuxième entraîne que l'on a unicité de la solution si

$$N \leq 3.$$

5.3. Fluide de Bingham

Comme nous l'avons vu au paragraphe 1, contrairement aux cas des deux exemples précédents, la fonctionnelle définissant un fluide de Bingham est non différentiable. On aura ainsi à résoudre l'inéquation variationnelle

$$(u', v-u) + \mu((u, v-u)) + b(u, u, v-u) + gj(v) - gj(u) \geq (f, v-u)$$

où j est la fonction convexe s.c.i. définie par

$$j(v) = 2\int_\Omega (D_{II}(v))^{1/2} dx.$$

On approche la fonctionnelle j par

$$j_\varepsilon(v) = \frac{2}{1+\varepsilon}\int_\Omega (D_{II}(v))^{\frac{1+\varepsilon}{2}} dx\ , \quad \varepsilon > 0.$$

L'équation approchée s'écrit donc

$$(u'_\varepsilon, v) + \mu(u_\varepsilon, v) + b(u_\varepsilon, u_\varepsilon, v) + g(j'_\varepsilon(u_\varepsilon), v) = (f, v)$$

avec

$$(j'_\varepsilon(u_\varepsilon), v) = \int_\Omega (D_{II}(u_\varepsilon))^{\frac{\varepsilon-1}{2}} D_{ij}(u_\varepsilon) D_{ij}(v) dx.$$

Comme pour un fluide newtonien, on a $p_1 = q_1 = 2$ et donc existence et unicité pour $N = 2$ dans $L^2(0,T;\mathcal{V}_1) \cap L^\infty(0,T;H)$ et seulement existence dans le même espace pour $N = 3$. Remarquons enfin qu'en dimension 2 on n'a pas besoin de birégularisation.

CHAPITRE 2 – VISCOSITE NON LINEAIRE BORNEE A L'INFINI

1 – Hypothèses — Formulation Variationnelle

On considère de nouveau une loi de comportement pour des fluides non newtoniens généralisés de la forme

$$\sigma = -pI + \varphi(s(u))D \tag{1.1}$$

où on a noté

$$s(u) = \sqrt{2D_{II}(u)}.$$

Supposons que φ sont des fonctions vérifiant les hypothèses suivantes:

H.1. μ est continûment différentiable de $\mathbb{R}_+$ dans $\mathbb{R}_+$.

H.2. Il existe $\mu_\infty > 0$ tel que

$$\lim_{z\to\infty} \mu(z) = \mu_\infty. \tag{1.2}$$

H.3. Il existe un ensemble $\mathcal{E} \subset \mathbb{R}_+$ tel que

$$\begin{cases} \mu'(z) \geq 0, & \forall z \in \mathcal{E} \\ z|\mu'(z)| \leq \mu(z), & \forall z \in \mathbb{R}_+ \setminus \mathcal{E}. \end{cases} \tag{1.3}$$

Remarque. L'hypothèse (1.4) n'exclut pas les cas où $\mathcal{E} = \mathbb{R}_+$ ou $\mathcal{E} = \emptyset$.

Exemples.

1. Fluide de Eyring-Prandtl. Pour ce fluide on a

$$\mu(z) = \mu_\infty + \frac{(\mu_0 - \mu_\infty)\,\mathrm{Log}[z\lambda + (z^2\lambda^2 + 1)^{\frac{1}{2}}]}{z\lambda}$$

où μ_0, μ_∞ et λ sont des constantes strictement positives.

2. Fluide de Cross. La loi est dans ce cas

$$\mu(z) = \mu_\infty + \frac{(\mu_0 - \mu_\infty)}{(1 + z^{\frac{2}{3}})}$$

où μ_0 et μ_∞ sont des constantes positives avec $\mu_\infty \geq \mu_0 > 0$. On vérifie que l'on a $\mathcal{E} = \mathbb{R}_+$.

3. Fluide de Williamson. Il est défini par

$$\mu(z) = \frac{\lambda(\mu_0 - \mu_\infty)}{\lambda + z} + \mu_\infty$$

où μ_0, μ_∞ et λ sont des constantes positives vérifiant $\mu_0 \geq \mu_\infty > 0$. Dans ce cas $\mathcal{E} = \emptyset$.

4. Fluide de Carreau. On a

$$\mu(z) = \mu_\infty + (\mu_0 - \mu_\infty)(1 + \lambda z^2)^{\frac{r-2}{2}}$$

où $\mu > \mu_\infty \geq 0,\ 1 < r \leq 2,\ \lambda > 0$. De nouveau, $\mathcal{E} = \emptyset$.

Dans la suite on pose

$$\text{(1.4)} \qquad \begin{cases} \sup\limits_{z \in \mathbb{R}_+} \mu(z) = M \\ \inf\limits_{z \in \mathbb{R}_+} \mu(z) = m. \end{cases}$$

Ces nombres existent grâce aux hypothèses H.1 - H.3 et vérifient, d'après (1.2)

$$0 \leq m \leq M < +\infty.$$

Les espaces fonctionnels dans lesquels on va travailler sont ceux classiques dans l'ètude des équations de Navier-Stokes que nous avons déjà introduits au Chapitre I, paragraphe 2.1

V = adhérence de $\mathcal{V}$ dans $[W^{1,2}(\Omega)]^N = \{v |\, v \in H_0^1(\Omega)\ \mathrm{div}\, v = 0\}$
H = adhérence de $\mathcal{V}$ dans $[L^2(\Omega)]^N = \{v |\, v \in [L^2(\Omega]^N,\ \mathrm{div}\, v = 0,\ v \cdot n|_{\partial\Omega} = 0\}$

où $\mathcal{V} = \{v \in \mathcal{D}(\Omega)^N, \mathrm{div}\, v = 0\}$. On a évidemment

$$V \subset H \subset V'$$

avec injection compacte.

Si on prend $v \in V$, on obtient à partir de (1.1), l'équation variationnelle

$$(1.5) \qquad \begin{cases} (u', v) + (\mathcal{A}u, v) + b(u,u,v) = (f, v), & \forall v \in V \\ u(0) = u_0, \end{cases}$$

où l'opérateur $\mathcal{A}$ est défini par

$$(\mathcal{A}v, w) = 2(\mu(s(v))Dv, Dw), \quad \forall v, w \in V$$

et u_0 est la condition initiale attachée au problème.

2 – Propriétés de l'Opérateur A

Il est facile à voir que , grâce aux hypothèses H.1 - H.3, l'opérateur $\mathcal{A}$ applique l'espace V dans V'. De plus

Proposition 2.1. *L'opérateur $\mathcal{A}$ est monotone hemicontinu.*

Démonstration:

(i). *Hemicontinuité.* Soient $u, v, w, \in V$. On doit montrer que la fonctionnelle réelle de variable réelle

$$h : \longmapsto (\mathcal{A}(u + \lambda v), w)$$

est continue. Soit $\{\lambda_n\}_{n \in \mathbb{N}}$ une suite réelle telle que $\lambda_n \to \lambda$. On a

$$(2.1) \qquad \begin{cases} (\mathcal{A}(u + \lambda_n v), w) = \int_\Omega 2\mu(s(u + \lambda_n v))\, D_{ij}(u) D_{ij}(w)\, dx + \\ \qquad + \lambda_n \int_\Omega 2\mu(s(u + \lambda_n v))\, D_{ij}(v) D_{ij}(w)\, dx. \end{cases}$$

Lorsque $\lambda_n \to \lambda$, puisque d'après H.1 μ est continue, on a

$$\mu(s(u + \lambda_n v)) \to \mu(s(u + \lambda v)) \quad \text{quand } n \to \infty.$$

Ainsi

$$2\mu(s(u + \lambda_n v)) D_{ij}(u) D_{ij}(w) \to \mu(s(u + \lambda v)) D_{ij}(u) D_{ij}(w)$$

ponctuellement. Par ailleurs

$$|2\mu(s(u + \lambda_n v)) D_{ij}(u) D_{ij}(w)| \leq M |D_{ij}(u) D_{ij}(w)|$$

avec $D_{ij}(u) D_{ij}(w) \in L^1(\Omega)$, car $u, v \in V$. On peut alors appliquer le théorème de convergence dominée de Lebesgue aux deux termes de (2.1). pour avoir le résultat.

(ii). *Monotonie.* On doit montrer que

$$(2.2) \qquad (Au - Av, u - v) \geq 0, \quad \forall u, v \in V.$$

où

$$(Au - Av, u - v) = \int_\Omega 2\mu(s(u))\, D_{ij}(u) D_{ij}(u - v)\, dx - \int_\Omega 2\mu(s(v))\, D_{ij}(u) D_{ij}(u - v)\, dx.$$

Soit $\alpha \in [0, 1]$. Considérons la fonction suivante:

$$\psi(\alpha) = \int_\Omega 2\mu(s(v + \alpha(u - v)))\, D_{ij}(v + \alpha(u - v)) D_{ij}(u - v)\, dx. \tag{2.3}$$

On remarque que ψ est continue sur $[0, 1]$ et de plus

$$(Au - Av, u - v) = 2[\psi(1) - \psi(0)]. \tag{2.4}$$

La démonstration de la monotonie s'effectue en deux étapes:

(ii$_1$). On prouve que ψ est une fonction dérivable sur l'intervalle $(0, 1)$.

(ii$_2$). On applique ensuite la formule de Taylor au second membre de (2.4) pour un $\lambda \in [0, 1]$, puis on utilise l'hypothèse (1.3) pour le minorer.

(ii$_1$). On pose pour $x \in \Omega$ et $\alpha \in [0, 1]$

$$f(x, \alpha) = \mu(s(v + \alpha(u - v)))\, D_{ij}(v + \alpha(u - v)) D_{ij}(u - v) \tag{2.5}$$

donc

$$\psi = \int_\Omega f(x, \alpha)\, dx.$$

On va montrer que $f(x, \alpha)$ est dérivable en α pour tout $x \in \Omega$ et que de plus, elle est majorée en module par une fonction intégrable sur Ω et indépendante de α. En appliquant de nouveau le théorème de Lebesgue on en déduit que ψ est dérivable.

On introduit les deux sous ensembles de Ω définis par

$$\Omega_1(\xi) = \{x \in \Omega | 0 < s(v + \xi(u - v)) < +\infty \text{ au point } x\}$$
$$\Omega_2(\xi) = \{x \in \Omega | s(v + \xi(u - v)) = 0 \text{ au point } x\}.$$

On a évidemment

$$\Omega_1(\xi) \cap \Omega_2(\xi) = \emptyset$$

et de plus l'ensemble $\Omega \backslash (\Omega_1(\xi) \cup \Omega_2(\xi))$ est de mesure nulle. On va vérifier que l'on a

$$\begin{cases} \dfrac{\partial f(x, \xi)}{\partial \xi} = \displaystyle\sum_{i,j=1}^{N} \mu(s(v + \xi(u - v)))[D_{ij}(u - v)]^2 + \\ \qquad + \dfrac{\mu'(s(v + \xi(u - v)))}{s(v + \xi(u - v))} \{\displaystyle\sum_{i,j=1}^{N} D_{ij}(v + \xi(u - v))\, D_{ij}(u - v)\}^2, & \text{si } x \in \Omega_1(\xi) \\ \dfrac{\partial f(x, \xi)}{\partial \xi} = \displaystyle\sum_{i,j=1}^{N} \mu(0)[D_{ij}(u - v)]^2, & \text{si } x \in \Omega_2(\xi) \end{cases} \tag{2.6}$$

En effet, soit $x \in \Omega_1(\xi)$. Alors

$$\begin{aligned} \frac{\partial f(x, \xi)}{\partial \xi} &= \sum_{i,j=1}^{N} \Big\{ \mu(s(v + \xi(u - v)))[D_{ij}(u - v)]^2 + \\ &+ \mu'(s(v + \xi(u - v))) \frac{\partial s(v + \xi(u - v))}{\partial \xi} D_{ij}(v + \xi(u - v))\, D_{ij}(u - v) \Big\}. \end{aligned} \tag{2.7}$$

Mais

$$s(u)=\sqrt{2D_{II}(u)},$$

donc

$$(2.8)\qquad \begin{aligned}\frac{\partial s(v+\xi(u-v))}{\partial\xi}=&\frac{\partial s(v+\xi(u-v))}{\partial D_{II}(v+\xi(u-v))}\frac{\partial D_{II}(v+\xi(u-v))}{\partial\xi}=\\ =&\frac{1}{s(v+\xi(u-v))}D_{ij}(v+\xi(u-v))\,D_{ij}(u-v)\end{aligned}$$

ce qui, remplacé dans (2.7), mène à la formule (2.6).

Soit maintenant $x\in\Omega_2(\xi)$. On ne peut plus appliquer (2.8) car $s(v+\xi(u-v))=0$. Mais cela implique que

$$\forall k,l\in\{1,...,N\}\quad \text{on a } D_{kl}(v)=-\xi D_{kl}(u-v).$$

Or pour que ceci soit vrai, on doit être dans une des situations suivantes:

$$\begin{aligned}&\text{(a)}\quad D_{kl}(u)=D_{kl}(v)=0,\quad \forall k,l\in\{1,...,N\}\\ &\text{(b)}\quad \exists k,l \text{ avec } D_{kl}(u-v)\neq 0.\end{aligned}$$

Dans le cas (a), on a d'après la définition (2.5) de la fonction f

$$f(x,\alpha)=0,\quad \forall\alpha\in(0,1)$$

donc $\frac{\partial f}{\partial\xi}=0$ et la formule (2.6) est encore valable. Dans le cas (b) la situation est plus compliquée. On a toujours $f(x,\alpha)=0$ mais quel que soit α dans un voisinage suffisamment petit de ξ, on a $f(x,\alpha)\neq 0$.

Soit $\{\alpha_k\}_{k\in\mathbb{N}}$ une suite d'éléments de $(0,1)$ tels que $\alpha_k\neq\xi$ pour tout $k\in\mathbb{N}$ et $\lim_{k\to\infty}\alpha_k=\xi$. D'après la remarque ci-dessus, la fonctin $\alpha\mapsto f(x,\alpha)$ est dérivable en α_k et sa dérivée est donnée par la formule (2.6). Ecrivons donc (2.6) au point (x,α_k) et soit $k\to\infty$. Compte tenu de la continuité de μ on obtient

$$\lim_{k\to\infty}\frac{\partial f(x,\xi)}{\partial\xi}=\mu(0)\sum_{i,j=1}^{N}[D_{ij}(u-v)]^2+\lim_{k\to\infty}I_k$$

où

$$I_k=\frac{\mu'(s(v+\alpha_k(u-v)))}{s(v+\alpha_k(u-v))}\{\sum_{i,j=1}^{N}D_{ij}(v+\alpha_k(u-v))\,D_{ij}(u-v)\}^2.$$

Si l'on montre que

$$\lim_{k\to\infty}I_k=O,$$

la formule (2.6) sera complètement établie. Or on a

$$\begin{aligned}|I_k|\leq&\frac{1}{4}\Big|\frac{\mu'(s(v+\alpha_k(u-v)))}{s(v+\alpha_k(u-v))}s^2(v+\alpha_k(u-v))\{\sum_{i,j=1}^{N}D_{ij}(u-v)\}^2\Big|\leq\\ \leq&|\mu'(s(v+\alpha_k(u-v)))\,s(v+\alpha_k(u-v))\{\sum_{i,j=1}^{N}D_{ij}(u-v)\}^2|.\end{aligned}$$

Si $k \to \infty$, $\alpha_k \to \xi$ et comme s est continue, $s(v+\alpha_k(u-v)) \to s(v+\xi(u-v)) = 0$ et de même $\mu'(s(v+\alpha_k(u-v))) \to \mu'(0)$, donc le membre de droite dans l'inégalité ci-dessus converge vers zéro, d'où le résultat.

Nous remarquons maintenant que $\dfrac{\partial f}{\partial \alpha}$ est majorée par une fonction intégrable. En effet, en utilisant (2.6) on a

$$\left|\frac{\partial f}{\partial \alpha}\right| \leq C|s^2(u-v)|$$

puisque μ, μ' et s sont des fonctions continues sur $[0,1]$, ce qui achève l'étape (ii_1).

(ii_2). On commence par appliquer à ϕ dans la formule (2.4) la formule de Taylor pour avoir

$$(\mathcal{A}u - \mathcal{A}v,\ u-v) = 2\frac{d\phi}{d\alpha}(\lambda) = 2\int_\Omega \frac{\partial f(x,\alpha)}{\partial \alpha}(\lambda)\,dx$$

pour un $\lambda \in (0,1)$. On a ainsi

$$(\mathcal{A}u - \mathcal{A}v,\ u-v) = 2\int_{\Omega_E} \frac{df}{d\xi}(x,\lambda)\,dx + 2\int_{\Omega\setminus\Omega_\mathcal{E}} \frac{df}{d\xi}(x,\lambda)\,dx$$

où $\Omega_\mathcal{E} = \{x \in \Omega |\, s(v+\lambda(u-v)) \in \mathcal{E}\}$. On considère d'abord la première intégrale. D'après l'hypothèse (1.3), on a

$$2\int_{\Omega_\mathcal{E}} \frac{df}{d\xi}(x,\lambda)\,dx \geq m\,\|u-v\|^2_{\Omega_\mathcal{E}} > 0 \tag{2.9}$$

puisque $\mu'(z) \geq 0$ si $z \in \mathcal{E}$. Il reste à minorer la deuxième intégrale. On a

$$\int_{\Omega\setminus\Omega_\mathcal{E}} \frac{df}{d\xi}(x,\lambda)\,dx = \int_{\Omega_1\cap(\Omega\setminus\Omega_\mathcal{E})} \frac{df}{d\xi}(x,\lambda)\,dx + \int_{\Omega_2\cap(\Omega\setminus\Omega_\mathcal{E})} \frac{df}{d\xi}(x,\lambda)\,dx.$$

On va montrer que chacune des intégrales est positive. Pour la deuxième ceci est evident d'après (1.4) et la formule (2.6). Par ailleurs,

$$\left\{\begin{aligned} &\int_{\Omega_1\cap(\Omega\setminus\Omega_\mathcal{E})} \frac{df}{d\lambda}(x,\lambda)\,dx \geq \sum_{i,j=1}^N \int_{\Omega_1\cap(\Omega\setminus\Omega_\mathcal{E})} \mu(s(v+\lambda(u-v)))[D_{ij}(u-v)]^2 - \\ &- \sum_{i,j=1}^N \int_{\Omega_1\cap(\Omega\setminus\Omega_\mathcal{E})} \frac{|\mu'(s(v+\lambda(u-v)))|}{s(v+\lambda(u-v))} s^2(v+\lambda(u-v))\,[D_{ij}(u-v)]^2 \geq 0 \end{aligned}\right. \tag{2.10}$$

où on a utilisé la formule (2.6) et le fait que

$$\mu(z) - z|\mu'(z)| \geq 0, \quad \forall z \in \mathbb{R}_+ \setminus \mathcal{E}$$

d'après l'hypothèse (1.3), ce qui achève la démonstration.

Remarque 2.2. On peut améliorer les estimations précédentes en faisant des hypothèses plus restrictives sur μ.

Supposons qu'au lieu de H.3, on ait

H.3′. Il existe un ensemble $\mathcal{E} \subset \mathbb{R}_+$ tel que

$$\begin{cases} \mu'(z) \geq 0, \quad \forall z \in \mathcal{E} \\ \mu(z) - z|\mu'(z)| \geq \mu_1 > 0, \quad \forall z \in \mathbb{R}_+ \setminus \mathcal{E}. \end{cases} \tag{2.11}$$

On peut noter que (2.11) est vérifiée par les fluides de Williamson ou de Carreau, définis par (1.6) et (1.7), avec $\mathcal{E} = \emptyset$.

Dans ce cas, en reprenant l'estimation (2.10), on obtient facilement

$$(\mathcal{A}u - \mathcal{A}v, u - v) \geq m\|u - v\|^2_{\Omega_\varepsilon} + m\|u - v\|^2_{\Omega_2\setminus(\Omega\setminus\Omega_\varepsilon)} + \mu_1\|u - v\|^2_{\Omega_1\setminus(\Omega\setminus\Omega_\varepsilon)} \geq C_0\|u - v\|^2_\Omega.$$

Remarque 2.3. Si μ vérifie H.1 - H.3 avec $\mathcal{E} = \mathbb{R}_+$, la démonstration ci-dessus conduit à

$$(\mathcal{A}u - \mathcal{A}v, u - v) \geq m\|u - v\|^2$$

3 – Résultats d'Existence

On a le résultat suivant:

Théorème 3.1. *Soit $f \in L^2(0,T;H)$ avec $0 < T < \infty, u_0 \in H$ et μ vérifiant les hypoth'eses H.1 - H.3. Alors il existe une solution u du problème (1.5) telle que*

$$\begin{cases} u \in L^2(0,T;V) \cap L^\infty(O,T;H) \\ u' \in L^1(O,T;V'). \end{cases} \tag{3.1}$$

Démonstration: On utilise la méthode de Faedo-Galerkin avec la base spéciale $\{w_j\}$ définie par

$$((w_j, v)) = \lambda_j\,(w_j, v) \quad \forall v \in V.$$

On cherche donc une solution approchée u_m définie par

$$u_m = \sum_{j=1}^{m} c_{jm}(t) w_j$$

satisfaisant à

$$\begin{cases} (u'_m, w_i) - (div[2\mu(s(u_m))]\, Du_m, w_i) + b(u_m, u_m, w_i) = (f, w_i)\,, \quad 1 \leq i \leq m \\ u_m(0) = u_{0m}. \end{cases} \tag{3.2}$$

avec

$$u_{0m} \in V,\ u_{0m} \to u_0 \text{ dans } V. \tag{3.3}$$

Ce système est un système de m équations différentielles ordinaires en c_{jm} de la forme

$$\frac{dc_{jm}}{dt} = G_j(t, c_{1m}, ..., c_{Nm})\,, \quad 1 \leq i \leq m$$

et, d'après le théorème de Cauchy, il a une solution locale u_m sur un intervalle $[0, T_m]$. On va obtenir des estimations a priori qui montrent que $T_m = T$.

3.1. Estimations a priori

On multiplie (3.2) par c_{jm}, on intègre par parties et on somme de 1 à m. On a

$$\frac{1}{2}\frac{d}{dt}|u_m(t)|^2 + 2\int_\Omega \mu(s(u_m))\, D_{ij}(u_m(t))D_{ij}(u_m(t))\, dx = \int_\Omega f(t)\, u_m(t)\, dx \tag{3.4}$$

où on utilise la convention de sommation des indices muets. D'après les notations introduites au début de ce chapitre, $D_{ij}(u_m(t))D_{ij}(u_m(t)) = s^2(u_m(t))$, on peut alors écrire

$$\int_\Omega \mu(s(u_m))\, s^2(u_m(t))\, dx = \int_\Omega [\mu(s(u_m)) - \mu_\infty]\, s^2(u_m(t))\, dx + \int_\Omega s^2(u_m(t))\, dx$$

où μ_∞ est donné par (1.2). On a ensuite, à partir de (3.4)

$$\frac{1}{2}\frac{d}{dt}|u_m(t)|^2 + 2\mu_\infty \|u_m(t)\|^2 \leq |f(t)|\, |u_m(t)| + 2\int_\Omega [\mu(s(u_m)) - \mu_\infty]\, s^2(u_m(t))\, dx. \tag{3.5}$$

On pose $\mu_1 = \mu(s) - \mu_\infty$. Par (1.2), $\mu_1(s) \to 0$ quand $s \to 0$, donc pour tout $\varepsilon > 0$, il existe $\beta > 0$ tel que si $|s(u_m(t))| > \beta$, on ait

$$|\mu_1(s(u_m))| < \varepsilon \text{ pour tout } (x,t) \in \Omega \times (O, T_m). \tag{3.6}$$

Soit $t \in [0, T_m]$ fixé, on considère les deux sous ensembles de Ω

$$\Omega_1^t = \{x \in \Omega | s(u_m(x,t)) \leq \beta\}$$
$$\Omega_2^t = \{x \in \Omega | s(u_m(x,t)) > \beta\}.$$

Avec ces notations et en utilisant (3.6), on a

$$\begin{aligned}\int_\Omega |\mu_1(s(u_m(x,t)))| s^2(u_m(x,t))\, dx &\leq \beta^2 \int_{\Omega_1^t} |\mu_1(s(u_m(x,t)))|\, dx + \varepsilon \int_{\Omega_2^t} s^2(u_m(x,t))\, dx \\ &\leq (\mu_\infty + M)\beta^2 \,\mathrm{mes}\Omega + \varepsilon \|u_m\|^2\end{aligned}$$

Ceci , remplacé dans (3.5) donne

$$\frac{1}{2}\frac{d}{dt}|u_m(t)|^2 + 2\mu_\infty \|u_m(t)\|^2 \leq |f(t)||u_m(t)| + 2(\mu_\infty + M)\beta^2 \,\mathrm{mes}\Omega + \varepsilon \|u_m(t)\|^2$$

On choisit $\varepsilon = \dfrac{1}{2}\mu_\infty$ et on pose

$$2(\mu_\infty + M)\beta^2 \,\mathrm{mes}\Omega = C(\Omega, \mu_\infty, M).$$

Alors

$$\frac{1}{2}\frac{d}{dt}|u_m(t)|^2 + 2\mu_\infty \|u_m(t)\|^2 \leq \frac{C_\Omega^2}{2\mu_\infty}|f(t)|^2 + \frac{1}{2}\mu_\infty \|u_m(t)\|^2 + C(\Omega, \mu_\infty, M)$$

où C_Ω est la constante de Poincaré de Ω. En intégrant en t on obtient les estimations a priori

$$\|u_m\|^2_{L^\infty(0,T;H)} + \mu_\infty \int_0^T \|u_m(t)\|^2\, dt \leq \frac{C_\Omega^2}{2\mu_\infty}|f|^2_{L^2(0,T;H)} + \|u_0\|^2 + TC(\Omega,\mu_\infty,M) \leq C_1(T)$$

pour tout T fini, la constante $C_1(T)$ dépendant évidemment de T. On a ainsi

$$u_m \in \text{ borné de } L^\infty(0,T;H) \cap L^2(0,T;V).$$

La suite de la démonstration suit pas à pas celle du théorème d'existence pour les équations de Navier-Stokes . On l'esquissera seulement ici et on renvoie le lecteur à J. L. Lions [2] pour les détails. Tout d'abord on vérifie que l'on a

$$b(u_m,u_m,v) = (Bu_m,v) \text{ avec } Bu_m \in \text{ borné } de L^1(O,T;V'),$$
$$\mathcal{A}u_m \in \text{ borné } de L^2(0,T;V')$$

et enfin

$$u'_m = P_m B u_m - P_m \mathcal{A} u_m + P_m f$$

où P_m est l'opérateur de projection de H sur W_m l'espace engendré par $\{w_1,...w_m\}$. On a par conséquent

$$u'_m \in \text{ borné } de L^1(0,T;V').$$

On a alors, après extraction eventuelle de sous suites

$$u_m \rightharpoonup \text{ dans } L^2(0,T;V) \text{ faible}$$
$$u_m \rightharpoonup \text{ dans } L^\infty(0,T;H) \text{ faible } \star$$
$$u_m \to \text{ dans } L^2(0,T;H) \text{ fort et p.p dans } \Omega.$$

De plus

$$\mathcal{A}u_m \rightharpoonup \chi \text{ dans } L^2(0,T;V') \text{ faible.}$$

On peut alors passer à la limite dans (3.3) pour obtenir

$$(u'(t),v) + (\chi(t),v) + b(u(t),u(t),v) = (f(t),v), \quad \forall v \in V.$$

Enfin, comme l'opérateur $\mathcal{A}$ est monotone, en utilisant le raisonnement de Minty, on a

$$(\chi(t),v) = (\mathcal{A}u(t),v), \quad \forall v \in V$$

ce qui achève la démonstration du théorème.

Dans le cas particulier où $m \neq 0$, on a existence globale de la solution. On prouve le résultat suivant:

Théorème 3.1. *Soient $f \in L^2(0,\infty;H), u_0 \in H$ et μ vérifiant les hypotèses H.1 - H.3. On suppose que*

(3.6) $$m \neq 0.$$

Alors il existe une solution u du problème (1.5) telle que

$$u \in L^2(0,\infty;V) \cap L^\infty(O,\infty;H).$$

Démonstration: On reprend la démonstration du théorème précédent. L'inégalité (3.6) ne permet toujours pas de prendre $T = +\infty$. On va estimer de manière plus fine le deuxième terme de (3.3). En effet, on a

$$\int_\Omega |\mu(s(u_m(x,t)))| s^2(u_m(x,t))\,dx \geq m \int_\Omega s^2(u_m(x,t))\,dx.$$

Il résulte alors de (3.3)

$$\frac{1}{2}\frac{d}{dt}|u_m(t)|^2 + 2m \int_\Omega s^2(u_m(x,t))\,dx \leq \int_\Omega f(t)\,u_m(x,t)\,dx.$$

Le même calcul que celui effectué dans la démonstration du Théorème 3.1 conduit ensuite aux estimations a priori

$$\|u_m\|_{L^\infty(0,\infty;H)} \leq C$$
$$\|u_m\|_{L^2(0,\infty;V)} \leq C$$

où la constante C dépend uniquement de μ_∞, de u_0, de m et de $\|f\|_{L^2(0,\infty;H)}$. Le passage à la limite se fait ensuite comme dans le cas précédent.

4 – Unicité et Régularité des Solutions

On donne ici quelques résultats concernant l'unicité, la régularité et la stabilité à l'infini des solutions. Il faut remarquer que les fluides introduits ici, définis par la loi (1.1) avec μ vérifiant les hypothèses H.1-H.3, se comportent " presque" comme les fluides newtoniens: on aura ainsi unicité dans le cas bidimensionnel pour $u \in L^2(0,T;V) \cap L^\infty(O,T;H)$. De même, dans le cas tridimensionnel l'unicité est prouvée dans l' espace $L^8(O,T;[L^4(\Omega)]^3$ qui n'est pas celui où on a existence des solutions. On commence par la dimension 2.

Théorème 4.1. *Soient $\Omega \subset \mathbb{R}^2$ et μ vérifiant H.1, H.2 et H.3'-(2.11) avec $m \neq 0$. Alors sous les hypothèses du Théorème 3.1 la solution faible u du problème (1.10) est unique et de plus*

$$u \in [L^4(\Omega \cdot (0,T))]^2 \qquad (4.1)$$
$$u' \in L^2(0,T;V').$$

Démonstration: Soient u_1 et u_2 deux solutions du problème (1.5). On pose $u_1 - u_2 = w$. Il vient alors

$$\frac{1}{2}\frac{d}{dt}|w(t)|^2 + (\mathcal{A}u_1 - \mathcal{A}u_2, w) + b(w,u_1,w) = 0. \qquad (4.2)$$

On a d'après (2.2)' ou (2.3)' et les hypothèses

$$(\mathcal{A}u_1 - \mathcal{A}u_2, w) \geq C_0\|w\|^2$$

avec $C_0 = m$ si $\mathcal{E} = \mathbb{R}_+$. La suite de la démonstration est identique à celle du théorème précédent. On majore tous les termes de (4.2) en utilisant en particulier l'inégalité de Ladyzhenskaya

$$\|u(t)\|^2_{[L^4(\Omega)]^2} \leq \sqrt{2}|u(t)|\,\|u(t)\| \tag{4.3}$$

pour aboutir à

$$\frac{1}{2}\frac{d}{dt}|w(t)|^2 + C_0\|w(t)\|^2 \leq \|w(t)\|\,|w(t)|\,\|u_1(t)\|$$

puis, en utilisant le lemme de Gronwall, on a $w = 0$.

Il reste à démontrer (4.1). De (4.3), compte tenu des propriétés (3.1) de u on obtient

$$\left(\int_0^T \|u(t)\|^4_{[L^4(\Omega)]^2}\right)^{\frac{1}{4}} \leq \sqrt{2}\,|u(t)|_{L^\infty(0,T;H)}\,\|u(t)\|_{L^2(0,T;V)}.$$

Ensuite, de (1.5) on a

$$u'(t) = f(t) - \mathcal{A}u(t) - Bu(t) \quad \text{dans } V' \tag{4.4}$$

Or, pour tout $v \in V$,

$$(\mathcal{A}u(t), v) \leq 2M\|u(t)\|\,\|v\|, \quad \forall t \in [0,T].$$

Ainsi $\mathcal{A}u \in L^2(0,T;V')$. Par ailleurs

$$b(u,u,v) = (Bu, v)$$

avec $Bu \in L^2(0,T;V')$. Avec toutes ces informations, de (4.4) on obtient (4.1).

Dans le cas de la dimension 3, on prouve

Théorème 4.2. *Soient $\Omega \subset \mathbb{R}^3$. On se donne f, u_0, et μ satisfaisant les hypothèses du Théorème 4.1. Si la solution u donnée par le Théorème 3.1 vérifie*

$$u \in L^8(0,T;[L^4(\Omega)]^3, \tag{4.4}$$

alors u est unique.

Démonstration: On ne la donne pas ici car elle est identique à celle pour les équations de Navier-Stokes en dimension 3.

Remarque 4.3. *Comportement pour $t \to \infty$.* L'unicité est démontrée dans le cas où $m \neq 0$. Remarquons que cette condition intervient aussi dans le Théorème 3.2 donnant l'existence globale. Si on suppose $f = 0$ et $\mathcal{E} = \mathbb{R}_+$, on a de (1.6)

$$\frac{1}{2}\frac{d}{dt}|u(t)|^2 + \frac{2m}{C_\Omega^2}|u(t)|^2 \leq 0$$

ce qui montre que $|u(t)|$ converge vers 0 exponentiellement lorsque $t \to \infty$, donc, tout comme pour les équations de Navier-Stokes, on a stabilité à l'infini.

BIBLIOGRAPHIE

[1] *K. Aziz, G. Govier – The Flow of Complex Mixtures in Pipes*, Van Nostrand Reinhold Company, 1972.

[2] *H. Brezis – Inéquations variationnelles relatives à l'opérateur de Navier-Stokes*, J. Math. Anal. Appl., **39**(1972).

[3] *S. Boujena – Étude d'une classe de fluides non newtoniens, les fluides newtoniens généralisés*, Thèse 3ème Cycle, Université Pierre et Marie Curie, 1986.

[4] *D. Cioranescu – Sur une classe de fluides non newtoniens*, Appl. Math. and Optimization, **3** (1977), 263–282.

[5] *D. Cioranescu, J. Saint Jean Paulin – Problèmes mathématiques en mécanique des fluides non newtoniens*, à paraître.

[6] *G. Duvaut, J.L. Lions – Les inéquations en mécanique des fluides*, Dunod, 1972.

[7] *O.A. Ladyzhenskaya – The Mathematical Theory of Viscous Incompressible Flow*, Gordon Breach, New York, 1969.

[8] *J.L. Lions – Quelques méthodes de résolution des problèmes aux limites non linéaires*, Dunod, 1969.

[9] *M. Reiner – A mathematical theory of dilatancy*, Amer. J. Math., **67** (1945), 350–362.

[10] *R.S. Rivlin – The hydrodynamics of non-newtonian fluids*, Proc. Roy. Soc. London (A), **193** (1948), 260–281.

[11] *P.E. Sobolevsky – Investigation of the asymptotically newtonian model of fluids*, Soviet Math. Dokl., **1** (1983), 95–98.

[12] *W.L. Wilkinson – Non-newtonian Fluids*, Pergamon Press, 1960.

Doina Cioranescu,
Laboratoire Analyse Numérique, Université Pierre et Marie Curie,
Tour 55-65, 5ème étage, 4 place Jussieu,
75252 PARIS Cedex 05 - FRANCE

Two Dimensional Incompressible Fluid Flow with Singular Initial Data

GEORGES-HENRI COTTET

Abstract

Significative new results have been obtained recently in the analysis of two dimensional incompressible fluids, in particular by Di Perna-Majda and Delort. This paper first reviews the classical analysis and then give some essential features in the concept of concentration and details the key steps in the proof of global existence for positive vortex sheet.

1 – Introduction

The purpose of these notes is to review some recent results concerning the mathematical theory of the two dimensional incompressible Euler and Navier-Stokes equations when the initial data are singular. More precisely we are going to focus on the case when the initial vorticity is a measure supported either on points or on curves. The motivation for that study is that in many interesting flows the generation of vorticity occurs on very small sets, typically at the surface of an obstacle; if the viscosity is small, as time goes on the support of the vorticity remains small, as it is the case for wakes or detached boundary layers. It is therefore natural to study the Navier-Stokes equations with an initial vorticity concentrated on a set of Lebesgue measure zero. As we will see this problem is relatively easy to deal with. However when the viscosity tends to zero we have to deal with a singular perturbation problem. The a *priori* estimates available for the Navier-Stokes equations obviously fail to be uniform with respect to the viscosity. If one has available a model for generating vorticity in the limit case of vanishing viscosity and expects that the Euler equations provide a good model for the evolution of this vorticity, the question of the well-posedness of these equations for singular initial data arises. In particular if one starts with a vorticity field concentrated on a curve, it is important to know whether more severe singularities can develop. This problem has received recently a lot of interest from applied mathematicians because it meets fundamental questions in fluid dynamics and requires sophisticated tools from applied analysis.

For the remainder of this discussion we will restrict our attention to the Euler equations with measure initial vorticity ω_0 supported on a curve Γ, to which we will refer to as the vortex sheet problem:

$$\frac{\partial \omega}{\partial t} + \nabla \cdot (u\omega) = 0 \tag{1.1}$$

$$\omega(\cdot, 0) = \omega_0 \tag{1.2}$$

$$\operatorname{div} u = 0 \tag{1.3}$$

$$\operatorname{curl} u = \omega \tag{1.4}$$

$$|u| \overset{\infty}{\to} 0, \tag{1.5}$$

where $\omega_0 = \alpha\delta_\Gamma$. C. Sulem, P.L. Sulem, C. Bardos and U. Frisch ([9]) have shown that in the analytic case (that is if Γ and the density α are analytic) a vortex sheet solution remains for small time. This critical time reflects the fact that in the linear stability analysis, the band width of analyticity decays linearly and therefore vanishes after sometime. This suggests that (1.1)-(1.5) is ill-posed in the classical framework of Sobolev spaces, a difficulty which has been confirmed in [5]. In this paper Duchon and Robert construct a vortex sheet solution which is analytic for all time $t < 0$ but with a possible breakdown in the curvature at time 0 (see also [1] for an independent construction). In particular L^∞ estimates on the velocity which, along with the straightforward compactness in L^1, would allow to pass to the limit in the equations and give a classical weak solution, although they are correct at time 0 if Γ is smooth enough, are probably out of reach for any positive time. On the other hand Krasny ([8]) has suggested by careful numerical experiments involving successive refinements in a vortex method that, for analytic initial perturbation of a flat vortex sheet, a weak solution might persist after the critical time derived from the linear stability analysis. However the meaning of this solution and how it could be obtained through a classical regularization limiting process was far from clear. In a series of papers (see [4] and the references therein) DiPerna and Majda have addressed this problem and recognized for the first time the importance of concentration in this limiting process. Concentration is a term to denote the possibility that when ω_0^ε is a sequence of C^∞ functions approaching ω_0, the related solutions u^ε tend to a function u in the sense of distributions while the energy of u is strictly smaller than the energy of u^ε. In other words, part of the energy has been lost in the limiting process. They give elementary steady state illustrations of this possibility. Moreover they show that this concentration phenomenon occurs only on small space-time sets, and the control they provide on the size of this set is sharp, as proved by examples. This gives a very important qualitative information on the kind of singularities which can appear and allows to prove that in the steady-state case the limit is a weak solution of the Euler equations. However this is not enough to prove existence of solutions for the time dependent problem.

More recently, using a more direct approach, Delort has shown that despite the phenomenon of concentration, it is still possible to pass to the limit inside the equation if the vorticity has a distinguished sign. He used the fact that only some quadratic quantities are present in the non-linear term of (1.1), and certainly not the energy itself. He also gives a counter example suggesting that if the vorticity changes sign this result is no longer true. This must be compared to a similar, but much simpler to establish, result concerning point vortices: if ω_0 is a linear combination of Dirac measures with positive weights, the hamiltonian structure of the system of underlying ODE combined with the conservation of the kinetic momentum allows to prove that the collapse of vortices cannot happen and thus there is a point vortices solution for all time.

Let us finally mention that, if there are still a lot of questions concerning the existence of solutions, the question of uniqueness is far behind since it is established only for bounded initial vorticity. This indicates that mathematical questions will remain open for long time in this field. The outline of this paper is as follows. In section 2 we give two results which can be obtained by classical compactness arguments, namely the existence of solutions to the Euler equations when the vorticity is in L^p for some $p > 1$ and the existence of solutions to the Navier-Stokes equations for any measure initial vorticity. In section 3 we describe the concentration approach by DiPerna-Majda and in section 4 we give the essential steps in the proof of Delort. Generally speaking, we tried to give a self contained introduction to the field; we will provide detailed proofs in particular when either they do not seem to appear in the litterature (in general because they are routine; the appendix is devoted to the proof of two such results) or when we think it is possible to simplify the original proofs, which in some instances is possible by combining the point of view of different authors. In this context we wish to mention the recent monograph by Evans ([6]) which is a detailed survey of functional analysis tools for dealing with weak convergence in non-linear PDE's.

2 – The simple cases for Euler and Navier-Stokes equations

To begin with let us rewrite the Euler equations with a more explicit dependance of u on ω. Let K be the kernel $K(x) = (2\pi|x|^2)^{-1}(-x_2, x_1)$ for $x = (x_1, x_2)$. Then we have

Lemma 2.1. *Assume either $\omega \in L^1(\mathbb{R}^2)$ and has compact support or $\omega \in L^1(\mathbb{R}^2) \cap L^p(\mathbb{R}^2)$ for some $p > 2$. Then (1.3)-(1.5) is equivalent to*

$$u = K \star \omega. \tag{2.1}$$

We give in the appendix a proof of the fact that (2.1) implies that u has the correct behavior at infinity.

From now on we are going to focus on the form (1.1),(1.2),(2.1) of the equation. However we must keep in minds that if $\omega_0 \in L^1(\mathbb{R}^2) \cap L^p(\mathbb{R}^2)$ for $1 < p < 2$ then it is not possible to make sure that the velocity vanishes at infinity. Even if ω_0 has compact support, since the velocity cannot be expected to be bounded, it is not possible to ensure that the support of the vorticity will remain bounded for positive time. We will use the classical notation $W^{k,p}$ for the space of functions which have their derivatives up to the order k in $L^p \equiv L^p(\mathbb{R}^2)$, and denote by $\|\cdot\|_{k,p}$ and $|\cdot|_{k,p}$ their usual norm and semi-norm. In the following lemma we summarize classical elliptic regularity results regarding (2.1).

Lemma 2.2. *If $f \in L^1 \cap L^p$ for some $p > 1$ then $K \star f \in L^q$ where $q \in]2, +\infty]$ satisfies $1/q = \min\{1/p - 1/2, 0\}$ and the derivatives of order 1 of $K \star f$ are in L^p. The following estimates hold:*

$$\|K \star f\|_{0,q} \leq C\|f\|_{0,p}, \tag{2.2}$$

$$|K \star f|_{0,p} \leq C\|f\|_{0,p}. \tag{2.3}$$

If $f \in L^1$ then for any bounded set Ω in $\mathbb{R}^2$ the following estimate holds

$$\int_\Omega |K \star f(x+y) - K \star f(x)|dx \leq C|y|[1 + \log|y|]\|f\|_{0,1} \tag{2.4}$$

The last assertion is proved in the appendix. However let us observe that this "quasi BV" estimate is just the integral equivalent of the more classical "quasi lipschitz" estimate. Note immediately that (2.4) implies that

$$\limsup_{y\to 0} \int_\Omega |K \star f(x+y) - K \star f(x)| dx = 0$$

which yields the compactness of the convolution by K as a mapping from L^1 into L^1_{loc}. With these preliminary estimates we can prove

Theorem 2.1. *Assume $\omega_0 \in L^1 \cap L^p$ for some $p > 1$. Then for all time $T > 0$ there exists a solution to (1.1),(1.2),(2.1) with $\omega \in L^\infty(0,T; L^1 \cap L^p)$.*

Proof: We focus here on the most singular case $1 < p < 2$. Let $\omega_0^\varepsilon \in C^\infty(\mathbb{R}^2)$ a regularization of ω_0 and $u^\varepsilon, \omega^\varepsilon$ the classical solution of (1.1),(1.2),(2.1) with this data (see the remark below on how to avoid to start with classical solutions). Writing for instance the conservation of ω^ε along the characteristics and using the fact that volumes are conserved along these characteristics we easily obtain the following a priori estimate, valid for all time:

$$\|\omega^\varepsilon(\cdot,t)\|_{0,p} = \|\omega_0^\varepsilon\|_{0,p} \leq \|\omega_0\|_{0,p}.$$

Let Ω a bounded domain in $\mathbb{R}^2$ and q such that $1/q = 1/p - 1/2$. By (2.2),(2.3) we have

$$u^\varepsilon \text{ is bounded in } L^\infty(0,T; W^{1,p}(\Omega)). \tag{2.5}$$

Using the equation we also get for $r = q/2$

$$\|\frac{\partial \omega^\varepsilon}{\partial t}\|_{-2,r} \leq \||u^\varepsilon|^2\|_{0,r} \leq \|u^\varepsilon\|_{0,q}^2 \leq \|\omega^\varepsilon\|_{0,p}^2.$$

By (2.3) we thus have

$$\frac{\partial u^\varepsilon}{\partial t} \text{ is bounded in } L^\infty(0,T; W^{-1,q/2}(\Omega)). \tag{2.6}$$

If now q' is such that $q/2 \leq q' < q$, we have the following diagram

$$W^{1,p}(\Omega) \subset L^{q'}(\Omega) \subset W^{-1,q/2}(\Omega)$$

where the first imbedding is compact. In view of (2.5) and (2.6) we are in a situation where classical compactness results allow us to extract a subsequence, still denoted by u^ε, satisfying

$$u^\varepsilon \to u \text{ in } L^2(0,T; L^{q'}_{loc}(\mathbb{R}^2))$$

To prove that u is solution of the Euler equation, we use the weak formulation of the equation (1.1): for all $\phi \in C_0^\infty(\mathbb{R}^2 \times [0,T[)$

$$\int_0^T \int u^\varepsilon \frac{\partial}{\partial t}(\operatorname{curl}\phi) + \sum_{j=1,2} u_j^\varepsilon u^\varepsilon \frac{\partial}{\partial x_j}(\operatorname{curl}\phi) dx dt = -\int u_0^\varepsilon \operatorname{curl}\phi(\cdot,0) dx$$

The only trouble could come for the quadratic terms $u_i^\varepsilon u_j^\varepsilon$. But we have $q > 2$ so we can choose $q' = 2$ and $u^\varepsilon \to u$ in $L^2(0,T;L^2_{loc}(\mathbb{R}^2))$. We can therefore pass easily to the limit in quadratic terms. ∎

Remark. It is actually not even necessary to assume the existence of classical solutions. If $\omega_0 \in L^1 \cap L^p$ for some $p > 4/3$ then we can consider a linearization of (1.1),(1.2),(2.1), for instance by an explicit time discretization of the equation, where the velocity is evaluated at the beginning of each time step. Then we can use the above arguments to pass to the limit. The only difference lies in the fact that in this situation we cannot write $\omega_\varepsilon = \operatorname{curl} u_\varepsilon$, because we do not solve exactly the Euler equation. Therefore we cannot reduce the nonlinear terms to a combination of terms like $u_i^\varepsilon u_j^\varepsilon$. Instead it is necessary to be able to pass to the limit in the term $u_\varepsilon \omega_\varepsilon$ which requires to have an a priori estimate fro the vorticity in some L^p for $p > 4/3$ (actually the value 4/3 is readily seen to be the minimal value for which it is possible to give sense to $u_\varepsilon \omega_\varepsilon$) . The general case can then be recovered by the above proof. For an approach through classical solutions when the vorticity is bounded, we refer to the seminal paper [10].

Let us now turn to the the Navier-Stokes equation:

$$\frac{\partial \omega}{\partial t} + \nabla \cdot (u\omega) - \nu \Delta \omega = 0 \tag{2.7}$$

$$\omega(\cdot, 0) = \omega_0 \tag{2.8}$$

$$u = K \star \omega \tag{2.9}$$

It is possible to prove the existence for any viscosity of a solution to this problem as soon as ω_0 is a Radon measure ([2]).

Theorem 2.2. *Assume ω_0 is a bounded Radon measure. Then (2.7)-(2.9) has at least one solution $\omega \in L^2(0,T;L^2(\mathbb{R}^2))$.*

Proof: As above we start by regularizing the initial condition and we look for a priori estimates on ω^ε. If $\|\cdot\|_M$ denotes the norm in the space of bounded measure, first observe that we have easily

$$\|\omega^\varepsilon(\cdot,t)\|_{0,1} \leq \|\omega^\varepsilon(\cdot,0)\|_{0,1} \leq \|\omega_0\|_M.$$

Let us now prove that there exists a constant C such that

$$\|\omega^\varepsilon(\cdot,t)\|_{0,2} \leq \frac{C}{\sqrt{\nu t}} \|\omega_0\|_M. \tag{2.10}$$

We classically multiply (2.10) by ω^ε and integrate by parts to find

$$\frac{d}{dt}\|\omega^\varepsilon(\cdot,t)\|_{0,2}^2 + 2\nu \|\nabla \omega^\varepsilon(\cdot,t)\|_{0,2}^2 = 0.$$

We introduce the Fourier transform

$$\hat{\omega}^\varepsilon(\xi,t) = \frac{1}{2i\pi}\int \exp\left(-i < x, \xi >\right) \omega^\varepsilon(x,t)dx,$$

so that

$$\|\nabla\omega^\varepsilon(\cdot,t)\|_{0,2}^2 = \int |\xi|^2 |\hat{\omega}^\varepsilon(\xi,t)|^2 d\xi.$$

To bound from below the above integral we split the Fourier space into low and high frequencies as follows

$$\mathbb{R}^2 = S_t \cup S_t^c \; ; \; S_t = \left\{\xi, |\xi| \le (\nu t)^{-1/2}\right\}$$

and we write

$$\int_{S_t^c} |\xi|^2 |\hat{\omega}^\varepsilon(\xi,t)|^2 d\xi \ge \frac{1}{\nu t}\int_{S_t^c} |\hat{\omega}^\varepsilon(\xi,t)|^2 d\xi.$$

Thus

$$\int_{\mathbb{R}^2} |\xi|^2 |\hat{\omega}^\varepsilon(\xi,t)|^2 d\xi \ge \frac{1}{\nu t}\int_{\mathbb{R}^2}\left[|\hat{\omega}^\varepsilon(\xi,t)|^2 d\xi - \int_{S_t} |\hat{\omega}^\varepsilon(\xi,t)|^2 d\xi\right]. \tag{2.11}$$

Since

$$\|\hat{\omega}^\varepsilon(\cdot,t)\|_{0,\infty} \le C\|\omega^\varepsilon(\cdot,t)\|_{0,1} \le C\|\omega_0\|_M,$$

this yields

$$\int_{\mathbb{R}^2} |\xi|^2 |\hat{\omega}^\varepsilon(\xi,t)|^2 d\xi \ge \frac{1}{\nu t}\int_{\mathbb{R}^2} |\hat{\omega}^\varepsilon(\xi,t)|^2 d\xi - \frac{C}{\nu^2 t^2}\|\omega_0\|_M \tag{2.12}$$

Combined with (2.11) and the Parseval identity this leads to the following differential inequality

$$\frac{d}{dt}\|\omega^\varepsilon(\cdot,t)\|_{0,2}^2 + \frac{2}{t}\|\omega^\varepsilon(\cdot,t)\|_{0,2}^2 \le \frac{C}{\nu t^2}\|\omega_0\|_M^2$$

If we set $y^\varepsilon(t) = \nu t^2\|\omega^\varepsilon(\cdot,t)\|_{0,2}^2$, we thus have $\dot{y}^\varepsilon \le C\|\omega_0\|_M^2$. Since in addition $y^\varepsilon(0) = 0$ we get the desired estimate (2.10). From (2.10) and lemma 2.1 we now have

$$u^\varepsilon \quad \text{is bounded in } \; L^p(0,T;W^{1,2}(\Omega)) \tag{2.13}$$

for all $p \in]1,2[$ and all bounded set Ω. Using the equation we also obtain

$$\frac{\partial u^\varepsilon}{\partial t} \text{ is bounded in } L^p(0,T;W^{-1,2}(\Omega)). \tag{2.14}$$

Classical compactness arguments allow then to extract a subsequence still denoted by u^ε such that, for instance

$$u^\varepsilon \to u \text{ in } L^{3/2}(0,T;L^2_{loc}(\mathbb{R}^2)),$$

from which it is possible to pass to the limit in the quadratic terms of the weak formulation of (2.7), as indicated in the proof of Theorem 2.1. ∎

Although this result has clearly a limited meaning (estimates are of course non uniform with respect to ν), it does check that the non linear terms do not interfere with the expected regularizing effect of the diffusion.

3 – Concentration and the problem of the vortex sheet

In this section and the following we are concerned with (1.1),(1.2),(2.1) in the case $\omega_0 = \alpha\delta_\Gamma$, where Γ is a smooth curve and α is a bounded compactly supported function on Γ. If $u_0 = K \star \omega_0$ then clearly $u_0 \in L^p$ for all $p \in]2, \infty]$. Let $u^\varepsilon, \omega^\varepsilon$ the sequence of approximate solutions produced either from a regularization ω_0^ε of the initial condition and the resolution of the Euler equation or by the resolution of the Navier-Stokes equations starting from the original initial condition with viscosity ε, the existence of which is guaranteed by Theorem 2.2. Let us first review the natural a *priori* estimates. We have obviously

$$\int |\omega_\varepsilon(x)| dx \leq \int_\Gamma |\alpha(x)| dx \tag{3.1}$$

Another expected invariant is the energy. However this one must be taken with care: since the velocity decays like $(\int \omega_0)/r$ at infinity the energy is not finite except if the vorticity has mean value 0. To overcome this technical difficulty, we introduce a smooth, radial function $\bar{\omega}$ and we split the initial vorticity into a smooth and singular part as follows

$$\omega_0 = (\int \omega_0)\, \bar{\omega} + \omega_0'.$$

We have $\int \omega_0' = 0$ and therefore $\int |u_0'|^2 < \infty$, where $u_0' = K \star \omega_0'$ (to check this assertion, observe that $u' \in L^2_{loc}$ and that u decays like r^{-2} at infinity). Moreover if we come back to the velocity formulation of the equations we observe that, if $\bar{u} = K \star \bar{\omega}$ then $(\bar{u}.\nabla)\bar{u} = 0$, for $\bar{\omega}$ is radial, and $u = \bar{u} + u'$ where u' is solution of

$$\frac{\partial u'}{\partial t} + (u.\nabla)u' = -(u'.\nabla)\bar{u} + \nabla p'.$$

We now multiply by u' and integrate to find , after cancellations resulting from the fact that u and u' are divergence free

$$\frac{d\|u'\|_{0,2}^2}{dt} \leq |\bar{u}|_{1,\infty} \|u'\|_{0,2}^2.$$

Due to the smoothness of $\bar{u}$ this implies that

$$\|u'(t)\|_{0,2} \leq C(T) \|u_0'\|_{0,2},$$

for all $t \leq T$. Recall now that $\bar{u}$ is bounded in L^2_{loc}. We thus get a bound for u in L^2_{loc}. This argument of course applies as well to the regularized solutions to give the following a priori estimate:

$$u_\varepsilon \text{ is bounded in } L^\infty(0, T; L^2_{loc}(\mathbb{R}^2)). \tag{3.2}$$

It is therefore possible to fin a subsequence still denoted by u_ε such that

$$u_\varepsilon \rightharpoonup u \text{ weakly in } L^\infty(0, T; L^2_{loc}(\mathbb{R}^2)).$$

However to pass directly to the limit in the equation (1.1) would require stronger convergence. DiPerna and Majda give the following very simple steady state example illustrating the fact that strong convergence in L^2 cannot be ensured in general. Let ω be a smooth non zero radial function with integral equal to 0, and let $\omega^\varepsilon(x) = \varepsilon^{-2}\omega(x/\varepsilon)$. Then $(u^\varepsilon = K \star \omega^\varepsilon, \omega^\varepsilon)$ is a steady state solution of the Euler equation, which clearly satisfies:

$$\omega^\varepsilon \rightharpoonup \Big(\int \omega\Big)\delta_{\{x=0\}} = 0.$$

Moreover since $\int \omega = 0$, $u = K \star \omega$ has finite energy and $u^\varepsilon(x) = \varepsilon^{-1}u(x/\varepsilon)$, so that

$$|u^\varepsilon|^2 \rightharpoonup \Big(\int |u|^2\Big)\delta_{\{x=0\}} \neq 0.$$

Let us first notice that the concentration compactness framework developped by P.L. Lions is not suitable here, because the logarithmic correction in (2.4) prevents the derivatives of the velocity from being integrable. To give a systematic account for the loss of energy in the limiting process that this example illustrates, DiPerna-Majda introduce instead the concept of reduced defect measure. Let u^ε be a sequence converging weakly in L^2_{loc} towards u. Let E a borel set in $\mathbb{R}^2 \times \mathbb{R}_+$. One defines the reduced defect measure of E by setting

$$\theta(E) = \limsup_{\varepsilon\to 0} \int_E |u^\varepsilon - u|^2 dx$$

Obviously θ is finite on all bounded sets and $\theta(E) = 0$ iff $u^\varepsilon \to 0$ strongly in $L^2(E)$. A concentration set for the sequence u^ε is a set which is a countable union of null sets of θ. To measure the size of concentration sets DiPerna and Majda use the concept of Hausdorff dimension:

Definition 1 [4], [7]. Let $\gamma > 0$. The sequence (v^ε) is said to concentrate on a set of Hausdorff dimension less than or equal to γ if for all $\gamma' > \gamma$ there exists $\delta > 0$ and a countable family of balls B_i with radius $r_i < \delta$ such that

$$\theta\Big(\big(\bigcup_i B_i\big)^c\Big) = 0, \quad \sum_i r_i^\gamma \leq \delta\ .$$

Then we have

Theorem 3.1. *Let (u^ε) a sequence satisfying (3.1)-(3.2). There is a subsequence such that either the convergence is strong in L^2_{loc} or it concentrates on a space time set of Hausdorff dimension* 1.

The second possibility corresponds somehow to a time-dependent generalization of the example seen above. In particular concentrations on sets with project for a fixed time onto a curve are forbidden. The proof of this essential result is twofold. In [4] it is proved that for a subsequence concentration occurs on a set of Hausdorff dimension ≤ 1. Then Greengard and Thomam observed that the dimension cannot be strictly less than 1, or else convergence is strong. Let us reproduce here their brief argument. Assume the sequence u^ε concentrates on a set of Hausdorff dimension strictly less than 1. Let Ω a bounded set in $\mathbb{R}^2$ and $\delta > 0$; there exists a countable family of balls B_i in $\Omega \times [0,T]$ with radius r_i such that $r_i < \delta$, $\sum_i r_i < \delta$ and, if $U = \cup_i B_i$:

$$\lim_{\varepsilon\to 0} \int_{U^c} |u^\varepsilon - u|^2 dx dt = 0.$$

If S denotes the projection of U onto the time axis we can write

$$\int_{\Omega\times[0,T]} |u^\varepsilon - u|^2 dxdt = \int_{\Omega\times S} |u^\varepsilon - u|^2 dxdt + \int_{\Omega\times S^c} |u^\varepsilon - u|^2 dxdt.$$

Next we observe that, from its definition, S can be covered by intervals with total length less than δ. Thus, by (3.2), the first integral in the right hand side above is bounded by $C\delta$. As for the second, since $\Omega \times S^c \subset U^c$ where convergence is strong, it is smaller than δ for ε small enough. So we finally get

$$\lim_{\varepsilon\to 0} \int_{\Omega\times[0,T]} |u^\varepsilon - u|^2 dxdt = 0.$$

Observe that in this proof we have used more information than for the definition of concentration sets, namely that L^2 estimates are uniform in time.

Unfortunately the control of the size of concentration sets given by Theorem 3.1, while it yields important qualitative informations on the nature of the singularities which can develop in this problem, is not enough to ensure that the weak limit is solution to the Euler equation, except in the steady state case. Actually, although we do not wish to give the full proof of this partial result, we are going to focus now on the steady state case because it allows to understand some essential features in the phenomenon of concentration. So we are going to prove the following particular case of Theorem 3.1:

Theorem 3.2. *Let (u^ε) be a steady state sequence satisfying (3.1)-(3.2). There is a subsequence which concentrates on a space set of Hausdorff dimension zero.*

The general case (Theorem 3.1) can then be proved by essentially tying together fixed time problems (at the expense of course of technicalities in the control of time derivatives). In the remainder of this section $u^\varepsilon, \omega^\varepsilon$ will denote a steady-state sequence satisfying (3.1)-(3.2). We will also use the stream functions $\psi^\varepsilon = G \star \omega^\varepsilon$, where $G(x) = -(2\pi)^{-1} \log |x|$. Let us first establish the following result

Lemma 3.1. *Let Ω be a bounded domain in $\mathbb{R}^2$ and F a closed subset in Ω such that $\psi^\varepsilon \to \psi$ in $L^\infty(F)$. Then $u^\varepsilon \to u = curl\,\psi$ in $L^2(F)$.*

Proof: First observe that since ω^ε is bounded in L^1, there is a subsequence which converges weakly in M. Let ς a smooth cut-off function satisfying $0 \leq \varsigma \leq 1$, $\varsigma \equiv 1$ in F and $\varsigma \equiv 0$ outside Ω. Let $\eta > 0$ and h_η the function defined by

$$h_\eta(v) = \begin{cases} v & \text{if } |v| \leq \eta \\ \eta \frac{v}{|v|} & \text{if not} \end{cases}$$

By the uniform convergence of ψ^ε to ψ we have $h_\eta(\psi^\varepsilon - \psi) = \psi_\varepsilon - \psi$ on F for ε small enough, which yields

$$\begin{aligned} \int_F |u^\varepsilon - u|^2 dx &= \int_F \varsigma(u^\varepsilon - u)\text{curl}\,(h_\eta(\psi_\varepsilon - \psi))dx \\ &\leq \int_\Omega \varsigma(u^\varepsilon - u)\text{curl}\,(h_\eta(\psi^\varepsilon - \psi))dx \end{aligned}$$

An integration by part thus gives (recall that $\varsigma = 0$ on $\partial\Omega$)

$$\begin{aligned}
\int_F |u^\varepsilon - u|^2 dx &\leq \int_\Omega \operatorname{curl}(\varsigma(u^\varepsilon - u))(h_\eta(\psi^\varepsilon - \psi))dx \\
&\leq \int_\Omega \varsigma(\omega^\varepsilon - \omega)h_\eta(\psi^\varepsilon - \psi)dx + \int_\Omega \operatorname{curl}\varsigma(u^\varepsilon - u)h_\eta(\psi^\varepsilon - \psi)dx \\
&\leq C\eta\|\omega^\varepsilon - \omega\|_M \leq C\eta
\end{aligned}$$

which ends the proof. ∎

If we now remember that under the assumptions (3.1)-(3.2) the sequence u^ε is compact in any L^p_{loc}, $1 < p < 2$, in view of the above lemma we just have to prove the following refinement of Egoroff's theorem

Theorem 3.3 [4]. *Assume that the sequence f_j converges to f in $W^{1,p}(\Omega)$ for all $p \in]1,2[$. Then there exists a subsequence f_k with the following property: for any $\gamma > 0$ and any $\delta > 0$ there exists a closed set F on which f_k converges uniformly to f and whose complement can be covered by a countable union of balls B_i with radius r_i satisfying $r_i \leq \delta$ and $\sum_i r_i^\gamma \leq \delta$.*

With the above theorem and Lemma 3.2, the proof of Theorem 3.2 is now clear. So we now proceed to give the essential steps in the proof of Theorem 3.3. The key ingredient is the following improvement of the Chebishev inequality

Lemma 3.2. *Let $f \in W^{1,p}(\Omega)$ for some $p \in]1,2[$ and $\lambda > 0$. Set $D_\lambda = \{x \in \Omega; |f(x)| \geq \lambda\}$ and fix $\gamma > 2 - p$. There exist a function $r_0(\lambda, |f|_{1,p})$ and a function $M(|f|_{1,p})$ satisfying*

$$\lim_{b \to 0} r(a,b) = 0, \text{ for fixed } a \tag{3.3}$$

$$M(a) > 0 \tag{3.4}$$

and such that, for some $r \leq r_0$, $D_\lambda \subset \{x \in \Omega; r^{-\gamma}\int_{B(x,r)} |\nabla f|^p dx \geq M\}$.

Proof: For the reader's convenience we reproduce here the proof given in [4]. Having in mind to use the Poincaré inequality which controls the oscillation of a function of $W^{1,p}$ around its average, we denote by $B_j(x)$ the ball of radius $r_j = 2^{-j}r_0$ around x and write

$$f = f_0 + \sum_{j \geq 1}(f_{j+1} - f_j)$$

where $f_j(x) = |B_j(x)|^{-1}\int_{B_j} f(y)dy$. We can always write

$$f_{j+1} - f_j = |B_{j+1}|^{-1}\int_{B_{j+1}}(f - f_j)dy,$$

whence

$$|f_{j+1} - f_j| \leq 4|B_j|^{-1}\int_{B_j}|f - f_j|dy$$

and by Hölder's and Poincaré inequalities

$$|f_{j+1} - f_j| \leq 4|B_j|^{-1/p}\left[\int_{B_j}|f - f_j|^p\right]dy \leq 4r_j\left[|B_j|^{-1}\int_{B_j}|\nabla f|^p dy\right]^{-1/p}.$$

Next, by Hölder's and Sobolev inequalities, if q denotes the Sobolev conjugate exponent of p we also have

$$|f_0| \leq \left[|B_0|^{-1}\int_{B_0}|f|^q dy\right]^{1/q} \leq C r_0^{-2/q}|f|_{1,p}.$$

So we obtain

$$|f| \leq C\left[r_0^{-2/q}|f|_{1,p} + \sum_j r_j^{1-2/p}\left(\int_{B_j}|\nabla f|^p dy\right)^{1/p}\right].$$

If the embedding claimed in the lemma was wrong at some point x, we would have

$$\int_{B_j}|\nabla f|^p dy \leq M r_j^{2-p+\delta}$$

for some $\delta > 0$ and all $j \geq 1$. Therefore

$$\lambda \leq C\left[r_0^{-2/q}|f|_{1,p} + M^{1/p}\sum_j r_j^{\delta}\right] \leq C'\left[r_0^{-2/q}|f|_{1,p} + M^{1/p}\right].$$

The choices

$$r_0 = \left(\frac{C'|f|_{1,p}}{3\lambda}\right)^{q/2}, \quad M = (3C')^{-p}\lambda^p$$

clearly lead to an impossibility. ∎

As a consequence of this lemma we can cover the sets D_λ with balls B_i of radius r_i satisfying

$$r_i \leq r_0 \tag{3.5}$$

and $\sum_i r_i^\gamma \leq M^{-1}\sum_i \int_{B_i}|\nabla f|^p dx$. A standard covering argument allows then to extract a family of balls with radius $5r_i$ which cover D_λ without overlapping. For these balls we have now

$$\sum_i r_i^\gamma \leq M^{-1}\int |\nabla f|^p dx \tag{3.6}$$

We can now give the proof of theorem 3.3. First we extract a subsequence such that the series converge:

$$\sum_k \|\nabla f_k - \nabla f\|_{0,p} < \infty \tag{3.7}$$

Then fix $\gamma > 0$, $\delta > 0$ and choose $p > 2 - \gamma$. We mimic the argument which proves Egoroff's theorem and set, for i, n integers

$$A_{n,i} = \{x; |f_n(x) - f(x)| \leq \frac{1}{i}\}\ ;\ B_{n,i} = \bigcap_{m \geq n} A_{m,i}$$

We must find a function $m(n)$ such that, if $F = \bigcap_{n \geq n_0} B_{m(n),n}$, F^c can be covered with balls satisfying the conclusions of theorem 3.3. Observe first that if $x \in F^c$, there exists $n \geq n_0$ and $m \geq m(n)$ such that $|f_m(x) - f(x)| \geq n^{-1}$, so that, in view of Lemma 3.2 and the comment which follows, any choice of $m(n)$ and n_0 gives a countable covering of F^c. Our specific choice of $m(n)$ will be to ensure that their radius along with the sum of their γ-power is smaller than δ. But, due to (3.3), for all n it is possible to find m(n) such that for $m \geq m(n)$, with the notations of Lemma 3.2,

$$r_0(n^{-1}, |f_m - f|_{1,p}) \leq \delta.$$

Moreover, by (3.7), we may also assume that $m(n)$ is such that

$$\sum_{m \geq m(n)} \frac{|f_m - f|_{1,p}}{M(n)} \leq \frac{\delta}{2^n}. \tag{3.8}$$

Thus, if $(B_{m,n,j})_j$ is a covering of $A_{m,n}$ given by Lemma 3.2, then F^c can be covered by the countable union of balls $(B_{m,n,j})_{m \geq m(n),n,j}$, with radius $r_{m,n,j}$ satisfying

$$r_{m,n,j} \leq \delta \quad \text{for } m \geq m(n)$$

and, by (3.6),(3.8)

$$\sum_{m \geq m(n),n,j} r_{m,n} \leq \sum_n \sum_{m \geq m(n)} \frac{|f_m - f|_{1,p}}{M(n)} \leq \sum_n \frac{\delta}{2^n} \leq \delta. \blacksquare$$

In closing this section, we wish to mention that the reference [4] contains much more than indicated here. In particular, to characterize concentration sets, the authors introduce the concept of maximal vorticity function and show how it can be used to recover, outside exceptional sets, elliptic regularity properties for the velocity field of the same nature as in (2.2).

4 – Global existence for positive vortex sheet

We now come to Delort's result:

Theorem 4.1 [3]. *Assume $\alpha \geq 0$. Then for all $T > 0$, equations (1.1),(1.2),(2.1) have at least one solution $u \in L^\infty(0,T; L^2_{loc}(\mathbb{R}^2))$.*

Of course the same results holds for negative vortex sheets. Before seeing where the sign restriction comes in the proof let us state the 2 basic observations on which it is based. Let as usual $(u^\varepsilon, \omega^\varepsilon)$ be a sequence of approximate solutions produced by a (positive) regularization of the initial condition, satisfying (3.1),(3.2). Let us come back to the weak formulation of the equations in the (u,p) formulation. If ϕ is a smooth test function on $\mathbb{R}^2 \times [0,T[$, we need only be able to pass to the limit in

$$I = \int (u^\varepsilon \otimes u^\varepsilon) \cdot \nabla(\operatorname{curl} \phi) dx dt.$$

If we set $\eta = \operatorname{curl} \phi$, we have

$$I = \int (u_1^\varepsilon)^2 \frac{\partial \eta_1}{\partial x_1} + u_1^\varepsilon u_2^\varepsilon \left(\frac{\partial \eta_1}{\partial x_2} + \frac{\partial \eta_2}{\partial x_1}\right) + (u_2^\varepsilon)^2 \frac{\partial \eta_2}{\partial x_2} dx dt.$$

But, since $\operatorname{div} \eta = 0$, this can be rewritten as

$$I = \int [(u_1^\varepsilon)^2 - (u_2^\varepsilon)^2] \frac{\partial \eta_1}{\partial x_1} + u_1^\varepsilon u_2^\varepsilon (\frac{\partial \eta_1}{\partial x_2} + \frac{\partial \eta_2}{\partial x_1}) dx dt.$$

It is therefore sufficient to pass to the limit in some specific non-linear terms, namely $u_1^\varepsilon u_2^\varepsilon$ and $(u_1^\varepsilon)^2 - (u_2^\varepsilon)^2$. In other words, strong convergence in L^2 is not required and concentration alone does not forbid to pass to the limit in the equations. In the sequel we focus on the term $u_1^\varepsilon u_2^\varepsilon$, for the other term can be reduced to this form after a rotation of the axis. Using the integral representation of the velocity in terms of the vorticity, we can write

$$\int u_1^\varepsilon u_2^\varepsilon \phi dx dt = \int H_\phi(x,y) \omega^\varepsilon(x,t) \omega^\varepsilon(y,t) dx dy dt, \tag{4.1}$$

where we have set

$$H_\phi(x,y) = -\frac{1}{4\pi^2} \int \frac{x_1 - z_1}{|x-z|^2} \frac{y_2 - z_2}{|y-z|^2} \phi(z,t) dz dt.$$

The crucial step in the proof is the following

Lemma 4.1. *Let $\phi \in C_0^\infty(\mathbb{R}^2 \times [0,T[)$. Then H_ϕ is continuous away from the diagonal $\{x = y\}$ and bounded in $\mathbb{R}^4$.*

We postpone the proof of this lemma to the end of this section. For now let us just mention that it results from careful estimates where the fact that the kernels have basically mean value zero is used to lower their singularity.

The next step is then to prove that the contribution of ω^ε in the integral (4.1) around the diagonal can be made small. From (3.1), there exists $\omega \in L^\infty(0,T;M)$ (where M still denotes the space of bounded Radon measure) such that when passing if necessary to a subsequence we have

$$\omega^\varepsilon \rightharpoonup \omega \text{ in } L^\infty(0,T;M) \text{ weak star.}$$

In addition we will use the fact, derived from the equation (1.1) itself, that, since $|u^\varepsilon|^2$ is bounded in L^1_{loc},

$$\frac{\partial \omega^\varepsilon}{\partial t} \text{ is bounded in } L^\infty(0,T;\mathcal{D}'(\mathbb{R}^\in)). \tag{4.2}$$

We can state

Lemma 4.2.

i) *The measure ω is for all time diffuse.*

ii) *Assume that $\omega^\varepsilon \geq 0$ and let Ω be a bounded domain in $\mathbb{R}^4$. For all $\delta > 0$, there exists a neighborhood $\mathcal{V}$ of the diagonal of Ω such that, for ε small enough*

$$\int_0^T \int_{\mathcal{V}} |\omega(x,t)||\omega(y,t)| dx dy dt \leq \delta \tag{4.3}$$

$$\int_0^T \int_{\mathcal{V}} |\omega^\varepsilon(x,t)|\omega^\varepsilon(y,t)| dx dy dt \leq \delta \tag{4.4}$$

(where in (4.4) integrals must be understood in the sense of measures).

Observe that the assertion i) above does not contradict the example of concentration in phantom vortices, since in this case the coefficient of the limit dirac mass is zero. If vorticity had not mean value 0, then energy coming from infinity in the scaling would make the L^2 estimate blow up. Lemma 4.1 and 4.2 easily combine to give the proof of Theorem 4.1 through the following argument. Given $\delta > 0$, let $\mathcal{V}$ a neighborhood of the diagonal provided by the above lemma, where Ω is the support of ϕ. We can split the integral in the right hand side of (4.1) in 2 pieces and obtain

$$\begin{aligned}\int_0^T \int (u_1^\varepsilon u_2^\varepsilon - u_1 u_2)\phi dx dt &= \int_{\mathcal{V}} [\omega^\varepsilon(x,t)\omega^\varepsilon(y,t) - \omega(x,t)\omega(y,t)] H_\phi(x,y) dx dy dt \\ &+ \int_{\mathbb{R}^4 - \mathcal{V}} [\omega^\varepsilon(x,t)\omega^\varepsilon(y,t) - \omega(x,t)\omega(y,t)] H_\phi(x,y) dx dy dt\end{aligned}$$

Due to the positivity of ω and ω^ε, the piece corresponding to the integration over $\mathcal{V}$ is easily bounded, for ε small enough, by $\|H_\phi\|_{0,\infty}\delta$. On the other hand, outside $\mathcal{V}$ the function H_ϕ is continuous so that we have immediately by (4.2)

$$\lim_{\varepsilon\to 0} \int_{\mathbb{R}^4 - \mathcal{V}} [\omega^\varepsilon(x,t)\omega^\varepsilon(y,t) - \omega(x,t)\omega(y,t)] H_\phi(x,y) dx dy dt = 0$$

and the proof of Theorem 4.2 is complete.

We now reproduce the proof of Lemma 4.2 given in [3]. For the first assertion we simply recall that all measure ω can be splitted as $\omega = \omega_d + \sum_j \omega_j \delta(x - x_j)$, where ω_d is diffuse and ω_j have a finite sum. But using the integral representation (2.1), it is readily seen that the velocity associated to this measure cannot be in L^2_{loc} unless all ω_j are zero. Let us now turn to (4.4). If it was wrong there would exist $\delta > 0$ such that for a sequence $\varepsilon_n \to 0$:

$$\int_0^T \int \omega^{\varepsilon_n}(x,t) dx \int_{\{|y-x|\le n^{-1}\}} |\omega^{\varepsilon_n}(y,t)| dy \ge \delta$$

where again we have used the positivity of ω^ε. Since $\int \omega^\varepsilon(x,t) = \int \omega_0^\varepsilon dx \neq 0$, this yields a sequence (x_n, t_n) in $\omega \times [0,T]$ and a sequence ε_n tending to 0 such that

$$\int_{\{|y-x_n|\le 1/n\}} \omega^{\varepsilon_n}(y,t_n) dy \ge \delta.$$

Passing to subsequences we have $(x_n, t_n) \to (x^*, t^*)$, and for n large enough

$$\int_{\{|y-x^*|\le 1/n\}} \omega^{\varepsilon_n}(y,t_n) dy \ge \delta.$$

If now ϕ is a smooth positive function equal to 1 around x^* we then get

$$\int \omega^\varepsilon(y,t_n)\phi(y) dy \ge \int_{\{|y-x_n|\le 1/n\}} \omega^\varepsilon(y,t_n)\phi(y) dy \ge \delta.$$

Passing now to the limit we obtain, using the additional estimate (4.2)

$$\int \omega(y,t^*)\phi(y) dy \ge \delta.$$

This bound, valid for all positive smooth function ϕ taking the value 1 at x^*, contradicts the fact that $\omega(\cdot, t^*)$ is diffuse and establishes (4.3). The estimate (4.4) follows from similar arguments.

The remainder of this section is now devoted to the proof of lemma 4.1. The continuity of H_ϕ outside the diagonal is clear, since the integrated function is smooth there, so let us prove that H_ϕ is bounded. Writing $v = x - z$, $w = x - y$, we have to estimate for a given smooth function ϕ and a ball B with center at the origin

$$\Lambda(x,w) = \int_B \frac{v_1}{|v|^2}\frac{v_2 + w_2}{|v+w|^2}\phi(x-v)dv.$$

Expanding the function ϕ around x yields

$$\Lambda(x,w) = \phi(x)\int_B \frac{v_1}{|v|^2}\frac{v_2 + w_2}{|v+w|^2}dv + \Lambda'(x,w),$$

with

$$|\Lambda'(x,w)| \le C\int_B \frac{|v_1|}{|v|}\frac{|v_2 + w_2|}{|v+w|^2}dv \le C\int_B |v+w|^{-1}dv \le C'.$$

So we really have to prove that

$$I(w) = \int_B \frac{v_1}{|v|^2}\frac{v_2 + w_2}{|v+w|^2}dv$$

is bounded. For a given w in $\mathbb{R}^2$ let us denote by B_1 the ball centered at 0 with radius $|w|/2$ and split the above integral between B_1 and B_1^c. Since for $v \in B_1$ we have $|v+w| \ge |w| - |v| \ge |w|/2$, the contribution of B_1 gives

$$|\int_{B_1} \frac{v_1}{|v|^2}\frac{v_2 + w_2}{|v+w|^2}dv| \le \frac{2}{|w|}\int_{B_1}|v|^{-1}dv \le C. \tag{4.5}$$

Next the integral over B_1^c can be rewritten as

$$J(w) = \int_{B_1^c} \frac{v_1}{|v|^2}\left[\frac{v_2 + w_2}{|v+w|^2} - \frac{v_2}{|v|^2}\right]dv.$$

This is precisely where sign cancellations in the integral due to the specific form of the quadratic terms under consideration (here the term $v_1 v_2$) came in the proof. To estimate J we again split the integral. We introduce the ball B_2 with center $-w$ and radius $|w|/2$. On the one hand we have $|v| \ge |w|/2$ and $|v+w| \le |w|/2$ for $v \in B_2 \cap B_1^c$, so that

$$\begin{aligned}\int_{B_1^c \cap B_2} \frac{v_1}{|v|^2}\left[\frac{v_2 + w_2}{|v+w|^2} - \frac{v_2}{|v|^2}\right]dv &\le \int_{B_1^c \cap B_2} \frac{1}{|v|}\left[\frac{1}{|v+w|} + \frac{1}{|v|}\right]dv \\ &\le \frac{4}{|w|}\int_{B_2} dv + \frac{1}{|w|}\int_{B_2}|v+w|^{-1}dv \le C'. \quad (4.6)\end{aligned}$$

On the other hand , for $v \in B_2^c \cap B_1^c$ we can write

$$\frac{v_2 + w_2}{|v+w|^2} - \frac{v_2}{|v|^2} = \frac{w_2|v|^2 - v_2|w|^2}{|v|^2|v+w|^2}$$

and thus

$$\left|\frac{v_2+w_2}{|v+w|^2}-\frac{v_2}{|v|^2}\right| \le \frac{|w|}{|v+w|^2}+\frac{|w|^2}{|v||v+w|^2}.$$

Moreover if $v \in B_1^c \cap B_2^c$ we have

$$3|v+w| = 2|v+w|+|v+w| \ge 2|v+w|+|v|-|w| \ge |w|+|v|-|w| = |v|$$

so that we can write

$$\begin{aligned}\int_{B_1^c\cap B_2^c} \frac{v_1}{|v|^2}\left[\frac{v_2+w_2}{|v+w|^2}-\frac{v_2}{|v|^2}\right] dv &\le C\left[|w|\int_{\frac{|w|}{2}\le|v|\le C}|v|^{-3}dv+|w|^2\int_{\frac{|w|}{2}\le|v|\le C}|v|^{-4}dv\right]\\ &\le C|w|^2 \le C'. \qquad (4.7)\end{aligned}$$

(notice that $\Lambda(x,w)$ is obviously bounded for large values of w so we need only consider values of $|w|$ smaller than 1 for instance. Combining (4.6),(4.7),(4.8) gives the desired bound for I and ends the proof of Lemma 4.2.

5 – Appendix

Let us start with a proof of Lemma 2.1 when $\omega_0 \in L^1(\mathbb{R}^2)\cap L^p(\mathbb{R}^2)$ for some $p>1$ (the case when ω_0 is in L^1 with compact support can be treated by obvious modifications in the argument below). As already noticed it is clear that (2.1) provides a divergence free vector field whose rotational is ω, so we only need to check that, with this formulation, we have $\lim_{|x|\to\infty} u(x)=0$. Let $\delta>0$. Since ω is integrable there exists $R>0$ such that, if B_R denotes the ball with center 0 and radius R

$$\int_{B_R^c}|\omega(y)|dy \le \delta. \qquad (5.1)$$

We then write

$$|u(x)| \le \int_{B_R}\frac{|\omega(y)|}{|x-y|}dy+\int_{B_R^c}\frac{|\omega(y)|}{|x-y|}dy. \qquad (5.2)$$

For $x\in\mathbb{R}^2$ such that $|x|>2R$ we first get

$$\int_{B_R}\frac{|\omega(y)|}{|x-y|}dy \le |x|^{-1}\int_{B_R}\frac{|\omega(y)|}{|1-|y||x|^{-1}|}dy \le 2|x|^{-1}\|\omega\|_{0,1}. \qquad (5.3)$$

Let us now consider $d>0$ to be determined later and split the second integral in the right hand side of (5.2) as follows

$$\begin{aligned}\int_{B_R^c}\frac{|\omega(y)|}{|x-y|}dy &= \int_{\{y\in B_R^c,|x-y|\le d\}}\frac{|\omega(y)|}{|x-y|}dy+\int_{\{y\in B_R^c,|x-y|\ge d\}}\frac{|\omega(y)|}{|x-y|}dy\\ &\le \|\omega\|_{0,p}\left[\int_{|x-y|\le d}|x-y|^{-p^*}dy\right]^{1/p^*}+d^{-1}\int_{B_R^c}|\omega(y)|dy\\ &\le C\|\omega\|_{0,p}d^{1-2/p}+d^{-1}\delta\end{aligned}$$

where p^* denotes the conjugate exponent to p and we have used (5.1) and the fact that $p > 2$. If we now take $d = \delta^{\frac{p}{2p-2}}$ we obtain

$$\int_{B_R^c} \frac{|\omega(y)|}{|x-y|} dy \leq C(1 + \|\omega\|_{0,p})\delta^{\frac{p-2}{2p-2}},$$

which, combined with (5.2) and (5.3) gives the desired result.

We now come to the proof of (2.3). Let Ω a bounded domain. We have to estimate in $L^1(\Omega)$ the function

$$I(x) = \int \left| \frac{x+y-z}{|x+y-z|^2} - \frac{x-z}{|x-z|^2} \right| |\omega(z)| dz.$$

We can of course assume $|y| \leq 1/2$ and we set $\Omega = \Omega_1 \cup \Omega_2 \cup \Omega_3$, where $\Omega_1 = \{z; |x-z| \leq 2|y|\}$ and $\Omega_2 = \{z; 2|y| \leq |x-z| \leq 1\}$. We split accordingly I into $I_1 + I_2 + I_3$. To estimate I_2 and I_3 we observe that the following inequality

$$\left| \frac{x+y-z}{|x+y-z|^2} - \frac{x-z}{|x-z|^2} \right| \leq C \frac{|y|}{|x-z|^2}$$

holds as soon as $2|y| \leq |x-z|$, so that we immediately get

$$I_3(x) \leq C|y| \int_{|x-z|\geq 1} \frac{|\omega(z)|}{|x-z|^2} dz \leq C|y| \|\omega\|_{0,1},$$

and therefore

$$\int_\Omega I_3(x) dx \leq C(\Omega)|y|. \tag{5.4}$$

For I_2, setting $\bar{z} = x - z$ yields

$$\int I_2(x) dx \leq C|y| \int \int_{2|y|\leq|\bar{z}|\leq 1} \frac{|\omega(z)|}{\bar{z}^2} dz d\bar{z} \leq C \log|y| \|\omega\|_{0,1}. \tag{5.5}$$

Finally to estimate I_1 we notice that, for $z \in \Omega_1$ we have $|x-y-z| \leq 3|y|$, and thus

$$\int I_1(x) dx \leq C \int \int_{\bar{z}\leq 3|y|} \frac{|\omega(z)|}{|\bar{z}|} dz d\bar{z} \leq C|y| \|\omega\|_{0,1}. \tag{5.6}$$

It remains now to combine (5.4),(5.5) and (5.6) to obtain (2.3).

ACKNOWLEDGEMENT – These notes correspond to a series of lectures given at the Summer School on Mathematical Topics in Fluid Mechanics held at the University of Lisbon in September 1991. We wish to thank the organizers, J.F. Rodrigues and A. Sequeira for giving us the opportunity to review some important recent results in the field.

REFERENCES

[1] *R. Caflish* and *O. Orellana* – *Singular solutions and ill-posedness for the evolution of a vortex sheet*, SIAM J. Math. Anal., **20**, (1989) 293–307.

[2] *G.-H. Cottet* – *Equations de Navier-Stokes dans le plan avec tourbillon initial mesure*, C.R. Acad. Sc. Paris, **303** (1986), 105–108.

[3] *J.-M. Delort*, – *Existence de nappes de tourbillon en dimension deux*, preprint.

[4] *R. DiPerna* and *A. Majda* – *Reduced Hausdorff dimension and concentration-cancellation for two dimensional incompressible flow*, J. Amer. Math. Soc., **1** (1988), 59–95.

[5] *J. Duchon* and *R. Robert* – *Solutions globales avec nappe de tourbillon pour les équations d'Euler dans le plan*, C.R. Acad. Sc. Paris, **302** (1986), 183–186.

[6] *L. Evans* – *Weak convergence methods for non linear partial differential equations*, in CBMS Regional Conference Series in Math., **74** (1990).

[7] *C. Greengard* and *E. Thomann* – *On DiPerna-Majda concentration sets for two dimensional incompressible flow*, Com. Pure ApplMath., **41** (1988), 295–303.

[8] *R. Krasny* – *Desingularization of periodic vortex sheet roll-up*, J. Comp. Phys., **65** (1986), 292–313.

[9] *C. Sulem* and *al.* – *Finite time analyticity for the two and three dimensional Kelvin Helmholtz instability*, Com. Math. Phys., **80** (1981), 485–516.

[10] *V.I. Yudovich* – *Non-stationnary flows of ideal incompressible liquid*, Zh. Vych. Mat., **3** (1963), 1032–1066 (in Russian).

G.-H. Cottet,
LMC-IMAG, Université Joseph Fourier,
BP 53x, 38041 GRENOBLE Cédex – FRANCE

On the Steady Stokes Flow in Exterior Domains

VIVETTE GIRAULT and ADÉLIA SEQUEIRA

Introduction

The mathematical theory of exterior stationary problems for flows past obstacles has been studied through different approaches by a large number of authors during the last years but there still remain many important open questions.

Among relevant early contributions are the works of Leray (1933) [31], Ladyzhenskaya (1959) [29], Cattabriga (1961) [8], Finn (1965) [10], concerning existence, uniqueness and asymptotic behaviour of solutions for exterior Stokes and Navier–Stokes problems. More recently we can refer for example to the work of Amick [1], Babenko [3], Borchers & Sohr [6], Galdi [11], [12], Galdi & Simader [13], Giga & Sohr [14], Girault [15], [16], Girault & Sequeira [20], Girault, Giroire & Sequeira [18], Guirguis [22], Guirguis & Gunzburger [23], Heywood [27], Kozono & Sohr [28], Ladyzhenskaya & Solonnikov [30], Masuda [33], Sequeira [36], [37], [38], Specovius-Neugebauer [39].

It is the purpose of this paper to review the variational approaches of [36], [20], [17] for the exterior Stokes problem in two and three dimensions in the framework of the weighted Sobolev spaces of Hanouzet [25] and Giroire [21]. One of the earliest formulations in the primitive variables was obtained by Sequeira in 1981 (cf. [36]) and by using isomorphisms established by Giroire [21] it was extended in a precise context in [20].

The present paper is organized as follows. In section 2 we recall the basic results of Girault & Sequeira [20] concerning the primitive variable (i.e. the velocity-pressure) formulation and we give in particular a simpler proof of the Ladyzhenskaya–Babušca–Brezzi (LBB) "inf-sup" condition. Section 3 is devoted to a mixed variational formulation in terms of the vorticity and stream function ([17], [18]). As a consequence of the previous results in Section 4 we give a Helmoltz decomposition of vector fields. Finally, in Section 5 we present some concluding remarks concerning numerical methods and a reference to open problems.

1 – Preliminaries

To be precise in describing our model problem, let Ω be a bounded domain of $\mathbb{R}^n$ ($n = 2$ or 3), simply connected with a Lipschitz-continuous and connected boundary Γ. Let Ω' be the complement (i.e. the exterior) of $\overline{\Omega}$. The unit normal to Γ pointing outside Ω is denoted by $\mathbf{n}$. Given two vector functions $\mathbf{f}$, $\mathbf{g}$ and a positive constant ν (viscosity) find a velocity field $\mathbf{u}$ and a scalar pressure p, such that

$$(S)\quad \begin{cases} -\nu\,\Delta\mathbf{u} + \nabla p = \mathbf{f} & \text{in } \Omega' \\ \operatorname{div}\mathbf{u} = 0 & \text{in } \Omega' \\ \mathbf{u} = \mathbf{g} & \text{on } \Gamma \end{cases}$$

with a certain decay at infinity described by means of weights obtained from Hardy's inequalities. Our essential tool are the weighted Sobolev spaces of Hanouzet and Giroire that we present briefly now. They are similar to the usual Sobolev spaces and we refer to [25] and [21] for more details.

Let $x = (x_1, ..., x_n)$ denote a general point of $\mathbb{R}^n$ ($n = 2$ or 3), $r = (x_1^2 + ... + x_n^2)^{\frac{1}{2}}$ its distance to the origin and consider the weights

$$\rho(r) = \sqrt{1 + r^2}\,, \qquad \ell(r) = 1 + \log \rho(r)\ .$$

For $\lambda \in \mathbb{N}^n$, denoting by ∂^λ the differential operator of order λ

$$\partial^\lambda = \frac{\partial^{|\lambda|}}{\partial x_1^{\lambda_1} \cdots \partial x_n^{\lambda_n}}, \qquad \text{with } |\lambda| = \lambda_1 + ... + \lambda_n\ ,$$

we define, for $m \in \mathbb{N}$ and $\alpha, \beta \in \mathbb{Z}$ the general weighted space

$$(1.1) \quad W^m_{\alpha,\beta}(\Omega') = \Big\{ v \in \mathcal{D}'(\Omega') : \forall \lambda \in \mathbb{N}^n,\ 0 \leq |\lambda| \leq k,\ \rho(r)^{\alpha - m + |\lambda|}\, \ell(r)^{\beta - 1}\, \partial^\lambda v \in L^2(\Omega');$$

$$k+1 \leq |\lambda| \leq m,\ \rho(r)^{\alpha - m + |\lambda|}\, \ell(r)^{\beta}\, \partial^\lambda v \in L^2(\Omega') \Big\}\ ,$$

where

$$k = \begin{cases} -1 & \text{if } \frac{n}{2} + \alpha \notin \{1, ..., m\} \\ m - (\frac{n}{2} + \alpha) & \text{otherwise.} \end{cases}$$

This is a Hilbert space for the norm

(1.2)

$$\|v\|_{W^m_{\alpha,\beta}(\Omega')} = \Big(\sum_{|\lambda|=0}^{k} \big\| \rho(r)^{\alpha-m+|\lambda|}\, \ell(r)^{\beta-1}\, \partial^\lambda v \big\|^2_{L^2(\Omega')} + \sum_{|\lambda|=k+1}^{m} \big\| \rho(r)^{\alpha-m+|\lambda|}\, \ell(r)^{\beta}\, \partial^\lambda v \big\|^2_{L^2(\Omega')} \Big)^{\frac{1}{2}}\ .$$

Its associated seminorm is

$$(1.3) \qquad |v|_{W^m_{\alpha,\beta}(\Omega')} = \Big(\sum_{|\lambda|=m} \big\| \rho(r)^{\alpha}\, \ell(r)^{\beta}\, \partial^\lambda v \big\|^2_{L^2(\Omega')} \Big)^{\frac{1}{2}}$$

and $(\cdot,\cdot)_{W^m_{\alpha,\beta}(\Omega')}$ denotes the corresponding scalar product.

It is easy to see that the functions of $W^m_{\alpha,\beta}(\Omega')$ are tempered distributions. For all bounded subsets $\mathcal{O}$ of Ω', we have $W^m_{\alpha,\beta}(\Omega') \subset H^m(\mathcal{O})$ and the usual trace theorems on Γ still hold for the functions of $W^m_{\alpha,\beta}(\Omega')$.

The space $\mathcal{D}(\overline{\Omega}')$ of indefinitely differentiable functions with compact support in $\overline{\Omega}$ is dense in $W^m_{\alpha,\beta}(\Omega')$. With this space we associate the two following spaces:

$$(1.4) \qquad \bullet \quad \overset{\circ}{W}{}^m_{\alpha,\beta}(\Omega') = \Big\{ v \in W^m_{\alpha,\beta}(\Omega') : \frac{\partial^j v}{\partial n^j} = 0 \text{ on } \Gamma,\ 0 \leq j \leq m-1 \Big\}$$

($\frac{\partial}{\partial n}$ = normal derivative) which is the closure of $\mathcal{D}(\Omega')$ in $W^m_{\alpha,\beta}(\Omega')$;

$$(1.5) \qquad \bullet \quad W^{-m}_{-\alpha,-\beta}(\Omega') \text{ the dual space of } \overset{\circ}{W}{}^m_{\alpha,\beta}(\Omega')\ .$$

The duality pairing between these two spaces is denoted by $\langle \cdot, \cdot \rangle$. If Ω has positive measure, the seminorm (1.3) is a norm on $\overset{\circ}{W}{}^m_{\alpha,\beta}(\Omega')$ equivalent to the norm $\|\cdot\|_{W^m_{\alpha,\beta}(\Omega')}$. This result follows from Hardy's inequality (see [26])

$$(1.6) \qquad \int_0^{+\infty} |f(r)|^p \, r^\beta \log^\gamma r \, dr \leq \left(\frac{2p}{|\beta+1|}\right)^p \int_0^{+\infty} \left|\frac{df}{dr}\right|^p r^{\beta+p} \log^\gamma r \, dr \ ,$$

with $p = 2$, $\beta \neq -1$.

Using spherical coordinates in $\mathbb{R}^n$ and relation (1.6), a standard compacteness argument derives in particular

$$(1.7) \qquad \|v\|_{W^m_{\alpha,\beta}(\Omega')} \leq C \, |v|_{W^m_{\alpha,\beta}(\Omega')} \, , \qquad \forall \, v \in \overset{\circ}{W}{}^m_{\alpha,\beta}(\Omega') \ .$$

(This is analogous to Poincaré's inequality valid in the case of connected open sets, bounded at least in one direction).

Let $\mathbb{P}_k$ denote the space of all polynomials (in two or three variables) of degree at most k. The weighted spaces contain $\mathbb{P}_k$, with k depending on the weights. The seminorm (1.3) is a norm on the quotient space $W^m_{\alpha,\beta}(\Omega')/\mathbb{P}_k$, equivalent to the quotient norm.

In the sequel we shall often use the basic Sobolev spaces $W^m_{\alpha,\beta}(\Omega')$, with $\beta = 0$. To simplify the notations, spaces $W^m_{\alpha,0}(\Omega')$ are denoted by $W^m_\alpha(\Omega')$, where

$$(1.8) \qquad \begin{aligned} W^m_\alpha(\Omega') = \Big\{ v \in \mathcal{D}'(\Omega') \colon \forall \, \lambda \in \mathbb{N}^n, \, 0 \leq |\lambda| \leq k, \ \rho(r)^{\alpha-m+|\lambda|} \, \ell(r)^{-1} \, \partial^\lambda v \in L^2(\Omega'); \\ k+1 \leq |\lambda| \leq m, \ \rho(r)^{\alpha-m+|\lambda|} \, \partial^\lambda v \in L^2(\Omega') \Big\} \ . \end{aligned}$$

In three dimensions these spaces contain no logarithmic weights and are easier to handle:

$$(1.9) \qquad W^m_\alpha(\Omega') = \Big\{ v \in \mathcal{D}'(\Omega') \colon \forall \, \lambda \in \mathbb{N}^3, \ 0 \leq |\lambda| \leq m, \ \rho(r)^{\alpha-m+|\lambda|} \, \partial^\lambda v \in L^2(\Omega') \Big\} \ .$$

In two and three dimensions we shall use mostly the following spaces:

- If $n = 2$,

$$(1.10) \qquad W^1_0(\Omega') = \Big\{ v \in \mathcal{D}'(\Omega') \colon \rho(r)^{-1} \, \ell(r)^{-1} \, v \in L^2(\Omega'); \ \nabla v \in L^2(\Omega') \Big\} \subset H^1_{\text{loc}}(\Omega')$$

which contains $\mathbb{P}_0$;

- If $n = 3$,

$$(1.11) \qquad W^1_0(\Omega') = \Big\{ v \in \mathcal{D}'(\Omega') \colon \rho(r)^{-1} \, v \in L^2(\Omega'); \ \nabla v \in L^2(\Omega') \Big\} \subset H^1_{\text{loc}}(\Omega')$$

which contains no polynomials.

According to (1.4), (1.5) with these spaces we associate

$$(1.12) \qquad \overset{\circ}{W}{}^1_0(\Omega') = \Big\{ v \in W^1_0(\Omega') \colon v|_\Gamma = 0 \Big\} = \overline{\mathcal{D}(\Omega')}^{W^1_0(\Omega')}$$

and the dual space

$$W_0^{-1}(\Omega') \subset H^{-1}(\Omega') \ .$$

Moreover the seminorm $v \mapsto \|\nabla v\|_{L^2(\Omega')}$ for $n = 3$ is a norm on $W_0^1(\Omega')$ equivalent to $\|v\|_{W_0^1(\Omega')}$ and for $n = 2$ is a norm on $\overset{\circ}{W}{}_0^1(\Omega')$ equivalent to $\|v\|_{W_0^1(\Omega')}$.

Observe that we have

$$W_0^0(\Omega') = L^2(\Omega') \; ; \tag{1.13}$$

we denote by $(\cdot,\cdot)$ the scalar product of $L^2(\Omega')$.

We close this section, considering also the particular spaces:

- For $n = 2$,

$$\begin{aligned} W_0^2(\Omega') = \Big\{ v \in \mathcal{D}'(\Omega') : \rho(r)^{-2}\, \ell(r)^{-1}\, v \in L^2(\Omega'); \\ \rho(r)^{-1}\, \ell(r)^{-1}\, \nabla v \in L^2(\Omega');\; \partial^2 v \in L^2(\Omega') \Big\} \supset \mathbb{P}_1 \, . \end{aligned} \tag{1.14}$$

- For $n = 3$,

$$\begin{aligned} W_0^2(\Omega') = \Big\{ v \in \mathcal{D}'(\Omega') : \rho(r)^{-2}\, v \in L^2(\Omega'); \\ \rho(r)^{-1}\, \nabla v \in L^2(\Omega');\; \partial^2 v \in L^2(\Omega') \Big\} \supset \mathbb{P}_0 \, . \end{aligned} \tag{1.15}$$

2 – The Velocity-Pressure Variational Formulation

In order to analyze the exterior Stokes problem (S) in a velocity-pressure formulation we need to introduce the following Hilbert spaces

$$V = \Big\{ \mathbf{v} \in \overset{\circ}{W}{}_0^1(\Omega')^n :\; \operatorname{div} \mathbf{v} = 0 \;\text{ in } \Omega' \Big\} \tag{2.1}$$

$$V^\perp = \Big\{ \mathbf{v} \in \overset{\circ}{W}{}_0^1(\Omega')^n :\; (\nabla \mathbf{v}, \nabla \mathbf{w}) = 0,\; \forall \mathbf{w} \in V \Big\} \tag{2.2}$$

$$V^0 = \Big\{ \mathbf{f} \in W_0^{-1}(\Omega')^n :\; \langle \mathbf{f}, \mathbf{w} \rangle = 0,\; \forall \mathbf{w} \in V \Big\} \, . \tag{2.3}$$

Theorem 2.1 (The "inf-sup" condition). *The following equivalent properties hold:*

i) *The divergence operator is an isomorphism from $V^\perp$ onto $L^2(\Omega')$;*

ii) *The gradient operator is an isomorphism from $L^2(\Omega')$ onto V^0;*

iii) *There exists a constant $\beta > 0$ such that*

$$\inf_{q \in L^2(\Omega')} \; \sup_{\mathbf{w} \in \overset{\circ}{W}{}_0^1(\Omega')^n} \frac{\displaystyle\int_{\Omega'} q \operatorname{div} \mathbf{w} \, dx}{\|q\|_{L^2(\Omega')} \, \|\mathbf{w}\|_{W_0^1(\Omega')^n}} \geq \beta \, . \tag{2.4}$$

Proof:

i) Let us take q in $L^2(\Omega')$. Its extension by zero in Ω belongs to $L^2(\mathbb{R}^n)$ and is still denoted by q. It is shown (Giroire [21]) that there exists a unique $\varphi \in W_0^2(\mathbb{R}^3)/\mathbb{R}$ (or $\varphi \in W_0^2(\mathbb{R}^2)/\mathbb{P}_1$) such that

$$\Delta\varphi = q \quad \text{in } \mathbb{R}^3$$

and the mapping $q \mapsto \varphi$ is continuous. Now, taking

$$\mathbf{v}_0 = \nabla\varphi \,,$$

then $\mathbf{v}_0 \in W_0^1(\mathbb{R}^n)^n$ satisfies

$$\operatorname{div} \mathbf{v}_0 = q$$

and

$$\|\mathbf{v}_0\|_{W_0^1(\mathbb{R}^n)^n} \leq C \, \|q\|_{L^2(\Omega')} \,.$$

It remains to find a divergence-free function with the same traces as $\mathbf{v}_0$ on Γ. Let B_R denote an open ball centered at the origin with radius R and boundary S_R, such that $\overline{\Omega} \subset B_R$, and set $\Omega_R = \Omega' \cap B_R$. Since S_R is smooth, Ω_R is a bounded open set with a Lipschitz-continuous boundary $\partial\Omega_R = \Gamma \cup S_R$. Moreover, since by construction $\operatorname{div} \mathbf{v}_0 = 0$ in Ω, it necessarily satisfies $\int_\Gamma \mathbf{v}_0 \cdot \mathbf{n} \, ds = 0$. Therefore, it follows (cf. Girault & Raviart [19]) that there exists a divergence-free lift $\mathbf{v} \in H^1(\Omega_R)^n$ with

$$\operatorname{div} \mathbf{v} = 0 \ \text{ in } \Omega_R \,, \qquad \mathbf{v}|_\Gamma = \mathbf{v}_0 \,, \qquad \mathbf{v}|_{S_R} = 0$$

and the mapping $\mathbf{v}_0 \mapsto \mathbf{v}$ is linear and continuous

$$\|\mathbf{v}\|_{H^1(\Omega_R)^n} \leq C \, \|\mathbf{v}_0\|_{W_0^1(\mathbb{R}^n)^n}$$

(with C depending on Ω_R).

Now extending $\mathbf{v}$ by zero outside Ω_R, the extended function still denoted $\mathbf{v}$ belongs to $W_0^1(\Omega')^n$ and has bounded support. The result follows by taking the function $\mathbf{w} = \mathbf{v} - \mathbf{v}_0 \in \mathring{W}_0^1(\Omega')^n$.

ii) The second property of the theorem is equivalent to the first one. In fact, Green's formula

$$\forall \mathbf{v} \in \mathring{W}_0^1(\Omega')^n, \ \forall q \in L^2(\Omega') \,, \quad (\operatorname{div} \mathbf{v}, q) = -\langle \mathbf{v}, \nabla q \rangle \tag{2.5}$$

shows that div: $V^\perp \to L^2(\Omega')$ and $-\nabla$: $L^2(\Omega') \to V^0$ are dual operators.

iii) The last statement is the LBB "inf-sup" condition. The equivalence to the other two properties is established in abstract terms as in Babuška [4], Brezzi [7] or Girault & Raviart [19]. This completes the proof. ∎

Remark 2.1. This "inf-sup" condition was proved by Girault & Sequeira [20], but the proof is less straightforward. The divergence-free lift operator is first constructed by imposing a constraint on the boundary data. This constraint is then removed assuming enough smoothness on the boundary. In the last step, the regularity assumption on the boundary is eliminated. The proof presented here is simpler because by extending the function to the whole space, the regularity restriction on the boundary becomes unnecessary. We refer to Girault [15] for a proof with more general weights. □

Let us now turn to the exterior Stokes boundary value problem (S). Suppose that $\mathbf{f} \in W_0^{-1}(\Omega')^n$, $\mathbf{g} \in H^{\frac{1}{2}(\Gamma)^n}$ and $\nu > 0$:

- *Find a pair* $(\mathbf{u}, p)$ *in* $W_0^1(\Omega')^n \times L^2(\Omega')$ *such that*

$$(S) \quad \begin{cases} -\nu\,\Delta\mathbf{u} + \nabla p = \mathbf{f} & \textit{in } \Omega' \\ \operatorname{div}\mathbf{u} = 0 & \textit{in } \Omega' \\ \mathbf{u}|_\Gamma = \mathbf{g}\,. \end{cases}$$

As in the case of a bounded domain (S) is equivalent to the variational problem

- *Find* $\mathbf{u} \in W_0^1(\Omega')^n$ *and* $p \in L^2(\Omega')$ *such that*

$$(Q) \quad \begin{cases} \forall \mathbf{v} \in \mathring{W}_0^1(\Omega')^n, & \nu(\nabla\mathbf{u}, \nabla\mathbf{v}) - (p, \operatorname{div}\mathbf{v}) = \langle \mathbf{f}, \mathbf{v} \rangle\,, \\ \forall q \in L^2(\Omega'), & (q, \operatorname{div}\mathbf{u}) = 0\,, \\ & \mathbf{u}|_\Gamma = \mathbf{g}\,. \end{cases}$$

According to theorem 2.1 ("inf-sup" condition) and considering that the bilinear form $(\mathbf{u}, \mathbf{v}) \mapsto \nu(\nabla\mathbf{u}, \nabla\mathbf{v})$ is elliptic on $\mathring{W}_0^1(\Omega')^n$, (S) is a well-posed problem (cf. Brezzi [7], Girault & Raviart [19]).

These arguments prove the main result in this section:

Theorem 2.2. *Let* $\mathbf{f}$ *be given in* $W_0^{-1}(\Omega')^n$ *and* $\mathbf{g}$ *given in* $H^{\frac{1}{2}}(\Gamma)^n$. *Then the exterior Stokes problem (S) is equivalent to the variational problem (Q). Moreover, problem (S) is well-posed (under the assumption that* Γ *is Lipschitz-continuous): it has a unique solution* $(\mathbf{u}, p) \in W_0^{-1}(\Omega')^n \times L^2(\Omega')$ *and the following estimate holds*

$$(2.6) \qquad \|\mathbf{u}\|_{W_0^1(\Omega')^n} + \|p\|_{L^2(\Omega')} \leq C\left(\|\mathbf{f}\|_{W_0^{-1}(\Omega')^n} + \|\mathbf{g}\|_{H^{\frac{1}{2}}(\Gamma)^n}\right). \blacksquare$$

In contrast to a bounded domain, there is no compatibility condition here on the boundary data $\mathbf{g}$, i.e. it is not necessary that $\int_\Gamma \mathbf{g}\cdot\mathbf{n}\,ds = 0$.

Assuming extra smoothness assumptions on the boundary and data, it is natural to expect that the solution $(\mathbf{u}, p)$ of the Stokes problem (S) is also smoother. Results in two and three dimensions must be treated separately because the weighted spaces involved are not of the same kind (it is more technical in $\mathbb{R}^2$). In each case the regularity is first established for a problem stated in the whole space and then it is extended to the exterior domain Ω', using suitable "cut-off" functions.

The precise regularity results for $(\mathbf{u}, p)$ are not given in this paper. All details can be found in Girault & Sequeira [20].

3 – A Stream Function-Vorticity Variational Formulation

This section is devoted to the theoretical analysis of a well-posed mixed variational formulation for the homogeneous exterior Stokes problem associated with (S)

$$(\mathrm{S}_0) \quad \begin{cases} -\nu\,\Delta \mathbf{u} + \nabla p = \mathbf{f} & \text{in } \Omega' \\ \operatorname{div} \mathbf{u} = 0 & \text{in } \Omega' \\ \mathbf{u} = \mathbf{0} & \text{on } \Gamma \end{cases}$$

written in terms of the vorticity ω and the stream function ψ of the flow, in two dimensions. The nonhomogeneous problem (S) can be easily derived from the homogeneous one.

Many authors starting with Ciarlet & Raviart [9] in 1974 (for $n = 2$) and Nedelec [34] in 1982 (for $n = 3$) studied ψ-ω formulations for the Stokes problem in bounded domains, but these formulations have the disadvantage of not satisfying the LBB "inf--sup" condition. As far as we know the first well-posed ψ-ω variational formulation was proposed in 1990 by Bernardi, Girault & Maday [5] when solving the Stokes problem in bounded domains, using a spectral method. Its extension to exterior domains in the context of the weighted Sobolev spaces $W^m_{\alpha,\beta}(\Omega')$, has the same properties as for bounded domains and is considered here.

For the sake of simplicity we only study the two-dimensional case we refer to Girault, Giroire & Sequeira [17] for the three-dimensional case.

Here again we assume that the domain is simply-connected but in contrast with the velocity-pressure formulation, this assumption is essential here.

For $n = 2$, recall that we have two different curl operators which are formal adjoint. One takes vectors into scalars and the other does the opposite(*):

$$\forall \mathbf{v} = (v_1, v_2) \in \mathbb{R}^2, \quad \operatorname{curl} \mathbf{v} = \frac{\partial v_2}{\partial x_1} - \frac{\partial v_1}{\partial x_2}, \tag{3.1}$$

$$\forall \varphi \in \mathbb{R}, \quad \mathbf{curl}\,\varphi = \left(\frac{\partial \varphi}{\partial x_2}, -\frac{\partial \varphi}{\partial x_1}\right). \tag{3.2}$$

Thus we define the vorticity ω to be the scalar field

$$\omega = \operatorname{curl} \mathbf{u} = \frac{\partial u_2}{\partial x_1} - \frac{\partial u_1}{\partial x_2} \tag{3.3}$$

(where $\mathbf{u} = (u_1, u_2)$ is the velocity of the flow) and the stream function ψ to be the scalar field

$$\mathbf{u} = \mathbf{curl}\,\psi = \left(\frac{\partial \psi}{\partial x_2}, -\frac{\partial \psi}{\partial x_1}\right). \tag{3.4}$$

Recall also that ψ-ω mixed formulations for two dimensional Stokes problems are based on the two identities

$$\begin{aligned} \mathbf{curl}(\operatorname{curl} \mathbf{v}) &= -\Delta \mathbf{v} + \nabla(\operatorname{div} \mathbf{v}) \\ \operatorname{curl}(\mathbf{curl}\,\varphi) &= -\Delta \varphi\,. \end{aligned} \tag{3.5}$$

(*) The notation used in this work to distinguish scalars from vectors is the same used now for the scalar curl and the vector **curl**.

The following theorem characterizes the stream function of a divergence-free vector field in $\overset{\circ}{W}{}_0^1(\Omega')^2$.

Theorem 3.1. *Let* $\mathbf{v}$ *be a vector field in* $\overset{\circ}{W}{}_0^1(\Omega')^2$ *such that*

$$\operatorname{div} \mathbf{v} = 0 \quad \textit{in } \Omega' .$$

Then $\mathbf{v}$ *has a unique stream function* $\varphi \in \overset{\circ}{W}{}_0^1(\Omega')$ *such that*

$$\mathbf{v} = \mathbf{curl}\,\varphi \quad \textit{in } \Omega' \tag{3.6}$$

and the following estimate holds

$$\|\varphi\|_{\overset{\circ}{W}{}_0^2(\Omega')} \le C \, \|\mathbf{v}\|_{\overset{\circ}{W}{}_0^1(\Omega')^2} . \tag{3.7}$$

Remark 3.1. We need the assumption that Ω is simply-connected in order that the stream function φ vanishes on the boundary. □

Sketch of the **proof:** Extending $\mathbf{v}$ by zero in Ω, the extended function belongs to $\overset{\circ}{W}{}_0^1(\mathbb{R}^2)^2$, $\operatorname{div} \mathbf{v} = 0$ in $\mathbb{R}^2$ and the result follows, constructing a stream function in the whole plane [17]. ■

The purpose of the ψ-ω mixed formulation is to reduce the regularity of ψ. We have seen that ψ belongs to $\overset{\circ}{W}{}_0^2(\Omega')$ (see formula (1.14)); but $\overset{\circ}{W}{}_0^2(\Omega') \subset \overset{\circ}{W}{}_{-1,-1}^1(\Omega')$, where

$$W_{-1,-1}^1(\Omega') = \left\{ v \in \mathcal{D}'(\Omega') \, ; \; \frac{v}{\rho(r)^2 \, \ell(r)} \in L^2(\Omega'), \; \frac{\nabla v}{\rho(r) \, \ell(r)} \in L^2(\Omega')^2 \right\} \tag{3.8}$$

and

$$\overset{\circ}{W}{}_{-1,-1}^1(\Omega') = \left\{ v \in W_{-1,-1}^1(\Omega') \, ; \; v = 0 \text{ on } \Gamma \right\} . \tag{3.9}$$

We want to write a formulation in which the stream function has no more regularity than $\overset{\circ}{W}{}_{-1,-1}^1(\Omega')$. Now, in the mixed formulation derived in [5], the curl of the right-hand side $\mathbf{f}$ belongs to the dual space of the stream function. This will be the case here if $\mathbf{f}$ belongs to $W_{1,1}^0(\Omega')^2$, where

$$W_{1,1}^0(\Omega') = \left\{ v \in \mathcal{D}'(\Omega') \, ; \; \rho(r) \, \ell(r) \, v \in L^2(\Omega') \right\} . \tag{3.10}$$

Therefore, we assume that

$$\mathbf{f} \in W_{1,1}^0(\Omega')^2 . \tag{3.11}$$

Now the derivation of our mixed formulation is very similar to that of [5]. We introduce the vorticity ω of the velocity as a new variable, $\omega = \operatorname{curl} \mathbf{u}$, and we take the curl of both sides of the Stokes equation (S_0):

$$-\nu \, \Delta \omega = \operatorname{curl} \mathbf{f} . \tag{3.12}$$

As expected, with the above assumption, $\mathrm{curl}\,\mathbf{f}$ belongs to $W_{1,1}^{-1}(\Omega')$ and therefore, $\Delta\omega$ also belongs to $W_{1,1}^{-1}(\Omega')$. This suggests to find the vorticity in the space:

(3.13) $$\mathcal{H} = \left\{\theta \in L^2(\Omega')\,;\ \Delta\theta \in W_{1,1}^{-1}(\Omega')\right\} .$$

Then, to set the state equation (3.12) in a variational form, it suffices to take the duality product of both sides by any function in the dual space of $W_{1,1}^{-1}(\Omega')$, that is $\overset{\circ}{W}{}^{1}_{-1,-1}(\Omega')$:

(3.14) $$\forall \varphi \in \overset{\circ}{W}{}^{1}_{-1,-1}(\Omega'), \quad -\nu\langle\Delta\omega,\varphi\rangle = \langle \mathrm{curl}\,\mathbf{f},\varphi\rangle = (\mathbf{f},\mathbf{curl}\,\varphi) .$$

Now, let ψ denote the stream function of $\mathbf{u}$; it is related to the vorticity by:

(3.15) $$\omega = -\Delta\psi .$$

Then the density of $\mathcal{D}(\Omega')$ in $\overset{\circ}{W}{}^{2}_{0}(\Omega')$ easily yields:

(3.16) $$\forall \theta \in \mathcal{H}, \quad (\omega,\theta) + \langle\psi,\Delta\theta\rangle = 0 ,$$

and this equation makes sense even when ψ belongs only to $\overset{\circ}{W}{}^{1}_{-1,-1}(\Omega')$.

Therefore, under the above assumption on $\mathbf{f}$, we see that the stream function ψ and the vorticity ω of the velocity solution $\mathbf{u}$ of the Stokes problem (S_0) also solve the mixed variational problem:

- *Find ψ in $\overset{\circ}{W}{}^{1}_{-1,-1}(\Omega')$ and ω in $\mathcal{H}$ such that:*

(3.14) $$\forall \varphi \in \overset{\circ}{W}{}^{1}_{-1,-1}(\Omega'), \quad -\nu\langle\Delta\omega,\varphi\rangle = (\mathbf{f},\mathbf{curl}\,\varphi) ,$$

(3.16) $$\forall \theta \in \mathcal{H}, \quad (\omega,\theta) + \langle\psi,\Delta\theta\rangle = 0 .$$

It is easy to prove that this mixed problem has no other solution. Furthermore, it is well-posed, in the sense that this unique solution depends continuously upon the data. Indeed, this problem can be put into the abstract framework of [4] and [7] with the following interpretation: the spaces are

$$X = \mathcal{H}, \quad M = \overset{\circ}{W}{}^{1}_{-1,-1}(\Omega') ;$$

the bilinear forms are

$$\forall \theta \text{ and } \mu \in X,\ \ a(\theta,\mu) = (\theta,\mu), \qquad \forall \theta \in X,\ \forall \varphi \in M,\ \ b(\theta,\varphi) = -\langle\Delta\theta,\varphi\rangle ;$$

and the element ℓ of M' is

$$\forall \varphi \in M, \quad \ell(\varphi) = \frac{1}{\nu}(\mathbf{f},\mathbf{curl}\,\varphi) .$$

The corresponding space V is:

$$V = \left\{\theta \in X\,;\ b(\theta,\varphi) = 0,\ \forall\varphi \in M\right\} .$$

Then, our problem is a particular case of:

For X given in X' and ℓ given in M', find ω in X and ψ in M such that:

$$\begin{cases} \forall \varphi \in M, & b(\omega,\varphi) = \ell(\varphi) \,, \\ \forall \theta \in X, & a(\omega,\theta) - b(\theta,\psi) = \chi(\theta) \,. \end{cases}$$

It is easy to see that the bilinear form a is elliptic on V:

$$\forall \theta \in V, \quad a(\theta,\theta) \geq \alpha \|\theta\|_X^2 \ \text{(with } \alpha > 0) \,;$$

in fact here $\alpha = 1$ and the only difficulty consists in showing that the bilinear form b satisfies the LBB "inf-sup" condition:

There exists a strictly positive constant β such that:

$$\forall \varphi \in M, \quad \sup_{\theta \in X} \frac{b(\theta,\varphi)}{\|\theta\|_X} \geq \beta \|\varphi\|_M \,.$$

The difficulty comes from the fact that, because of the weights at infinity, $M = \overset{\circ}{W}{}^1_{-1,-1}(\Omega')$ is not a subspace of $X = \mathcal{H}$. The proof of the LBB condition, which here reads:

$$\forall \varphi \in \overset{\circ}{W}{}^1_{-1,-1}(\Omega'), \quad \sup_{\theta \in \mathcal{H}} \frac{\langle \Delta\theta,\varphi\rangle}{\|\theta\|_{\mathcal{H}}} \geq \beta \|\varphi\|_{W^1_{-1,-1}(\Omega')}$$

is derived by a duality argument. Then the well-posedness of the mixed problem follows from the theorem of Babuška and Brezzi.

The above results are summarized in the following theorem.

Theorem 3.2. *Let $\mathbf{f}$ be given in $W^0_{1,1}(\Omega')^2$. Then problem (3.14), (3.16) has the unique solution (ψ,ω), where ψ is the unique stream function in $\overset{\circ}{W}{}^1_{-1,-1}(\Omega')$ of the velocity solution $\mathbf{u}$ of the Stokes problem (S_0) and $\omega = \operatorname{curl}\mathbf{u}$ is the vorticity of $\mathbf{u}$.*

4 – A Helmholtz Decomposition

We can derive a Helmholtz decomposition of the space $\overset{\circ}{W}{}^1_0(\Omega')^2$ by an easy combination of Theorems 2.2 and 3.1. Indeed, let $\mathbf{f}$ be given in $\overset{\circ}{W}{}^1_0(\Omega')^2$ and let us solve the Stokes problem with right-hand side $\Delta\mathbf{f}$. Since $\Delta\mathbf{f}$ belongs to $W_0^{-1}(\Omega')^2$, it follows from Theorem 2.2 that there exists a unique $\mathbf{w}$ in $\overset{\circ}{W}{}^1_0(\Omega')^2$ and q in $L^2(\Omega')$ such that

$$-\Delta\mathbf{w} + \nabla q = \Delta\mathbf{f} \quad \text{and} \quad \operatorname{div}\mathbf{w} = 0 \ \text{in } \Omega' \,.$$

Set $\mathbf{v} = \mathbf{f} + \mathbf{w}$; then $\mathbf{v}$ is the unique solution of the Dirichlet problem for the Laplace operator:

(4.1) $$\mathbf{v} \in W_0^1(\Omega')^2 \,, \quad \Delta\mathbf{v} = \nabla q \ \text{in } \Omega' \,, \quad \mathbf{v} = \mathbf{0} \ \text{on } \Gamma \,,$$

with q in $L^2(\Omega')$. On the other hand, applying Theorem 3.1, there exists a unique φ in $\mathring{W}_0^2(\Omega')$ such that $\mathbf{w} = \mathbf{curl}\,\varphi$. Moreover, this is an orthogonal decomposition, for

$$\int_{\Omega'} \nabla \mathbf{w} \cdot \nabla \mathbf{v}\, d\mathbf{x} = -\langle \mathbf{w}, \Delta \mathbf{v} \rangle = -\langle \mathbf{w}, \nabla q \rangle = 0 .$$

Thus, we have derived the following Helmholtz decomposition:

Theorem 4.1. *The space* $\mathring{W}_0^1(\Omega')^2$ *has the decomposition*

$$\mathring{W}_0^1(\Omega')^2 = V \oplus V^\perp ,$$

where V *and* $V^\perp$ *are defined by (2.1) and (2.2) respectively and have the following characterization*

$$V = \left\{ \mathbf{curl}\,\varphi;\ \varphi \in \mathring{W}_0^2(\Omega') \right\}, \quad V^\perp = \left\{ (\Delta^{-1}) \nabla q;\ q \in L^2(\Omega') \right\} , \tag{4.2}$$

and $(\Delta^{-1})\nabla q$ *is the unique solution of problem (4.1).*

5 – Concluding Remarks and Open Problems

In this short survey on the Stokes problem in exterior domains, we have seen that the weighted Sobolev spaces of Hanouzet in three dimensions and those of Giroire in two dimensions are a good framework for obtaining well-posed variational formulations, either in the primitive variables or in terms of the stream function and vorticity. As far as the theory is concerned, several extensions are in preparation.

First, we can seek a solution with a different growth or decay at infinity. In three dimensions, this can be achieved by looking for the velocity in $W_\alpha^1(\Omega')^3$ and the pressure in $W_\alpha^0(\Omega')$ for some integer α other than zero, which is the exponent of the weight $\rho(r)$ that we have used here. The approach is then no longer variational and the problem can be handled by first constructing the null space of the Stokes operator for strictly negative values of α and from there by deriving by duality necessary orthogonality conditions that the data must satisfy in order that the solution exists for stictly positive α. This technique is inspired from that used by Giroire for the Laplace operator. Of course, since we are dealing here with a Stokes problem, the "inf-sup" condition must also be established for those values of α, but in contrast to the case of a bounded domain, this is not a major difficulty. All this concerns the velocity-pressure formulation. If we want to derive a stream function-vorticity formulation for arbitrary values of the integer α, we must construct the stream-function that corresponds to such values of α. These two aspects are the object of a forthcoming paper [16].

Next, instead of setting the problem in Hilbert spaces (observe that all the spaces used here are defined with respect to the L^2-norm), we can try to solve the Stokes problem in the Banach spaces $W_{\alpha,\beta}^{m,p}(\Omega')$, which are weighted Sobolev spaces defined in terms of the L^p norm. These results are more difficult to obtain, because they rely on good and complete isomorphism theorems for the Laplace operator in those spaces. They are in preparation and some of them are announced in [2].

As far as the numerical analysis is concerned, once the problem is set in variational form, an important application is the coupling of this variational formulation with an integral formulation in order to be able to solve the problem, by some standard numerical method in a bounded domain. We refer to [36], [37], [38] for such a coupling of the Stokes problem in the primitive variables, in the case of a finite-element approximation. In a recent work by L. Halpern, this is extended to the coupling with a spectral approximation in polar coordinates. A coupling with a spectral approximation is also now being studied, in the case of a piecewise rectangular exterior domain. In the same direction, it would be interesting to couple the stream function-vorticity formulation of the Stokes problem with an integral formulation. This extension is now being studied, but we do not know yet whether such a coupling is possible or not.

Many problems remain opened in this area. For example, we have not examined at all the time-dependent Stokes problem in weighted spaces, whether from a theoretical or from a practical point of view. But for us, the most interesting open problem is the extension of the above results to nonlinear problems, and in particular to the Navier-Stokes system. Of course, many results have been established on the existence and regularity of classical solutions. Now, it would be very useful to derive a well-posed variational formulation for the stationary Navier-Stokes problem in weighted Sobolev spaces. But, because the domain is unbounded, the difficulties arise from the lack of adequate Sobolev imbedding theorems. It may be that the weighted spaces we have used are not adapted to the Navier-Stokes problem (or even to the Oseen equation) and that the answer lies in the choice of other spaces. So there is still a wide area to explore.

REFERENCES

[1] *C. Amick – On Leray's problem of steady Navier–Stokes flow past a body in the plane*, Acta Math., **161** (1988), 71–130.

[2] *C. Amrouche, V. Girault* and *J. Giroire – Espaces de Sobolev avec poids et équation de Laplace dans* $\mathbb{R}^n$ *(parties I et II)*, to appear in C.R.A.S., Paris (1992).

[3] *K.I. Babenko – On stationary solutions of the problem of flow past a body of a viscous incompressible fluid*, Mat. Sbornik, **91**(133) (1973).

[4] *I. Babušca – The finite element method with Lagrangian multipliers*, Numer. Math., **20** (1973), 179–192.

[5] *C. Bernardi, V. Girault* and *Y. Maday – Mixed spectral element approximation of the Navier-Stokes equations in the stream-function and vorticity formulation* (to appear in IMA Journal of Num. Anal. (1992).

[6] *W. Borchers* and *H. Sohr – On the semigroup of the Stokes operator for exterior domains in* L^q *spaces*, Math. Z., **196** (1987), 415–425.

[7] *F. Brezzi – On the existence, uniqueness and approximation of saddle-point problems arising from Lagrange multipliers*, RAIRO, Anal. Num., **R2** (1974), 129–151.

[8] *L. Cattabriga – Su un problema al contorno relativo al sistema di equazioni di Stokes*, Rend. Sem. Mat. Univ. Padova, **31** (1961), 308–340.

[9] *P.G. Ciarlet* and *P.A. Raviart* – *A mixed finite element method for the biharmonic equation* in "Mathematical Aspects of Finite Elements in Partial Differential Equations" (*C. de Boor*, ed.), 125–145, Academic Press, New York, 1974.

[10] *R. Finn* – *On the exterior stationary problem of the Navier–Stokes equations and associated perturbation problems*, Arch. Rat. Mech. Anal., **19** (1965), 363–406.

[11] *G.P. Galdi* – *An Introduction to the Mathematical Theory of the Navier–Stokes Equations*, Vol. I, Chapter V, Springer Tracts in Natural Phylosophy, 1992.

[12] *G.P. Galdi* – *On the existence of steady motions of a viscous flow with non-homogeneous boundary conditions*, Le Matematiche, **XLVI**, Fasc. I (1991).

[13] *G.P. Galdi* and *C.G. Simader* – *Existence, uniqueness and L^q-estimates for the Stokes problem in an exterior domain*, Arch. Rat. Mech. Anal., **112** (1990), 291–318.

[14] *Y. Giga* and *H. Sohr* – *On the Stokes operator in exterior domains*, J. Fac. Sci. Univ. Tokyo, Sect. IA, **36** (1989), 103–130.

[15] *V. Girault* – *The gradient, divergence, curl and Stokes operators in weighted Sobolev spaces of $\mathbb{R}^3$*, to appear in Journal of Fac. of Sci. of Tokyo (1992).

[16] *V. Girault* – *The Stokes problem and vector potential in three-dimensional exterior domains. An approach in weighted Sobolev spaces* (in preparation).

[17] *V. Girault, J. Giroire* and *A. Sequeira* – *Formulation variationnelle en fonction courant-tourbillon du problème de Stokes extérieur dans les espaces de Sobolev à poids*, C.R.A.S. Paris, **313**, Série I (1991), 499–502.

[18] *V. Girault, J. Giroire* and *A. Sequeira* – *A stream function-vorticity variational formulation for the exterior Stokes problem in weighted Sobolev spaces*, to appear in Math. Meth. in Appl. Sci. (1992).

[19] *V. Girault* and *P.A. Raviart* – *Finite Element Methods for Navier–Stokes Equations*, SCM **5**, Springer-Verlag, Berlin, 1986.

[20] *V. Girault* and *A. Sequeira* – *A well-posed problem for the exterior Stokes equations in two and three dimensions*, Arch. Rat. Mech. and Anal., **114** (1991), 313–333.

[21] *J. Giroire* – *Étude de quelques problèmes aux limites extérieurs et résolution par équations intégrales*, Thèse UPMC, Paris, 1987.

[22] *G. Guirguis* – *On the existence, uniqueness and regularity of the exterior Stokes problem in $\mathbb{R}^3$*, Com. in Partial Diff. Equat., **11** (1986), 567–594.

[23] *G. Guirguis* and *M. Gunzburguer* – *On the approximation of the exterior Stokes problem in three dimensions* M2AN, **21** (1987), 445–464.

[24] *M. Gunzburguer* – *Finite element methods for viscous incompressible flows*, Academic Press, London, 1989.

[25] *B. Hanouzet* – *Espaces de Sobolev avec poids. Application au problème de Dirichlet dans un demi-espace*, Rend. Sem. Univ. Padova, **XLVI** (1971), 227–272.

[26] *G.G. Hardy, D.E. Littlewood* and *G. Polya* – *Inequalities*, Cambridge Univ. Press, 1959.

[27] *J.G. Heywood* – *The Navier–Stokes equations: On the existence, regularity and decay of solutions*, Indiana Univ. Math. J., **29** (1980), 639–681.

[28] *H. Kozono* and *H. Sohr* – *L^q-regularity theory of the Stokes equations in exterior domains*, Fachbereich Mathematik, Universität – Gesamthochschule, Paderborn, 1990 (preprint).

[29] *O.A. Ladyzhenskaya* – *Investigation of the Navier–Stokes equations for a stationary flow of an incompressible fluid*, Upp. Mat. Nauk, **3** (1959), 75–97.

[30] *O.A. Ladyzhenskaya* and *V.A. Solonnikov* – *On the solution of boundary and initial value problems for Navier–Stokes equations in domains with non-compact boundaries*, Vestnik Leningrad, **13** (1977), 39–47.

[31] *J. Leray - Étude de diverses équations intégrales non linéaires et de quelques problèmes que pose l'hydrodynamique*, J. Math. Pures Appl., **9** (1933), 1–82.
[32] *J. Leray - Sur le mouvement d'un liquide visqueux emplissant l'espace*, Acta Math., **63** (1934), 193–249.
[33] *K. Masuda - On the stability of incompressible viscous fluid motions past objects*, J. Math. Soc. Japan, **27** (1975), 294–327.
[34] *J.C. Nedelec - Elements finis mixtes incompressibles pour l'équation de Stokes dans* $\mathbb{R}^3$, Num. Math., **39** (1982), 97–112.
[35] *J.C. Nedelec - Équations Intégrales* in "Analyse Mathématique et Calcul Numérique pour les Sciences et les Techniques (*R. Dautray & J.L. Lions*, eds.) 2, Ch. XI, collection C.E.A., Masson, Paris, 1985.
[36] *A. Sequeira - Couplage entre la méthode des éléments finis et la méthode des équations intégrales. Application au problème extérieur de Stokes stationnaire dans le plan*, Thèse 3ème Cycle UPMC, Paris, 1981.
[37] *A. Sequeira - The coupling of boundary integral and finite element methods for the bidimensional exterior steady Stokes problem*, Math. Meth. in Appl. Sci., **5** (1983), 356–375.
[38] *A. Sequeira - On the computer implementation of a coupled boundary and finite element method for the bidimensional exterior steady Stokes problem*, Math. Meth. in Appl. Sci., **8** (1986), 117–133.
[39] *M. Specovius-Neugebauer - Exterior Stokes problem and decay at infinity*, Math. Meth. in Appl. Sci., **8** (1986), 351–367.

Vivette Girault,
Analyse Numérique - Université Pierre et Marie Curie,
Tour 55-65, 5ème étage, 4 Place Jussieu,
F-75252 PARIS Cedex 05
FRANCE

Adélia Sequeira,
CMAF/University of Lisbon,
Av. Prof. Gama Pinto, 2
1699 LISBOA Codex
PORTUGAL

Mathematical Problems arising in Differential Models for Viscoelastic Fluids

C. GUILLOPÉ and J.C. SAUT

Abstract

This paper is a survey of recent mathematical results concerning flows of viscoelastic fluids which obey a constitutive law of differential type: initial boundary value problem, steady solutions, stability issues. As far as possible we try to present robust results, i.e. not too dependent on the model. Its contents is the following:

1 – Introduction. The models

Viscoelastic fluids are fluids like polymer melts or solutions which display memory effects. The stress tensor at present time t depends on the history of the deformation tensor. Such effects explain experimental facts (Weissenberg effect, die swell, stress relaxation, ...) which cannot be predicted by the standard Newtonian constitutive law or its "non-elastic" generalizations. The reader is referred to the books [1] [4] [23] [41] [64] for a precise description of these experimental facts. On the other hand polymer processing industry is faced to severe instabilities problems (extrusion defects, melt fracture, ...) which limit the quality of the products and the efficiency of the processes (see e.g.

(*) This work was partly supported by the GDR 901 "Rhéologie et transformation des polymères fondus" of CNRS.

Tordella [71], El Kissi and Piau [12] [42] for a striking description of the phenomena and for some possible physical explanations).

The complexity of viscoelastic flows requires a multidisciplinary approach including modelling, computational and mathematical aspects. We shall restrict to the latter and refer to [4] [32] [41] for modelling, and to the recent surveys [9] [28] for numerical simulations.

For incompressible fluids, the stress tensor is given by $-pI + \underline{\tau}$, p being the hydrostatic pressure (determined by the flow), and $\underline{\tau}$ the extra-stress tensor. The extra-stress tensor is given by a constitutive law. Contrary to Newtonian fluids, there is no universal constitutive equation for viscoelastic fluids, and deriving reasonable constitutive laws is still the object of several researches (see Larson [32] for an up to date exposition). There are two classes of constitutive laws, both derived by molecular or continuum mechanics considerations, the integral and the differential ones.

To introduce a popular example of the first category, we need to introduce a few kinematics notions.

Let $\chi_t(x, \tau)$ be the position at time $\tau \leq t$ of the fluid particle which at present time t is at x. The relative deformation gradient is the tensor

$$F_t(\tau) = \nabla \chi_t(x, t)$$

where the gradient is taken with respect to x.

The strain is measured by two relative strain tensors, the right relative Cauchy-Green tensor

$$C_t(\tau) = F_t^T(\tau) F_t(\tau)$$

and the Finger tensor $C_t^{-1}(\tau)$.

The KBKZ (Kaye, Bernstein, Kearsley and Zapas) model is an integral model motivated by the theory of elasticity and supported by thermodynamics arguments. It is defined in term of a "strain energy" $W(t - \tau, \mathrm{I}, \mathrm{II})$, where I and II are the two invariants of C, $(\mathrm{I}, \mathrm{II}) = (\operatorname{Tr} C(\tau), \operatorname{Tr} C^{-1}(\tau))$. More precisely, the extra-stress is given by

$$\underline{\tau}(t) = \int_{-\infty}^{t} \left\{ -\frac{\partial W}{\partial \mathrm{II}}(t - \tau, \mathrm{I}, \mathrm{II},) C_t^{-1}(\tau) + \frac{\partial W}{\partial \mathrm{I}}(t - \tau, \mathrm{I}, \mathrm{II},) C_t(\tau) \right\} d\tau \tag{1.1}$$

This model contains many classical ones, for instance the upper convected Maxwell model (UCM)

$$\underline{\tau}(t) = \frac{\eta}{\lambda^2} \int_{-\infty}^{t} \exp[-(t - \tau)/\lambda] \left[C_t^{-1}(\tau) - I \right] d\tau, \tag{1.2}$$

or the lower convected Maxwell model (LCM)

$$\underline{\tau}(t) = \frac{\eta}{\lambda^2} \int_{-\infty}^{t} \exp[-(t - \tau)/\lambda] \left[I - C_t(\tau) \right] d\tau, \tag{1.3}$$

which we shall encounter later on in their differential forms.

We shall focus on differential models from now on, because most mathematical and numerical studies deal with them. Actually they display (at least qualitively) most of the striking phenomena observed in viscoelastic flows.

These models obey the constitutive law

$$\underset{\sim}{\tau} = \underset{\sim}{\tau}^s + \underset{\sim}{\tau}^p \quad , \quad \underset{\sim}{\tau}^p = \sum_{i=1}^{N} \underset{\sim}{\tau}_i, \tag{1.4}$$

$$\underset{\sim}{\tau}_i + \lambda_i \frac{\mathcal{D}_a \underset{\sim}{\tau}_i}{\mathcal{D}t} + \beta_i(\underset{\sim}{\tau}_i, \nabla v) = 2\eta_i D, \qquad \eta_i, \lambda_i > 0, i = 1, \cdots, N, \tag{1.5}$$

$$\underset{\sim}{\tau}^s = 2\eta_s D, \eta_s \geq 0, \tag{1.6}$$

where $\underset{\sim}{\tau}$ is the (symmetric) extra-stress tensor. The coefficients η_i, $1 \leq i \leq N$, and η_s are viscosities ; $\lambda_i, 1 \leq i \leq N$, are relaxation times. The stress $\underset{\sim}{\tau}^s$ corresponds to a Newtonian contribution or to a fast relaxation mode. It plays a fundamental role for mathematical and numerical analysis: models with $\eta_s = 0$ will be called "of Maxwell type", those with $\eta_s > 0$ "of Jeffreys type".

In (1.5) $\frac{\mathcal{D}_a}{\mathcal{D}t}$ represents an objective derivative of a tensor, more precisely

$$\frac{\mathcal{D}_a}{\mathcal{D}t}\underset{\sim}{\tau} = \left(\frac{\partial}{\partial t} + v.\nabla\right)\underset{\sim}{\tau} + \underset{\sim}{\tau}W - W\underset{\sim}{\tau} - a(D\underset{\sim}{\tau} + \underset{\sim}{\tau}D) \tag{1.7}$$

where v is the velocity field, $D = \frac{1}{2}\left(\nabla v + \nabla v^T\right), W = \frac{1}{2}\left(\nabla v - \nabla v^T\right), -1 \leq a \leq 1$; $\beta_i\left(\underset{\sim}{\tau}_i, \nabla v\right)$ is a tensor valued function, smooth and at least quadratic, submitted to certain restrictions due to objectivity.

A few remarks on the principle of frame indifference (or objectivity) are in order here. This principle states roughly speaking that the constitutive law should not depend on the frame of reference (or on the observer).

A *change of frame* is a transformation

$$x^* = Q(t)x + b(t)$$

where $Q(t)$ is an orthogonal tensor and $b(t)$ a vector. A tensor valued function $A(x,t)$ is *indifferent* if any change of frame transforms it according to the formula

$$A^*(x^*, t) = Q(t)A(x,t)Q(t)^T.$$

For instance, the tensor $L = \nabla v$ is not indifferent since it transforms as

$$L^*(x^*, t) = Q(t)L(x,t)Q(t)^T + \dot{Q}(t)Q(t)^T,$$

through the transformation $x \to x^*$.

On the other hand, the rate of deformation tensor $D = \frac{1}{2}\left(\nabla v + \nabla v^T\right)$ is indifferent.

The extra-stress tensor, which we denotes temporarily by $T(x,t)$, is assumed to be indifferent,

$$T^*(x^*, t) = Q(t)T(x,t)Q(t)^T.$$

If it has to satisfy a differential equation like

$$\lambda \frac{\mathcal{D}T}{\mathcal{D}t} + T = 2\eta D,$$

the derivative $\frac{\mathcal{D}}{\mathcal{D}t}$ has to be *objective*, i.e. it must verify

$$\frac{\mathcal{D}^*}{\mathcal{D}t}T^* = Q(t)\frac{\mathcal{D}T}{\mathcal{D}t}Q^T(t).$$

Then

$$\lambda\frac{\mathcal{D}^*T^*}{\mathcal{D}t} + T^* - 2\eta D^*[v^*] =$$

$$Q(t)\left(\lambda\frac{\mathcal{D}T}{\mathcal{D}t} + T - 2\eta D[v]\right)Q(t)^T = 0,$$

and all observers will be in agreement about the form of the constitutive equation.

Now, the convected derivative $\dot{A} \underset{\text{def}}{=} \frac{\partial A}{\partial t} + (v\cdot\nabla)A$ is not objective when applied to a tensor-valued function, since

$$\dot{T}^* = Q\dot{T}Q^T + \dot{Q}TQ^T + QT\dot{Q}^T.$$

This rules out all "constitutive laws" based on this derivative and invalidates from a physical viewpoint all results obtained when using them (see e.g. Oskolkov *et al* [7]).

The idea of objective derivatives is to take the time derivative with respect to a reference frame suitably fixed to the body (Oldroyd [39]). Different choices of body-fixed frames will yield different objective derivatives. Here are a few classical examples (see e.g. Joseph [23]).

Upper convected invariant derivative. The body-fixed frame is built from a convected coordinate system $\left(\xi^i\right)$, i.e. a system which deforms with the body so that the coordinates (ξ^1, ξ^2, ξ^3) associated with a particular material point do not change with time. Base vectors g_i at a material point are defined by

$$g_i\left(\xi^1, \xi^2, \xi^3, t\right) = \frac{\partial}{\partial \xi^i}x\left(\xi^1, \xi^2, \xi^3, t\right),$$

where $x\left(\xi^1, \xi^2, \xi^3, t\right)$ is the position vector at time t of the material point with convected coordinates $\left(\xi^1, \xi^2, \xi^3\right)$. Using the frame at the material point with basis $\{g_1, g_2, g_3\}$ leads to the upper convected derivative

$$\overset{\triangledown}{A} \underset{\text{def}}{=} \frac{\partial A}{\partial t} + (v\cdot\nabla)A - LA - AL^T, \qquad L = \nabla v. \tag{1.8}$$

Lower convected invariant derivative. Using the dual basis $\left(g^i\right)$ of (g_i) leads to the upper convected derivative

$$\overset{\triangle}{A} \underset{\text{def}}{=} \frac{\partial A}{\partial t} + (v\cdot\nabla)A + L^TA + AL. \tag{1.9}$$

Corotational (Jaumann) invariant derivative. It corresponds to the time derivative of A with respect to a basis which shares the rotation of the neighborhood of the material point but not its deformation. It reads

$$\overset{\circ}{A} = \frac{1}{2}\left(\overset{\triangle}{A} + \overset{\triangledown}{A}\right) = \frac{\partial A}{\partial t} + (v.\nabla)A - WA + AW, \quad W = \frac{1}{2}(\nabla v - \nabla v^T). \tag{1.10}$$

"Interpolate" (Oldroyd) invariant derivatives. They are defined a priori as combinations of the upper convected and lower convected derivatives,

$$\frac{\mathcal{D}_a A}{\mathcal{D} t} = \frac{\partial A}{\partial t} + (v \cdot \nabla)A - WA + AW - a(DA + AD), \tag{1.11}$$

with $-1 \leq a \leq 1$. The values $a = 1, 0$, and -1 correspond respectively to the upper convected, corotational and lower convected derivative. The parameter a has to satisfy $|a| \leq 1$ as could be seen in simple viscometric flows such as Couette flows.

Most differential models of viscolastic fluids reduce to (1.5) with an appropriate choice of β_i. Here are some examples where for simplicity we only take one relaxation time, $N = 1$, $\beta_1 \equiv \beta$.

(i) $\beta \equiv 0$ corresponds to a version of Oldroyd models. When $\eta_s = 0$, the particular values $a = 1, 0$, and -1 are respectively the upper convected, corotational and lower convected Maxwell models. Note that (1.2) (1.3) are the integral form of the (UCM) and (LCM) models.

(ii) $\beta(\underline{\tau}, D) = 2\,\mathrm{Tr}(\underline{\tau} D)\alpha(\mathrm{Tr}\,\underline{\tau})(\underline{\tau} + I)$ and $a = 1$: Larson's model. Here $\alpha(\mathrm{Tr}\,\underline{\tau})$ is a scalar function of $\mathrm{Tr}\,\underline{\tau}$.

(iii) $\beta(\underline{\tau}, D) = \alpha\underline{\tau}\,\mathrm{Tr}\,\underline{\tau}, \alpha =$ constant : the Phan-Thien – Tanner model. Larson generalized it in $\beta(\underline{\tau}, D) = a\underline{\tau}^2 + b\underline{\tau}$ where a and b are scalar functions of $\mathrm{Tr}\,\underline{\tau}$ and $\underline{\tau}$.

(iv) $\beta(\underline{\tau}, D) = \alpha\underline{\tau}^2, \alpha =$ constant: Giesekus'model. It is a particular case of (iii).

(v) $\beta(\underline{\tau}, D) = \mu_0(\mathrm{Tr}\,\underline{\tau})D + \nu_1(\underline{\tau} D)I + \mu_2 D^2 + \nu_2\,\mathrm{Tr}\,(D^2)\,I$, where $\mu_0, \nu_1, \mu_2, \nu_2$ are some constants: 8-constant Oldroyd model.

(vi) White-Metzner's models correspond to $\beta \equiv 0$ and λ, η functions of the second invariant $\mathrm{II} = \mathrm{II}_D = \frac{1}{2}\,\mathrm{Tr}\,(D^2)$.

Typically, $\eta = \eta_0 \left\{1 + (\lambda_0 \mathrm{II})^{2\alpha}\right\}^{\frac{n-1}{2\alpha}}$ (Carreau's law).

Of course the previous list is not limitative. Other popular models (Leonov, Bird, De Aguiar, ...) can be put within the same framework. This is also the case of models with "internal variable" (order parameter) as those of Kwon and Shen [30], which seem to be quite realistic. All these models are derived by molecular or continuum mechanics considerations.

Equations (1.5) (1.6) are to be solved together with the equations of conservation of momentum and mass

$$\rho\left(\frac{\partial v}{\partial t} + (v \cdot \nabla)v\right) + \nabla p = \mathrm{div}\,\underline{\tau}, \tag{1.12}$$

$$\mathrm{div}\ v = 0, \tag{1.13}$$

and appropriate initial and boundary conditions.(1)

(1) Compressible effects cannot always be neglected in polymer flows (see for instance Edwards and Beris [15] for some "compressible models"), but we will restrict here to incompressible flows.

Solving problem (1.5) (1.6) (1.12) (1.13) is (mathematically) a formidable task, and a lot of issues are still unanswered. This is not surprising because of the complexity of the equations, and because of their recent derivation (around 1950 for Oldroyd models). On the other hand, the mathematical theory for the Euler and the Navier–Stokes equations of incompressible fluids is not still complete though these equations were derived respectively in 1755 and 1821!

The rest of the paper will be organized as follows. Section 2 will recall some basic and important facts on Maxwell type models. Section 3 will survey the existence theory for stationary flows. In Section 4 we will study unsteady flows and discuss stability issues.

2 – Maxwell type models : loss of evolution, change of type

Maxwell type models ($\eta_s = 0$) display two striking phenomena which are not present in Jeffreys type models ($\eta_s > 0$) and which we will describe now (see the books [20] and [40] for a more extensive treatment).

2.1. Loss of evolution (Hadamard instabilities, instabilities to short waves)

This concerns the Cauchy problem (initial value problem). Let us consider a system

$$A_0(x,q)\frac{\partial q}{\partial t} + \sum_{\ell=1}^{n} A_\ell(x,q)\frac{\partial q}{\partial x_\ell} = g(x,q). \tag{2.1}$$

$q = (q_1, \cdots, q_k)\,, A_0, A_\ell(1 \le \ell \le n)$ are smooth $k \times k$ matrices, $x = (x_1, \cdots, x_n)\,.$

The method of frozen coefficients and a plane wave analysis ($q = q_0 \exp[i(\alpha \cdot x + \nu t)]$) leads to the following.

Definition. (2.1) is *evolutionary* (of evolution type) if for every choice of the parameter $\alpha \in \mathbb{R}^n$, the roots ν of

$$\det\left(\nu A_0 + \sum_{\ell=1}^{n} \alpha_\ell A_\ell\right) = 0$$

are real.

If this condition is violated, the Cauchy problem for (2.1) is not well posed in any good but analytic class, which is not physical: highly oscillating initial data will grow exponentially in space at any prescribed time. This phenomenon was first described by Hadamard in his famous example on the ill posedness of the Cauchy problem for the Laplace equation. An ill-posed problem leads to catastrophic instabilities in numerical simulations. Even if one initiates the solution in a "stable" region, one can get arbitrarily close to an "unstable" region. We refer to Joseph-Saut [27] for a survey of ill-posed problems in fluid mechanics.

It was noticed by Rutkevich [60] [61] and systematized by Joseph-Renardy-Saut [24], Joseph-Saut [25], Dupret-Marchal [11] that Maxwell type models can present Hadamard instabilities.

For instance, the Maxwell model

$$\left.\begin{aligned}&\lambda\frac{\mathcal{D}_a\tau}{\mathcal{D}t}+\tau=2\eta D\\&\rho\left(\frac{\partial v}{\partial t}+(v\cdot\nabla)v\right)=\operatorname{div}\tau-\nabla p\\&\operatorname{div}\ v=0\end{aligned}\right\}\tag{2.2}$$

is Hadamard ill-posed provided

$$\frac{1-a}{2}\wedge_{\max}-\frac{1+a}{2}\wedge_{\min}>\frac{\eta}{\lambda},\tag{2.3}$$

where $\wedge_{\max}$ (resp. $\wedge_{\min}$) is the largest (resp. smallest) eigenvalue of τ.

Actually (2.3) is equivalent to the fact that the (second order in t and x) equation for the vorticity is not of evolution type.

It can also be shown ([24]) that Hadamard instabilities are possible for admissible motions if $a \in (-1,1)$, e.g. in extensional flows.

On the other hand, restrictions on the eigenvalues of τ prevent Hadamard instabilities for $a = \pm 1$. This is immediately seen from the integral from (1.2) (1.3) of the (UCM) and (LCM) models since the tensors $C_t(\tau)$ and $C_t^{-1}(\tau)$ are symmetric and positive definite, implying that (2.3) is never fullfilled in this case.

2.2. Change of type in steady flows: a transonic phenomenon in viscoelastic fluids

The P.D.E. system for *steady* flows of Maxwell models (i.e. (2.2) with $\frac{\partial}{\partial t} = 0$) is of *composite type*, neither elliptic nor hyperbolic. This is not surprising, the same being true for instance for the stationary system of perfect incompressible fluids. The new feature, discovered in [24], is that some *change of type* may occur. In fact an easy but tedious computation shows that three type of characteristics are present:

- complex characteristics (elliptic part) associated with incompressibility;
- real characteristics (hyperbolic part) associated with the propagation of information along streamlines (double in 2D, of multiplicity 4 in 3D);
- characteristics which change type: *complex* if and only if the equation for the vorticity is *elliptic*, *real* if it is *hyperbolic*.

The change of type (analogous to the well known situation in gas dynamics) occurs when the modulus $|v|$ of the velocity exceeds $\sqrt{\frac{\eta}{\rho\lambda}}$, which is the speed of propagation of shear waves in the fluid at rest. In other words, if one introduce a viscoelastic Mach number $M = \text{Re We}$ (see Section 3), the flow goes from sub- to supercritical as M crosses 1. Then the vorticity from elliptic becomes hyperbolic, and there are waves of vorticity. This of course implies a qualitative change in the nature of the flow, which is supported by some experiments (e.g. Joseph, Matta, Chen [26]). This change of type leads to drastic changes in the boundary conditions (see Section 3). Very few (mathematical or numerical) results are known in the super-critical case (see Renardy [57], Fraenkel [16], Crochet-Delvaux [10], Bazin [3]).

Among differential models without Newtonian contribution, Maxwell like models possess the nice feature that the change of type is associated with a change of type in the vorticity. A general class of differential models sharing this property was exhibited in [24]. This paper (and [23] [25]) contains also a classification of classical flows (Couette, Poiseuille, extensional, ...) for various Maxwell models with regard to type. To illustrate the generality of change of type in elastic fluids (i.e. fluids without Newtonian contribution) we present here a brief analysis of the linearization around a uniform flow $v = \begin{pmatrix} U \\ 0 \\ 0 \end{pmatrix}$, using the Coleman-Noll theory of fading memory [7]. Steming from the rather general concept of simple fluid (Noll), this theory provides a systematic way to derive constitutive laws for *special flows* (e.g. perturbations of rigid motions). The idea is to choose a L^2-weighted space for the history of deformations (see [62] for other choices of function spaces), and to use the Riesz theorem and isotropy to express the Fréchet derivative of the stress at a given motion.

One gets in this fashion constitutive laws of the type

$$\underline{\tau} = \int_0^\infty \kappa(s)\,[C_t(t-s) - I]\,ds$$

where κ is a scalar kernel (the relaxation kernel) satisfying $\frac{\kappa}{h^2} \in L^2(0,\infty)$, and h is the weight function associated to the space of histories of deformations.

For example, the (UCM) model corresponds to $\kappa(s) = \eta/\lambda\ \exp(-s/\lambda)$.

It can then be shown that the vorticity ξ of the linearized flow around the uniform flow v satisfies the equation

$$\rho U^2 \frac{\partial^2 \xi}{\partial x^2} - \kappa(0)\Delta_\perp \xi = \text{lower order terms}, \tag{2.4}$$

where $\Delta_\perp = \frac{\partial^2}{\partial y^2} + \frac{\partial^2}{\partial z^2}$, $M = \frac{U}{c}$, $J \quad c = \sqrt{\frac{\kappa(0)}{\rho}}$ (speed of shear waves). Then (2.4) reads

$$\left(M^2 - 1\right) \frac{\partial^2 \xi}{\partial x^2} - \Delta_\perp \xi = \text{lower order terms},$$

and the vorticity changes type from elliptic to hyperbolic when M crosses 1.

For the Maxwell (UCM) model, the full equation for ξ reads

$$\rho U^2 \frac{\partial^2 \xi}{\partial x^2} - \frac{\eta}{\lambda}\Delta_\perp \xi + \frac{\rho U}{\lambda}\frac{\partial \xi}{\partial x} = 0$$

(see Ultman and Denn [59]).

A change of type analysis for more complicated models is performed in [63].

3 – Steady flows

First we write (1.5) (1.6) (1.12) (1.13) in a nondimensional form (see [17] for the details) to get

$$\mathrm{Re}\left(\frac{\partial v}{\partial t} + (v \cdot \nabla)v\right) - (1-\varepsilon)\Delta v + \nabla p = \operatorname{div} \underline{\tau} + f, \tag{3.1}$$

$$\mathrm{We}\left(\frac{\partial \underline{\tau}}{\partial t} + (v\cdot\nabla)\underline{\tau} + \beta(\nabla v, \underline{\tau})\right) + \underline{\tau} = 2\varepsilon D, \tag{3.2}$$

$$\mathrm{div}\ v = 0. \tag{3.3}$$

The parameters in (3.1) (3.2) (3.3) are the Reynolds number $\mathrm{Re} = \rho UL/\eta$ (U and L are a typical velocity and length of the flow), the Weissenberg number $\mathrm{We} = \lambda U/L$, and the retardation parameter $\varepsilon = \eta_p/(\eta_p + \eta_s)$. Obvioulsy, $0 \le \varepsilon < 1$; $\varepsilon = 1$ corresponds to Maxwell type fluids, and $0 < \varepsilon < 1$ corresponds to Jeffreys type fluids. Observe the change of notation in (3.2) where $\beta(\nabla v, \underline{\tau})$ denotes now all the nonlinear terms in ∇v and $\underline{\tau}$ other than $(v\cdot\nabla)\underline{\tau}$. Finally f is a given body force.

To start with, we consider steady flows of Maxwell type fluids with simple boundary conditions, namely the system, for a bounded smooth domain of $\mathbb{R}^N, N = 2, 3$,

$$\left.\begin{aligned} &\mathrm{Re}(v\cdot\nabla)v + \nabla p = \mathrm{div}\,\underline{\tau} + f,\\ &\mathrm{div}\ v = 0,\\ &\mathrm{We}((v\cdot\nabla)\underline{\tau} + \beta(\nabla v, \underline{\tau})) + \underline{\tau} = 2\varepsilon D,\\ &v_{|\partial\Omega} = v_0, \qquad v\cdot n_{|\partial\Omega} = 0. \end{aligned}\right\} \tag{3.4}$$

In (3.4), n is the unit exterior normal vector to the boundary $\partial\Omega$ of Ω, and v_0 is a prescribed velocity field.

The following result is due to Renardy [47].

Theorem 3.1. (Existence of slow flow)(1) Let $\|f\|_2, \|v_0\|_{H^{5/2}(\partial\Omega)}$ be sufficiently small. Then there exist $v \in \mathbf{H}^3(\Omega), \underline{\tau} \in \mathbf{H}^2(\Omega), p \in H^2(\Omega)$, solution of (3.4), unique among the small solutions of (3.4).

If moreover $f \in \mathbf{H}^s, v_0 \in \mathbf{H}^{s+1/2}(\partial\Omega)$, $s \ge 2$, then $v \in \mathbf{H}^{s+1}(\Omega), \underline{\tau} \in \mathbf{H}^s(\Omega), p \in \mathbf{H}^s(\Omega)$.

Remark. A similar result holds for models with several relaxation times and/or a Newtonian contribution (Jeffreys models), see [47], and with traction boundary conditions [53].

Sketch of the proof: The idea is to use an iterative scheme which alternates between a perturbed Stokes system (corresponding to the elliptic part of (3.4)) and a hyperbolic equation whose characteristics are streamlines. To make things simpler, let us consider the (UCM) model, $\beta(\nabla v, \underline{\tau}) = -(\nabla v)\underline{\tau} - \tau(\nabla v)^T$; we have to solve

$$\mathrm{Re}(v\cdot\nabla)v + \nabla p = \mathrm{div}\,\underline{\tau} + f, \qquad \mathrm{div}\ v = 0, \tag{3.5}$$

$$(v\cdot\nabla)\underline{\tau} - (\nabla v)\underline{\tau} - \underline{\tau}(\nabla v)^T + \frac{1}{\mathrm{We}}\underline{\tau} = \frac{\varepsilon}{\mathrm{We}}(\nabla v + \nabla v^T). \tag{3.6}$$

Assuming that we have a smooth solution in hands we "extract" the elliptic part of (3.5) (3.6). First we apply the div operator to (3.6), to get

$$(v\cdot\nabla)\,\mathrm{div}\,\underline{\tau} - (\nabla v)\,\mathrm{div}\,\underline{\tau} + \frac{1}{\mathrm{We}}\,\mathrm{div}\,\underline{\tau} = \frac{\varepsilon}{\mathrm{We}}\Delta v + \underline{\tau} : \partial^2 v, \tag{3.7}$$

(1) Throughout the paper $\|\cdot\|_s$ will stand for the norm in the Sobolev space H^s or $\mathbf{H}^s$ (Sobolev space of vector or tensor-valued functions).

where

$$\underline{\tau} : \partial^2 \underset{\text{def}}{=} \sum_{j,k} \underline{\tau}^{jk} \frac{\partial^2}{\partial x_j \partial x_k}.$$

We express $\operatorname{div} \underline{\tau}$ by (3.5) and report in (3.7) to obtain

$$\begin{aligned} \underline{\tau} : \partial^2 v + \frac{\varepsilon}{\mathrm{We}} \Delta v - \mathrm{Re}(v \cdot \nabla)(v \cdot \nabla) v + \mathrm{Re}(\nabla v)(v \cdot \nabla) v - \frac{\mathrm{Re}}{\mathrm{We}}(v \cdot \nabla) v \\ = \nabla q - \left(\nabla v + \nabla v^T\right) \nabla p - (v \cdot \nabla) f + (\nabla v) f - \frac{1}{\mathrm{We}} f \end{aligned} \tag{3.8}$$

where $q = (v \cdot \nabla) p + \frac{1}{\mathrm{We}} p$.

This construction allows to define the following iterative scheme.

Start with $v^0 = 0, p^0 = q^0 = 0, \tau^0 = 0$. Then solve

$$\left.\begin{aligned} \underline{\tau}^n : \partial^2 v^{n+1} & + \frac{\varepsilon}{\mathrm{We}} \Delta v^{n+1} - \mathrm{Re}\,(v^n \cdot \nabla)\,(v^n \cdot \nabla)\, v^{n+1} - \nabla q^{n+1} \\ & = -\left[\nabla v^n + (\nabla v^n)^T\right] \nabla p^n - (v^n \cdot \nabla)\, f + (\nabla v^n)\, f - \frac{1}{\mathrm{We}} f \\ & - \mathrm{Re}\,(\nabla v^n)\,(v^n \cdot \nabla)\, v^n + \frac{\mathrm{Re}}{\mathrm{We}}\,(v^n \cdot \nabla)\, v^n, \\ \operatorname{div}\ v^{n+1} = 0, & \quad \textstyle\int_\Omega q^{n+1} dx = 0, \\ v^{n+1} = v_0 & \quad \text{on } \partial\Omega, \end{aligned}\right\} \tag{3.9}$$

$$\left(v^{n+1} \cdot \nabla\right) p^{n+1} + \frac{1}{We} p^{n+1} = q^{n+1}, \tag{3.10}$$

$$\left.\begin{aligned} (v^{n+1} \cdot \nabla)\, \underline{\tau}^{n+1} - (\nabla v^{n+1})\, \underline{\tau}^{n+1} - \underline{\tau}^{n+1} (\nabla v^{n+1})^T + \frac{1}{\mathrm{We}} \underline{\tau}^{n+1} \\ = \frac{\varepsilon}{\mathrm{We}} \left[\nabla v^{n+1} + (\nabla v^{n+1})^T\right]. \end{aligned}\right\} \tag{3.11}$$

Note that (3.9) is a perturbation of a Stokes system,

$$\left.\begin{aligned} & \frac{\varepsilon}{\mathrm{We}} \Delta v^{n+1} + \nabla q^{n+1} = g^n, \\ & \operatorname{div}\ v^{n+1} = 0, \\ & v^{n+1}{}_{|_{\partial\Omega}} = v_0, \qquad v_0 \cdot n_{|_{\partial\Omega}} = 0. \end{aligned}\right\} \tag{3.12}$$

Actually the perturbation $\underline{\tau} : \partial^2 v^{n+1} - \mathrm{Re}\,(v^n \cdot \nabla)\,(v^n \cdot \nabla)\, v^{n+1}$ is "small" if v^n and $\underline{\tau}^n$ are small enough.

On the other hand, (3.10) and (3.11) amount to inverting $(v^{n+1} \cdot \nabla) + \frac{1}{\mathrm{We}} I$ which is easy since v^{n+1} is divergence free and tangent to the boundary.

The first step consists in showing that the iterates are well defined and remain small in the norms indicated in Theorem 1, for f and v_0 small enough. A key result for that is the

Lemma 3.2. *Let $s \geq 1$. There exists $\varepsilon_1 > 0, \varepsilon_2 > 0$ such that the conditions $\|\underline{\tau}^n\|_{s+1} \leq \varepsilon_1, \|v^n\|_{s+1} \leq \varepsilon_2$ imply that (3.9) has a unique solution (v^{n+1}, q^{n+1}) such that*

$$\left\|v^{n+1}\right\|_{s+2} + \left\|q^{n+1}\right\|_{s+1} \leq c_1 \Big\{ \|p^n\|_{s+1} \|v^n\|_{s+2}$$

$$+ \|v^n\|_{s+1} \|f\|_{s+1} + \|f\|_s + \|v^n\|_{s+2}^2 \|v^n\|_s + \|v^n\|_{s+1}^2 \Big\}.$$

The estimates on (3.10) (3.11) are obtained by energy methods.

The second step consists now in showing that the algorithm defines a contraction (from which the convergence follows) in weaker norms ($\mathbf{H}^s \times \mathbf{H}^{s-1} \times \mathbf{H}^{s-1}$). ∎

The existence Theorem 1 has several extensions which we briefly survey.

3.1. White-Metzner models

A. Hakim [21] has shown that Theorem 1 is valid for White-Metzner models

$$\underset{\sim}{\tau} = \underset{\sim}{\tau}^s + \underset{\sim}{\tau}^p,$$

$$\underset{\sim}{\tau}^s = 2\eta_\infty D, \qquad \eta_\infty \geq 0 \quad \text{(constant)},$$

$$\underset{\sim}{\tau}^p + \tilde{\lambda}_{\text{II}} \frac{\mathcal{D}_a \underset{\sim}{\tau}^p}{\mathcal{D}t} = 2\tilde{\eta}_{\text{II}} D$$

provided $\tilde{\lambda}_{\text{II}}$ and $\tilde{\eta}_{\text{II}}$ are smooth functions of $\text{II} = \frac{1}{2}\,\text{Tr}\,(D^2)$ satisfying $\tilde{\lambda}_{\text{II}} > 0$ and $\tilde{\eta}_{\text{II}} > 0$.

3.2. Weakly elastic fluids (see Guillopé-Saut [19])

Interest is focused here on Jeffreys models:

$$\left.\begin{aligned} &\text{Re}(v \cdot \nabla)v - (1-\varepsilon)\Delta v + \nabla p = \text{div}\,\underset{\sim}{\tau} + f, \\ &\text{We}((v \cdot \nabla)\underset{\sim}{\tau} + \beta(\nabla v, \underset{\sim}{\tau})) + \underset{\sim}{\tau} = 2\varepsilon D, \\ &\text{div } v = 0, \\ &v_{|\partial\Omega} = 0, \qquad 0 < \varepsilon < 1. \end{aligned}\right\} \tag{3.13}$$

Problem (3.13) is to be solved in a smooth bounded domain Ω of $\mathbb{R}^N, N = 2, 3$.

Let $\tilde{v} \in X_3 = \mathbf{H}^2 \cap V$, where $V = \{v \in \mathbf{H}^1(\Omega);\ \text{div } v = 0, \quad v_{|\partial\Omega} = 0\}$, be a solution of the steady Navier-Stokes equation,

$$\begin{aligned} &\text{Re}(\tilde{v} \cdot \nabla)\tilde{v} - \Delta\tilde{v} + \nabla\tilde{p} = f, \\ &\text{div}\,\tilde{v} = 0, \quad \tilde{v}_{|\partial\Omega} = 0. \end{aligned} \tag{3.14}$$

Note that $\tilde{v}$ is not assumed to be "small". We are asking the following question : does there exist a solution $(v, \underset{\sim}{\tau})$ of (3.13) which is close to $(\tilde{v}, 0)$?

Let $L[\tilde{v}]$ be the linearization of (3.14) around $\tilde{v}$, that is

$$L[\tilde{v}]v = \text{Re}\,[(\tilde{v} \cdot \nabla)v + (v \cdot \nabla)\tilde{v}] - \Delta v + \nabla p.$$

Theorem 3.3. *Let $\tilde{v} \in X_3$ be a solution of (3.14) such that $L[\tilde{v}]$ is an isomorphism. Assume moreover that (for some constant c_β depending only on β and Ω)*

$$c_\beta\,\text{We}\,\|\tilde{v}\|_3 < 1. \tag{3.15}$$

Then there exists $\varepsilon_0 > 0$ such that for any ε, $0 < \varepsilon \leq \varepsilon_0$, (3.13) admits a unique solution $(v_\varepsilon, \underset{\sim}{\tau}^\varepsilon) \in X_3 \times \mathbf{H}^2$. Moreover the sequence $(v_\varepsilon, \underset{\sim}{\tau}^\varepsilon) \to (\tilde{v}, 0)$ in $X_3 \times \mathbf{H}^2$ as $\varepsilon \to 0$.

Idea of the proof: It uses an iterative scheme inspired by (3.9) (3.10) (3.11). We look for a solution of (3.13) of the form $(\tilde{v} + v, \underset{\sim}{\tau})$, where v and $\underset{\sim}{\tau}$ are small. The main difference lies in the elliptic part (see (3.8) (3.9)). This elliptic part is now

$$\mathcal{L}w = L[\tilde{v}]w - \varepsilon\Delta w - \left(\mathrm{We}(\tilde{v} + v)\cdot\nabla + I\right)^{-1}\varepsilon\Delta w$$

$$- \left(\mathrm{We}(\tilde{v} + v)\cdot\nabla + I\right)^{-1}\mathrm{We}\,\ell_2\left(\partial^2 w, \underset{\sim}{\tau}\right),$$

where $\ell_2\left(\partial^2 w, \underset{\sim}{\tau}\right)$ is at least quadratic. $\mathcal{L}$ is an isomorphism provided that $L[\tilde{v}]$ is an isomorphism, (3.15) holds, and v and $\underset{\sim}{\tau}$ are small enough. ∎

3.3. Problems with inflow boundaries

It is well known that for Navier-Stokes equations, the prescription of the velocity field or of the traction on the boundary leads to a well posed problem. On the other hand, viscoelastic fluids have memory: the flow inside the domain depends on the deformations that the fluid has experimented before it entered the domain, and one needs conditions at the inflow boundary. For integral models, an infinite number of such conditions are required. For differential models, only a finite number of conditions are required (more and more as the number of relaxation times increases ...). The number of boundary conditions is model dependent (Maxwell/Jeffreys) and flow dependent (subcritical/supercritical). Determining the nature of the boundary conditions is by no way a trivial matter and no complete answer is known so far. We treat briefly two different approaches.

3.3.1. Perturbation of a uniform flow.

The first one, due to Renardy [51] [57] deals with a special nonlinear situation, namely the small perturbations of a uniform flow transverse to a strip, for a Maxwell fluid.

Then we consider a perturbation of a uniform flow transversal to a strip, with periodicity in the directions orthogonal to the flow. More precisely, let us consider a flow transverse to the strip $0 \leq x \leq 1$ with periodic boundary conditions in y and z (with periods L_y and L_z). In the absence of body forces, $(v = (U,0,0)^T, p = 0, \underset{\sim}{\tau} = 0)$ is a solution. We look for flows which are perturbations of such a flow. The aim of [51] [57] is to derive an existence theory for such flows, with appropriate boundary conditions, in case of Maxwell or Jeffreys models. The idea is to use suitable modifications of the algorithm described in the proof of Theorem 1, which lead to a well-posed problem. Instead of giving a precise statement of the results in [51] [57] we will just describe the inflow boundary conditions. One has first to impose the natural conditions:

$$\left.\begin{array}{lll} v = (U,0,0)^T + v_0(y,z) & \text{at} & x = 0 \\ v = (U,0,0)^T + v_1(y,z) & \text{at} & x = 1 \end{array}\right\} \qquad (3.16)$$

with the compatibility condition

$$\int_0^{L_z}\int_0^{L_y} v_0(y,z)dydz = \int_0^{L_z}\int_0^{L_y} v_1(y,z)dydz.$$

We write the total extra-stress as

$$\underline{\tau} = \underline{\tau}^s + \underline{\tau}^p \quad J, \quad \underline{\tau}^s = 2\eta_s D \quad \eta_s \geq 0,$$

$\underline{\tau}^p$ satisfies an equation of type (1.5).

Jeffreys models ($\eta_s > 0$). In this case all components of the elastic part $\underline{\tau}^p$ of the extra stress can be prescribed at $x = 0$.

Maxwell models ($\eta_s = 0$).

(i) Subcritical case (i.e. $U < \sqrt{\frac{\eta}{\rho\lambda}}$).

- In 2 space dimension, one can prescribe the diagonal components τ_{11}^p and τ_{22}^p.

- In 3-D, a correct choice of boundary conditions for $\underline{\tau}^p$ is not so simple. A possible choice of 4 boundary conditions is as follows. First expand each stress component in a Fourier series,

$$\tau_{11}^p(0,y,z) = \sum_{k,\ell} \tau_{11}^{k\ell} \exp\left(2i\pi\left(ky/L_y + \ell z/L_z\right)\right).$$

Then, one can prescribe

$$\begin{aligned}
&\tau_{11}^{k\ell}, \tau_{22}^{k\ell}, \tau_{13}^{k\ell}, \tau_{33}^{k\ell}, \quad \text{if} \quad |k| >> |\ell|, \\
&\tau_{11}^{k\ell}, \tau_{22}^{k\ell}, \tau_{12}^{k\ell}, \tau_{33}^{k\ell}, \quad \text{if} \quad |\ell| >> |k|, \\
&\tau_{11}^{k\ell}, \tau_{13}^{k\ell}, \tau_{23}^{k\ell}, \tau_{33}^{k\ell}, \quad \text{if} \quad |k| \text{ and } |\ell| \text{ are comparable,}
\end{aligned} \tag{3.17}$$

$$\tau_{11}^{k\ell}, \tau_{22}^{k\ell}, \tau_{23}^{k\ell}, \tau_{33}^{k\ell} \text{ if } k = \ell = 0. \tag{3.18}$$

An alternative approach leading to first order differential boundary conditions at the inflow boundary is presented in [54].

(ii) Supercritical case ($U > \sqrt{\frac{\eta}{\rho\lambda}}$).

The choice of the boundary conditions (3.17) (3.18) leads in this case to an ill-posed problem (as the Dirichlet problem for a hyperbolic equation), as shown by Renardy [56]. In addition to the normal velocities at both boundaries and to the previous inflow conditions on the stresses, one can prescribed the vorticity and its normal derivative in two dimensions, or the second and third components of the vorticity and their normal derivatives in three dimensions.

3.3.2. Absorbing boundary conditions for viscoelastic fluids.

We describe briefly here some results from Tajchman's Doctoral Thesis [69]. Interest is focused here on models described by (see (3.1)-(3.3))

$$\text{Re}\left[v_t + (v\cdot\nabla)v\right] + \nabla p + (1-\varepsilon)\Delta v = \text{div}\,\underline{\tau} + f, \tag{3.19}$$

$$\mathrm{We}\,\frac{\mathcal{D}_a \underline{\tau}}{\mathcal{D}t} + \underline{\tau} = 2\varepsilon D[v], \tag{3.20}$$

$$\mathrm{div}\;\; v = 0, \tag{3.21}$$

or

$$p_t + \beta\,\mathrm{div}\;\; v = 0, \tag{3.21$'$}$$

according to whether we consider the incompressible case ($\beta = +\infty$) or the slightly compressible case ($\beta > 0$).

We assume that the geometry of the flow is "infinite" or very large (flow around an obstacle, "infinite" die, ...). The flow at infinity is assumed to be known (uniform, Poiseuille flow, ...). For computational purposes, one introduce artificial boundaries, at a finite, hopefully not too large, distance. The problem is to know what to impose on the artificial boundaries in order to obtain a solution of the truncated problem which is as close as possible to the solution of the original problem.

Such considerations were first carried out by Engquist and Majda [14] for wave equations (linear hyperbolic equations), and developed by many authors later on. In particular, Halpern [22] considered the case of parabolic perturbations of hyperbolic systems.

From physical considerations (plane waves travelling through the fluid), one gets boundary conditions which make the artificial boundaries transparent to the waves leaving the computational domain and absorb the waves entering the domain (other than those generated by the solution at infinity).

We shall make the hypothesis that the artificial boundaries are far enough in order to justify the linearizations around the solution at infinity. The linearized problem reads

$$\left.\begin{aligned}
&\mathrm{Re}\left[v_t + (v_\infty \cdot \nabla)\, v + (v \cdot \nabla) v_\infty\right] + \nabla p - (1-\varepsilon)\Delta v = \mathrm{div}\;\; \underline{\tau},\\
&\mathrm{We}\left(\frac{\mathcal{D}_a \underline{\tau}}{\mathcal{D}t}\right)_{\text{linearized}} + \underline{\tau} = 2\varepsilon D[v],\\
&\mathrm{div}\;\; v = 0 \quad \text{or} \quad p_t + \beta\,\mathrm{div}\;\; v = 0.
\end{aligned}\right\} \tag{3.22}$$

Looking for plane waves solutions amounts to testing non trivial solutions of the type

$$U(s,\psi)\exp\, i\left[st + \xi(s,\psi)x_1 + \sum_{k=2}^{N}\psi_k x_k\right], \qquad \psi = (\psi_2, \cdots, \psi_N)$$

where N is the space dimension, and x_1 is the normal direction to the artificial boundary, outwards pointing.

The sign of the real part of these solutions determines the direction of propagation of the corresponding wave. For each wave entering the domain of computation one imposes a boundary condition which eliminates it:

$$\iint V(s,\psi)^T \hat{v}\; ds\; d\psi = 0,$$

when V is the corresponding left eigenvector, and $\hat{v}$ denotes the Laplace transform of v with respect to t and the Fourier transform with respect to $x' = (x_2, \cdots, x_n)$.

These conditions are not local (they are integral relations difficult to incorporate into a numerical code). By perturbation techniques, one gets *local* approximations in terms of partial differential equations on the boundary. Higher order approximations can be obtained at the price of an increasing difficulty in the computations.

To be of any practical use, the artificial boundary conditions have to be stable, in particular they should not depend on rounding errors. The stability analysis is made on the linearized problem, by computing the time evolution of the solution norms.

Finally one can compare the absorbing conditions which are obtained in different cases, e.g. the limit $\varepsilon \to 1$ or $\beta \to +\infty$.

The results depend crucially on the subcritical or supercritical nature of the flow ($\mathrm{Re}\,\mathrm{We} \gtrless 1$). We refer to [69] for a precise description.

4 – Unsteady flows

4.1. Existence results

We first consider Jeffreys type models, namely (3.1) (3.2) (3.3) with $\varepsilon < 1$, which we complete with the Dirichlet boundary condition

$$v_{|\partial\Omega} = 0, \tag{4.1}$$

and the initial values

$$\left.\begin{aligned} v_{|t=0} &= v_0 \\ \underline{\tau}_{|t=0} &= \underline{\tau}_0 . \end{aligned}\right\} \tag{4.2}$$

To start with, we state a local existence theorem. In what follows, we shall denote as previously the Sobolev spaces of vector or tensor valued functions by $\mathbf{H}^k = \mathbf{H}^k(\Omega)$.

Theorem 4.1. *[17] Let Ω be a bounded domain of $\mathbb{R}^N$, $N = 2, 3$ with C^3 boundary. Let $f \in L^2_{\mathrm{loc}}(\mathbb{R}_+; \mathbf{H}^1)$, $f' \in L^2_{\mathrm{loc}}(\mathbb{R}_+; \mathbf{H}^{-1})$, $v_0 \in \mathbf{H}^2 \cap \mathbf{H}^1_0$, div $v_0 = 0$, $\underline{\tau}_0 \in \mathbf{H}^2$.*

Then there exist $T_* > 0$ and a unique solution $(u, \underline{\tau}, p)$ of (3.1)-(3.3), (4.1), (4.2) which satisfies

$$v \in L^2\left(0, T^*; \mathbf{H}^3\right) \cap C\left([0, T^*); \mathbf{H}^2 \cap \mathbf{H}^1_0\right),$$

$$v' \in L^2\left(0, T^*; \mathbf{H}^1_0\right) \cap C\left([0, T^*); \mathbf{L}^2\right),$$

$$p \in L^2\left(0, T^*; \mathbf{H}^2_0\right),$$

$$\underline{\tau} \in C\left([0, T^*); \mathbf{H}^2\right).$$

The proof of Theorem 4.1 is obtained by Schauder's fixed point theorem.

A similar result has been proved by Hakim [20] for a class of White-Metzner models (still with $\varepsilon < 1$) under the following assumptions on the constitutive functions $\lambda(\mathrm{II})$,

η(II):

$$\left.\begin{aligned}
&\lambda(x) > 0, \quad \eta(x) > 0 && \forall\, x \in \mathbb{R}_+,\\
&\max\left\{\left|\frac{\lambda'(x)}{\lambda^2(x)}\right|, \left|\frac{\lambda''(x)}{\lambda^2(x)}\right|, \left|\frac{\lambda'^2(x)}{\lambda^3(x)}\right|\right\} \le M && \forall\, x \in \mathbb{R}_+,\\
&\left|\tfrac{1}{\lambda(x)} - \tfrac{1}{\lambda(y)}\right| \le M|x-y| && \forall\, x \in \mathbb{R}_+,\\
&|\mu'(x)| \le M, \quad |\mu''(x)| \le M(1+x) && \forall\, x \in \mathbb{R}_+.
\end{aligned}\right\} \tag{4.3}$$

These assumptions are satisfied in particular by the Carreau or Gaidos-Darby laws.

Remarks.

1. The result in Theorem 4.1 does not depend on the precise nature of the term $\beta(\nabla v, \underline{\tau})$ and thus is model independent.

2. If more regularity is assumed on the data, then more regularity of the solution is obtained provided an additional compatibility condition is imposed.

We now turn to local existence for the Maxwell type models. The situation is much more tricky here since these models can display Hadamard instabilities (see Section 2.1), and no general result seem to be known so far. One has anyway to restrict initial data to "Hadamard stable ones". A possible way to overcome the difficulty is to consider models satisfing an ellipticity condition, which will imply well-posedness. This approach was followed by Renardy [55], whose results are briefly described below.

The extra-stress tensor $\underline{\tau} = (\tau_{ij})$ is assumed to satisfy an equation of the form

$$\left(\frac{\partial}{\partial t} + v\cdot\nabla\right)\tau_{ij} = A_{ijk\ell}(\underline{\tau})\frac{\partial v_k}{\partial x_\ell} + g_{ij}(\underline{\tau}), \tag{4.4}$$

where (due to frame-indifference)

$$A_{ijk\ell}(\underline{\tau}) = \frac{1}{2}\left[\delta_{ik}\,\tau_{\ell j} - \delta_{i\ell}\,\tau_{kj} - \tau_{ik}\,\delta_{\ell j} + \tau_{i\ell}\,\delta_{kj}\right] + B_{ijk\ell}(\underline{\tau}),$$

$B_{ijk\ell}$ is symmetric in k and ℓ (and of course in i and j), and satisfies

$$B_{ijk\ell} = B_{k\ell ij}.$$

We set

$$C_{ijk\ell} = B_{ijk\ell} + \frac{1}{2}\left[\delta_{ik}\,\tau_{\ell j} - \delta_{i\ell}\,\tau_{kj} - \tau_{ik}\,\delta_{\ell j} + \tau_{i\ell}\,\delta_{kj}\right],$$

so that

$$A_{ijk\ell} = \tau_{i\ell}\,\delta_{kj} + C_{ijk\ell},$$

and

$$C_{ijk\ell} = C_{k\ell ij}.$$

The crucial hypothesis is the strong ellipticity condition,

$$C_{ijk\ell}(\underline{\tau})\,\varsigma_i\,\varsigma_k\,\eta_j\,\eta_\ell \ge \kappa(\underline{\tau})|\varsigma|^2|\eta|^2 \qquad \forall\, \varsigma, \eta \in \mathbb{R}^3, \tag{4.5}$$

where $\kappa(\underset{\sim}{\tau}) > 0$.

Note that, for Maxwell models with $-1 \leq a \leq 1$ (4.5) is satisfied locally in time provided it is satisfied at $t = 0$. For $a = \pm 1$, (4.5) is equivalent to (2.3), and this is a natural condition to impose on the stress. On the other hand, for $a \neq \pm 1$, condition (4.5) is a demonstration of the fact that the model is not always of evolution type.

Under hypothesis (4.5) and some smoothness conditions on the (bounded) domain of the flow, and on the functions $A_{ijk\ell}$ and g_{ij}, Renardy proves the local existence and uniqueness of a $\mathbf{H}_0^1 \cap \mathbf{H}^4$ solution of (1.12), (1.13), (4.4) provided that the initial data v_0 and $\underset{\sim}{\tau}_0$ are smooth, and satisfy a compatibility condition at $t = 0$.

Concerning global (in time) existence of solutions, the only general result seems to be the following (Guillopé-Saut [17]) which is established for Jeffreys models with a sufficiently small retardation parameter.

Theorem 4.2. *Assume that Ω in Theorem 4.1. is of class C^4. There exists $\varepsilon_0, 0 < \varepsilon_0 < 1$ (depending on Ω) such that if $0 < \varepsilon < \varepsilon_0$ and if $v_0 \in \mathbf{H}^2(\Omega) \cap \mathbf{H}_0^1(\Omega)$, div $v_0 = 0, f \in L^\infty(\mathbb{R}_+; \mathbf{H}^1(\Omega)), f' \in L^\infty(\mathbb{R}_+; \mathbf{H}^{-1}(\Omega))$ are small enough, then (3.1)-(3.3), (4.1), (4.2) admit a unique solution*

$$v \in C_b\left(\mathbb{R}_+; \mathbf{H}^2(\Omega) \cap \mathbf{H}_0^1(\Omega)\right) \cap L^2_{\text{loc}}\left(\mathbb{R}_+; \mathbf{H}^3(\Omega)\right),$$

$$v' \in C_b\left(\mathbb{R}_+; \mathbf{L}^2(\Omega)\right) \cap L^2_{\text{loc}}\left(\mathbb{R}_+; \mathbf{H}_0^1(\Omega)\right),$$

$$\underset{\sim}{\tau} \in C_b\left(\mathbb{R}_+; \mathbf{H}^2(\Omega)\right), \quad \underset{\sim}{\tau}' \in C_b\left(\mathbb{R}_+; \mathbf{H}^1(\Omega)\right).$$

Idea of the proof: Using energy methods, one shows that the quantity

$$Y(t) = \frac{(1-\varepsilon)^2}{19\ 2\ \text{Re}^2} \|v(t)\|_1^2 + \text{Re}\,\|v'(t)\|_0^2 + \frac{\text{We}}{2\varepsilon} \|\underset{\sim}{\tau}'(t)\|_0^2$$

$$+ \frac{\text{We}}{\text{Re}^3} \frac{(1-\varepsilon)^3}{3c_0\varepsilon^3} \|\underset{\sim}{\tau}(t)\|_2^2$$

satisfies

$$Y' = \lambda Y \leq \alpha\left(Y^2 + Y^3 + Y^6\right) + \beta, \quad \lambda > 0, \alpha > 0$$

where $\beta > 0$ is small when the data are small.

Remarks.

1. The restriction $0 < \varepsilon < \varepsilon_0$ is due to the treatment of the boundary conditions in the linear coupled terms.

2. No result such as Theorem 4.2 seems to be known for Maxwell models. We have however to mention Kim [29] where the upper-convected Maxwell model in the whole space $\mathbb{R}^3$ is considered.

3. A similar result for White-Metzner-Jeffreys model is proven by Hakim [20] under the hypothesis (4.3) and $\lambda(x) \leq M, \forall\, x \in \mathbb{R}_+$.

One can also get a stability result, in the context of Theorem 4.1.

Theorem 4.3. *Let ε and Ω be as in Theorem 4.1. Let $v_0^i \in \mathbf{H}^2(\Omega) \cap \mathbf{H}_0^2(\Omega)$, $\operatorname{div} v_0^i = 0$, $\underline{\tau}_0^i \in \mathbf{H}^2(\Omega)$, $i = 1, 2$, $f \in L^\infty(\mathbb{R}_+; \mathbf{H}^1(\Omega))$, $f' \in L^\infty(\mathbb{R}_+; \mathbf{H}^{-1}(\Omega))$ be small enough. Then the global corresponding solutions $\left(v^i, \underline{\tau}^i\right)$ of (3.1)-(3.3), (4.1), (4.2) satisfy*

$$\begin{aligned} &\|v^1(t) - v^2(t)\|_0^2 + \|\underline{\tau}^1(t) - \underline{\tau}^2(t)\|_0^2 \le \\ &C\left(\|v_0^1 - v_0^2\|_0^2 + \|\underline{\tau}_0^1 - \underline{\tau}_0^2\|_0^2\right) e^{-\delta t}, \quad \delta > 0,\ \forall t > 0. \end{aligned} \tag{4.6}$$

By classical arguments (see Serrin [65], Valli [73]) one obtains from Theorem 4.3 :

Corollary 4.4. *Let ε and Ω be as in Theorem 4.1.*

(i) *Let $f \in L^\infty(\mathbb{R}_+; \mathbf{H}^1(\Omega))$, $f' \in L^\infty(\mathbb{R}_+; \mathbf{H}^{-1}(\Omega))$ be time periodic of period $T > 0$, and small enough. Then there exists a T-periodic solution of (3.1)-(3.3), (4.1), (4.2), unique among the small T-periodic solutions.*

(ii) *If moreover f is time independent, there exists a steady solution, which is unique among the small steady solutions.*

(iii) *The aforementioned solutions are stable in the Lyapunov sense.*

Remarks and open problems.

1. Corollary 4.3 obviously implies the stability of the null solution (the rest state). This fact is not known without assuming the restriction on ε (and a fortiori for Maxwell models.)

2. The existence of (small) time periodic solutions for Maxwell models is unknown, as is that of arbitrary (not small) periodic solutions for Jeffreys models.

3. Nothing is known concerning the global existence of solutions of unsteady flows for differential models in 2 or 3 space dimensions (weak solutions, singularities in finite time, ...). See below for some situations in one dimension.

More specific results can be obtained in some 1-D situations which we describe now.

Following [18] we consider shearing motions and Poiseuille flows of an Oldroyd fluid. The dimensionless equations are easily reduced to a system for the velocity $v(x,t)$, the shear stress $\tau(x,t)$ and a linear combination of normal stresses $\alpha(x,t)$, $x \in \mathrm{I}$, $t \ge 0$,

$$\left.\begin{aligned} &\mathrm{Re}\ v_t - (1-\varepsilon)v_{xx} = \tau_x - f, \\ &\alpha_t + \frac{\alpha}{\mathrm{We}} = (1 - a^2)\tau v_x, \\ &\tau_t + \frac{\tau}{\mathrm{We}} = \left(\frac{\varepsilon}{\mathrm{We}} - \alpha\right) v_x. \end{aligned}\right\} \tag{4.7}$$

For Couette flow, one has $\mathrm{I} = (0,1)$ and $f = 0$, while for Poiseuille flow $\mathrm{I} = (-1,1)$, and $f = 1$ is the (constant) pressure gradient in the flow direction. The boundary conditions are

$$\left.\begin{aligned} &v(t,-1) = v(t,1) = 0 \qquad t \ge 0 \quad \text{for the Poiseuille flows,} \\ &\qquad\qquad\text{and} \\ &v(t,0) = 0, \quad v(t,1) = 1 \quad t \ge 0 \qquad \text{for Couette flow.} \end{aligned}\right\} \tag{4.8}$$

The parameter a will be assumed to satisfy $-1 < a < 1$. The crucial observation is the following a priori bound on the stress components

$$\left.\begin{aligned}&\left(\alpha(x,t)-\frac{\varepsilon}{\mathrm{We}}\right)^2+(1-a^2)\tau^2(x,t)\\&\le\left[\left(\alpha(x,0)-\frac{\varepsilon}{\mathrm{We}}\right)^2+(1-a^2)\tau^2(x,0)\right]e^{-t/\mathrm{We}}\\&+\left(\frac{\varepsilon}{\mathrm{We}}\right)^2\left(1-e^{-t/\mathrm{We}}\right),\quad (x,t)\in I\times\mathbb{R}_+.\end{aligned}\right\}\tag{4.9}$$

Using the bound (4.9), we can then prove the

Theorem 4.5. *Let* $0 < \varepsilon < 1$.

(i) *Uniqueness. There exists at most one solution* (v, α, τ) *of (4.7) (4.8) in the space* $[L^\infty(\mathbb{R}_+; L^2) \cap L^2_{\mathrm{loc}}(\mathbb{R}_+; H^1)] \times [L^\infty(I \times \mathbb{R}_+)]^2$.

(ii) *Existence. Let* $v_0, \alpha_0, \tau_0 \in H^1_0(I)$. *Then for any* $T > 0$, *there exists a unique solution* (v, α, τ) *of (4.7) (4.8) with initial data* (v_0, α_0, τ_0) *in the space* $[C([0,T]; H^1) \cap L^2(0,T; H^2)] \times [C([0,T]; H^1)]^2$. *Moreover* $v \in C_b(\mathbb{R}_+; L^2)$ *and the bound (4.9) holds true.*

Remarks.

1. In the case $\varepsilon = 1$ (Maxwell models), system (4.7) is not always of evolution type (see Section 2.1). However, Hrusa, Nohel and Renardy [49] have constructed initial data in the hyperbolic domain, with steep gradients such that the velocity and the stress develop singularities in the first space derivatives in finite time. The idea is to reduce by a clever change of variable to a degenerate system of three nonlinear hyperbolic equations.

2. The result in Theorem 4.5 depends crucially on the model (an Oldroyd model). It would be interesting to know what happens for 1-D flows of general differential models with a Newtonian contribution.

3. Another proof of Theorem 4.5 has been obtained by Malkus, Nohel and Plohr [35].

4.2. Stability issues

They are one of the main challenging problems in the theory of viscoelastic fluids. We shall consider only bounded geometries. Let us briefly review the goals and main difficulties. One wants to explain (predict, avoid, ...) the instabilities occuring in polymer processing. When polymer melts are extruded from a pipe, instabilities often occur at a critical value of the wall shear stress. They are known as spurt flows, shark skin defects, melt fracture,.... They manifest themselves in an increase of the flow rate for a given pressure gradient, irregularities on the surface of the extrudate, pressure oscillations, chaotic behavior,... (see Tordella [71], El Kissi-Piau [12] [42]). Their physical explanation is not well understood. Possible causes could be slip at the wall (interaction of the fluid with the wall of the pipe), change of type, constitutive instabilities (non monotone shear stress / shear rate curve), propagation of defects in the pipe,... . In any case they differ drastically from the classical "hydrodynamic instabilities" which occur in Newtonian flows at high Reynolds numbers. They happen at moderate Reynolds numbers and are basically due to the elastic effects. Note that the viscoelastic analogue of the classical hydrodynamic instabilities (Bénard and Taylor problems) can in principle be studied, at least formally, by the methods of bifurcation theory. (See for instance the recent work of Renardy and Renardy [59] on the Bénard problem, and of Avgousti and Beris [2] on the Taylor problem.).

From a fundamental point of view, none of the theoretical results necessar to justify rigorously bifurcation or stability studies have been established so far for the equations governing viscoelastic fluids (contrarily to the Newtonian case). Such results concern for instance the relation between linear stability and spectrum, between linear and nonlinear stability, the reduction of the dynamics to a center manifold,...

The difficulties are due to the partial hyperbolicity of the problem: non-Newtonian fluids with memory do not generally lead to evolution equations associated with analytic semi-groups. (The difficulty is the same for Hamiltonian systems or for compressible viscous fluids.)

Consider for instance the linearized equations of motion around a given solution $(\tilde{u},\tilde{\underline{\tau}}) \in \mathbf{W}^{1,\infty}(\Omega) \times \mathbf{W}^{1,\infty}(\Omega)$, with div $\tilde{v} = 0$, and $\tilde{v} \cdot n_{|\partial\Omega} = 0$. They read (see [19])

$$\left.\begin{aligned} &\mathrm{Re}\,[(\tilde{v}.\nabla)v + (v\cdot\nabla)\tilde{v}] - (1-\varepsilon)\Delta v + \nabla p - \operatorname{div}\,\underline{\tau} = 0,\\ &\mathrm{We}\,[(\tilde{v}\cdot\nabla)\underline{\tau} + (v\cdot\nabla)\tilde{\underline{\tau}}] + \beta(\nabla v,\tilde{\underline{\tau}}) + \beta(\nabla\tilde{v},\underline{\tau}) + \underline{\tau} = 2\varepsilon D[v],\\ &\operatorname{div}\ v = 0,\\ &v_{|\partial\Omega} = 0. \end{aligned}\right\} \tag{4.10}$$

We first introduce the classical spaces (see [70])

$$\begin{aligned} \mathcal{V} &= \left\{u \in C_0^\infty(\Omega)^3 \ \ \operatorname{div}\ u = 0\right\},\\ V &= \text{closure of } \mathcal{V} \text{ in } \mathbf{H}^1,\\ H &= \text{closure of } \mathcal{V} \text{ in } \mathbf{L}^2. \end{aligned}$$

P denotes the orthogonal projection on H in $\mathbf{L}^2$. We also set $\mathcal{H} = H \times \mathbf{L}^2_{\mathrm{sym}}$, where $\mathbf{L}^2_{\mathrm{sym}}(\Omega)$ denotes the $\mathbf{L}^2$ valued symmetric tensors.

Lemma 4.6.

(i) *Jeffreys models* ($0 < \varepsilon < 1$). *Let $\mathcal{A}$ be the unbounded operator in $\mathcal{H}$, defined by its domain*

$$D(\mathcal{A}) = \left\{(u,\underline{\tau}) \in V \times \mathbf{L}^2_{\mathrm{sym}}, P((1-\varepsilon)\Delta u + \operatorname{div}\underline{\tau}) \in H, (\tilde{v}\cdot\nabla)\underline{\tau} \in \mathbf{L}^2\right\},$$

and $\mathcal{A}(u,\underline{\tau})$ by (4.10). Then $-\mathcal{A}$ is the infinitesimal generator of a C^0 semi-group in $\mathcal{H}$.

(ii) *Maxwell models* ($\varepsilon = 1$). *We assume moreover that $\tilde{v} \in \mathbf{W}^{2,\infty}(\Omega)$. Let us define $\mathcal{B}$ by its domain*

$$D(\mathcal{B}) = \left\{(u,\underline{\tau}) \in V \times \mathbf{L}^2_{\mathrm{sym}}, \qquad P\ \operatorname{div}\ \underline{\tau} \in H, (\tilde{v}\cdot\nabla)\underline{\tau} \in \mathbf{L}^2\right\},$$

and $\mathcal{B}(u,\underline{\tau})$ by (4.10) with $\varepsilon = 1$.

Then, assuming that $\mathrm{We}\left(\mathrm{Re}\,\|\tilde{v}\|_\infty^2 + C_\beta \|\tilde{\underline{\tau}}\|_\infty\right) < 1$, $-\mathcal{B}$ *is the generator of a* C^0 *semi-group in* $\mathcal{H}$.

Remark. The smallness condition in the Maxwell case, although not optimal, is not surprising in view of the possible Hadamard instabilities inherent to Maxwell models (see Section 2.1).

To illustrate the fact that the semi-groups $\mathcal{A}$ or $\mathcal{B}$ are not in general analytic, we consider the case where $\tilde{u} = 0$ and $\tilde{\underline{\tau}} = 0$, and show that $\mathcal{B}$ is not sectorial. In fact the spectrum of $\mathcal{B}$ is as follows. It contains the eigenvalues $\xi = 1/\mathrm{We}$ (with eigenspace $= \{(0, \underline{\tau}), \mathrm{div}\ \underline{\tau} = \nabla p\}$, and the sequence of eigenvalues

$$\xi_k = \frac{1 \pm i\sqrt{4\Lambda_k \mathrm{We}/\mathrm{Re} - 1}}{2\,\mathrm{We}}, \qquad k = 1, 2, \cdots,$$

if $\Lambda_1 > \mathrm{Re}/4\,\mathrm{We}$. On the other hand, if $\Lambda_1 < \mathrm{Re}/4\,\mathrm{We}$, the spectrum of $\mathcal{B}$ is

$$\tilde{\xi}_k = \frac{1 \pm \sqrt{1 - 4\Lambda_k \mathrm{We}/\mathrm{Re}}}{2\,\mathrm{We}} \qquad k = 1, 2, \cdots, k_0 - 1,$$

where $k_0 = \mathrm{Inf}\{k, \Lambda_k > \mathrm{Re}/4\,\mathrm{We}\}$, and

$$\xi_k = \frac{1 \pm i\sqrt{4\Lambda_k \mathrm{We}/\mathrm{Re} - 1}}{2\,\mathrm{We}}, \qquad k \geq k_0.$$

In the previous formulas, $0 < \Lambda_1 < \Lambda_2 < \cdots$ indicates the increasing sequence of eigenvalues of the Stokes operator.

In any case the spectrum of $\mathcal{B}$ contains an unbounded set of eigenvalues lying on a parallel to the imaginary axis wich implies that $\mathcal{B}$ is not sectorial and cannot generate a analytic semi-group [40].

We now turn to describing rigorous results concerning stability issues for viscoelastic flows.

4.2.1. Nonlinear stability.

We mean here stability in the Lyapunov sense. Due to the lack of connection between linearized and nonlinear stability, the only known results are obtained by energy methods [18], or by a singular perturbation analysis [38].

Stability of small steady or time-periodic solutions have been already mentioned in Section 4.1 for Jeffreys models with a sufficiently large Newtonian contribution. We shall focus now on the stability of special one dimensional flows of Oldroyd type models [18] [38].

To start with, we describe the steady plane Couette, and Poiseuille flows i.e. the steady solutions of (4.7) with the boundary conditions (4.8).

The case of the plane Couette flow is easy. One finds

$$\left.\begin{aligned} &v(x) = x \qquad 0 \leq x \leq 1, \\ &\tau = \frac{\varepsilon}{1 + k^2}, \quad \alpha = \frac{\varepsilon k^2}{\mathrm{We}(1 + k^2)} \quad \text{where } k^2 \underset{\mathrm{def}}{=} \mathrm{We}^2(1 - a^2). \end{aligned}\right\} \tag{4.11}$$

The situation for plane Poiseuille flow is not as simple as shown by the following results.

Lemma 4.7.

(i) *Let* $\varepsilon \in [0, 8/9]$. *Then there exists a unique steady solution* $v = v_\varepsilon(x)$, *which is* C^∞ *on* $(0, 1)$.

(ii) *Let* $\varepsilon \in (8/9, 1)$. *There exists a critical Weissenberg parameter* $k_c^\varepsilon > 0$ *such that*

(a) *if* $k \leq k_c^\varepsilon$, *the conclusion of (i) holds;*

(b) *if* $k > k_c^\varepsilon$, *there does not exist any* C^1 *solution, but a continuum of* C^0 *solutions which are* C^∞ *except at a finite number of points.*

Here is a typical half profile of Poiseuille flows.

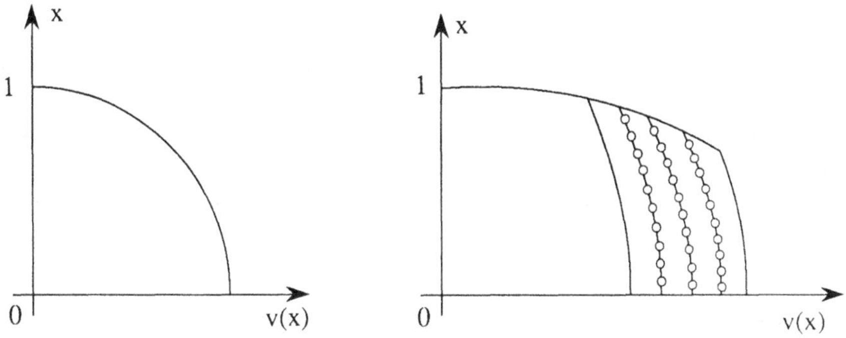

$0 < \varepsilon < 8/9$ $\qquad$ $8/9 < \varepsilon < 1, k > k_c^\varepsilon$

Fig. 4.1

Concerning the Lyapunov stability of the Couette flow under 1-D perturbations, we have for instance [18].

Theorem 4.8.

(i) *Let* $\varepsilon \in (0, 8/9)$. *Then the Couette flow is* H^1*-stable.*

(ii) *Let* $\varepsilon \in (8/9, 1)$. *Then the Couette flow is* H^1*-stable for* $k < k_1(\varepsilon)$.

Remark. As we mentioned before, Theorem 4.8 is obtained by energy methods. We refer to [18] for a proof, and other related results, e.g. sufficient conditions on ε and k for unconditional stability in L^2.

Results of stability for nonregular Poiseuille flows, described in Lemma 4.6, at small Reynolds number are obtained by Nohel and Pego [38].

More precisely the steady Poiseuille flows for which the velocity takes its values in the increasing part of the shaped curve (see Figure 4.2) only (excluding a neighborhood of the max and the min) are nonlinearly stable. Stability holds when the perturbations from a steady state are small in H^1 for the total shear stress, small in L^1 and bounded pointwise by some large constant for the normal stresses.

Moreover, if Re is small enough, every unsteady Poiseuille flow converges, as $t \to \infty$, to some steady state, possibly nonregular, possibly unstable.

The proofs of these results rely on the geometric study of the approximated dynamical system obtained at zero Reynolds number. This system has been thoroughly studied in [36], while [37] was devoted to a model system with only one equation for the stress.

4.2.2. Linearized stability.

We mean here that the solutions of the linearized unsteady equations around the given steady solution decay to zero as $t \to +\infty$. For Newtonian flows in bounded geometries this holds true provided the spectrum of the (steady) linearized operator lies in the left hand side of the imaginary axis. Due to the lack of analyticity of the underlying semi-groups the link between the spectrum and linearized stability is not clear for viscoelastic flows. (See Pazy [40] for an example of a C^0 semi-group with "good" spectrum, and unstable associated linear equation.)

There are very few rigorous results concerning the linear stability of viscoelastic flows. One of them deals with 1-D perturbations of the plane Couette flow. The rather special form of the flow and perturbations lead to a kind of degeneracy which induces an *analytic* underlying semi-group.

The result are better understood if we plot the curve of the total shear stress $\tilde{\tau}^*(\gamma)$ versus the shear rate $\gamma = U/L$ (U is the velocity of the upper plate and L the distance between the two parallel plates).

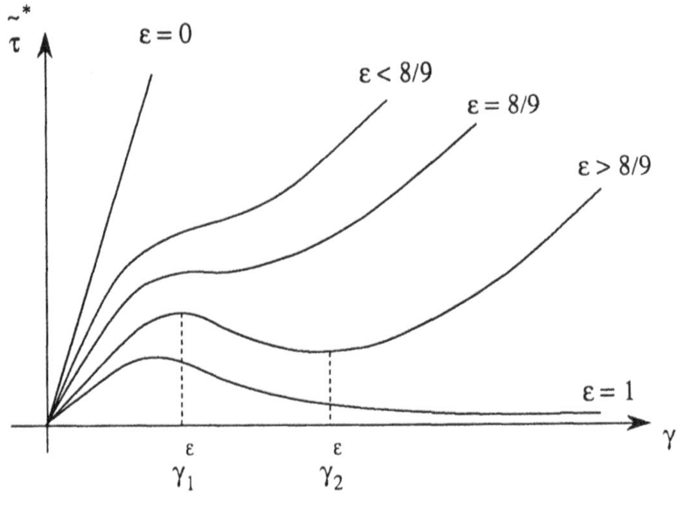

Fig. 4.2.

Theorem 4.9. [18]

(i) *If $0 < \varepsilon < 8/9$, the Couette solution (v_s, τ_s, α_s) is linearly stable for all k.*

(ii) *If $8/9 < \varepsilon < 1$, (v_s, τ_s, α_s) is linearly stable if and only if $0 \le k < k_-$ or $k > k_+$*

The stability regions are exactly those where the curve in Figure 4.2 is monotone (i.e. they correspond to $0 < \gamma < \gamma_1^\varepsilon$ or $\gamma > \gamma_2^\varepsilon$)

The stability is lost when the spectrum of the linearized operator crosses the imaginary axis at 0, eigenvalue of infinite multiplicity.

The proof of Theorem 4.9 consists in proving that the underlying semi-group is analytic (because of the degeneracy of the equation for α and τ in (4.7)), and then in localizing the spectrum by the Routh-Hurwitz criterion.

Remarks.

1. It seems natural to conjecture that the linear stability regions in Theorem 4.8 coincide with the *nonlinear* stability regions. (Theorem 4.7 gives the answer for part of the first increasing portion of the S-shaped curve in Figure 4.2.) No complete proof seems available.

2. The instability in Theorem 4.9 is due to the aforementioned S-shaped curve. This constitutive instability (inherent to the model) has been questioned as unphysical by some authors.

Another result we know of is due to Renardy [58]. It concerns the linear stability of the Couette flow for two-dimensional disturbances, in the case of the Maxwell upper convected (UCM) model, when *inertia is neglected*. Using the special form of the equations and a stream function formulation, Renardy proves that the (linearized) two-dimensional disturbances decay exponentially to 0 as $t \to +\infty$. The problem is open for other models and/or without neglecting the inertia terms.

Finally we comment briefly on weakly elastic fluids (see [19] and Section 3.2).

Let c_1 be the constant of continuity defined by

$$|((u \cdot \nabla)v, u)| \leq c_1 \|v\|_1 \|u\|_1^2 \quad u, v \in V.$$

We assume that the given Newtonian solution $\tilde{v}$ satisfies

$$\|\tilde{v}\|_1 < (c_1 \operatorname{Re})^{-1}. \tag{4.12}$$

Note (see e.g. Prodi [43]) that this condition insures that $\tilde{v}$ is asymptotically (Lyapunov) stable. Concerning the linear stability of the viscoelastic solution $(v, \underline{\tau})$ close to $(\tilde{v}, 0)$, we have

Theorem 4.10. *Assume (3.15) and (4.12). Then, for any $\varepsilon > 0$ small enough, the solution $(v_\varepsilon, \underline{\tau}_\varepsilon)$ of (3.13) is linearly asymptotically stable.*

The proof of Theorem 4.10 lies on the following result of Prüss [44] and on the localization of the spectrum of a perturbation of a Stokes-like operator

Lemma 4.11. [44] *Let $\mathcal{A}$ be the infinitesimal generator of a C^0 semi-group in a Hilbert space H. The zero solution of the equation*

$$\frac{dv}{dt} + \mathcal{A}v = 0$$

is linearly asymptotically stable if and only if the resolvent set of $\mathcal{A}$ contains the set $\{\xi \in \mathbb{C}, \operatorname{Re}\xi \leq 0\}$, and there exists a constant $M \geq 1$ such that

$$\|R(\xi; \mathcal{A})\|_{\mathcal{L}(H)} \leq M \qquad \forall\, \xi \in \mathbb{C}, \operatorname{Re}\xi \leq 0.$$

We refer to [19] for an application of Theorem 4.10 to the Bénard problem for a weakly viscoelastic fluid.

4.2.3. Spectral studies.

As we mentioned before no theoretical results exist which would justify the usual procedure of calculating the eigenvalues of the linearization to asserts the stability of viscoelastic flows. Despite this fact, a lot of studies have been devoted to such computations, and it is impossible to quote them all. We have therefore selected significant examples, and quoted a few others.

The rough idea is to look for plane waves perturbations $e^{\sigma t}e^{ik\cdot x}$ leading to spectral problems in σ, which can be solved numerically. We emphasize once again that this tells nothing about the nonlinear (even the linear !) stability of the flow.

M. Renardy and Y. Renardy [48] [52] have investigate the stability of the plane Couette flow for Maxwell models involving the derivative (1.7).

The flow lies between parallel plates at $y = \pm 1$ which are moving in the x direction with velocities ± 1. The basic Couette solution is $v = (y, 0, 0)$ which yields the following stresses

$$\tau_{12} = \frac{1}{1+k^2}, \quad \tau_{11} = \frac{(a+1)\,\mathrm{We}}{1+k^2}, \quad \tau_{22} = -\frac{(1-a)\,\mathrm{We}}{1+k^2},$$

with $k^2 = \mathrm{We}^2(1-a^2)$. One adds to the basic flow a perturbation which has a velocity only in the z-direction :

$$v = ye_x + \varepsilon w(y) e^{i\alpha x + \sigma t} e_z.$$

The equations of motion and the constitutive relation are then linearized with respect to ε. To avoid some possible unphysical instabilities, the range of the parameter a, which a priori belongs to $(-1, 1)$ is restricted to $a \geq 1/2$, and $k \leq 1$. The first restriction is to insure that the model is consistent with rod climbing (a well known viscoelastic effect), the later insures that one stays on the increasing part of the curve of the shear stress as a function of the shear rate (this function has a maximum for $k = 1$).

The eigenvalue problem is solved numerically by the Chebishev-tau method [52]. It can be determined analytically that there is a continuous spectrum given by

$$\mathrm{Re}\,\sigma = \frac{1}{\mathrm{We}}\left[\frac{(1-a^2)\,\mathrm{We}^2}{2\left(2+(1-a^2)\,\mathrm{We}^2\right)} - 1\right], -\alpha \leq \mathrm{Im}\,\sigma \leq \alpha.$$

In addition to this continuous spectrum there is an infinite number of discrete eigenvalues which are essentially lined up along a parallel to the imaginary axis (an other illustration of the fact that the underlying semi-group is not analytic).

Various computations, involving different values of the parameters Re, We, α and a are performed in [52]. None of them leads to instability.

Many studies have been devoted to the Couette-Taylor problem (flow between concentric cylinders with radii R_1 and R_2, $R_1 \leq R_2$, of infinite length, and rotating with angular velocities Ω_1 and Ω_2 respectively) for viscoelastic flows. For instance Demay and Zielinska [74] considered the general Maxwell models $(-1 < a < 1)$. They showed that the axisymmetric steady-state solution (the Couette flow) does not exist for all values of a and the relaxation time (see also Bouidi and Guillopé [5]). In the region of parameters where the steady state exists, all models (except for a very close to -1) predict stabilization of the Couette flow (in the spectral sense) for small enough values of the relaxation time λ_1.

On the other hand, Larson, Muller and Shaqfeh [33] [34] [36] [66] have reported on a purely elastic Taylor-Couette instability for models with or without Newtonian contribution (Maxwell or Oldroyd). The conclusion of their studies is that negative second normal stresses are stabilizing, especially for very small gap ratios, and that the Newtonian relative contribution has a stabilizing influence.

5 – Remarks on integral models

We have tried to survey the mathematical issues and problems arising from the theory of viscoelastic fluids. We have restrict ourselves to the *differential models* which are widely used in numerical simulations. On the other hand, *integral models* such as the KBKZ model (1.1) are very popular among polymer scientists, and are likely to be more and more used for both numerical and theoretical purposes. They lead to very complicated integro-differential systems, and their mathematical study is relatively lacunary.

We list below a few significant papers. Renardy proves the local well-posedness for the initial boundary value problem associated to KBKZ type models [46], and for integral models with Newtonian viscosities [45] in 2 or 3 dimensions. The existence of slow steady flows was proven in [50].

Slemrod [67] [68] and Laklalech [31] study various properties, including stability, of the linear semi-groups associated to simple fluids (see Section 2.2).

Various results, including global existence, for one dimensional (e.g. shear flows) of the KBKZ models are proven by Engler [13], Brandon and Hrusa [6].

We refer to the book [49] for many other references.

REFERENCES

[1] Astarita, G. and Marrucci, G., *Principles of Non-Newtonian Fluid Mechanics*, McGraw-Hill, New York, 1974.

[2] Avgousti, M., and Beris, A.N., Viscoelastic Taylor-Couette flow: bifurcation analysis in the presence of symmetry, Preprint, University of Delaware, Newark, 1992.

[3] Bazin, V., Contributions à l'analyse mathématique d'écoulements de fluides viscoélastiques dans un canal ondulé, Thèse de Doctorat, Université de Paris-Sud, 1990.

[4] Bird, R.B., Armstrong, B., and Hassager, O., *Dynamics of Polymeric Liquids*, J. Wiley, New York, 1977.

[5] Bouidi A., and Guillopé, C., Stability of viscoelastic Couette-Taylor flows, in preparation.

[6] Brandon, D., and Hrusa, W.J., Global existence of smooth shearing motions of a nonlinear viscoelastic fluid, J. Integr. Eq. Appl. 2, (1990), 333-351.

[7] Coleman, B.D., and Noll, W., Foundations of linear viscoelasticity, Rev. Modern Phys. 33, (1961), 239-249.

[8] Cotsiolis, A.A., and Oskolkov, A.P., J. Soviet Math. 46, (1989), 1595-1599.

[9] Crochet, M.J., Numerical simulation of viscoelastic flow : a review, Rubber Chemistry and Technology 62, 3, (1989).

[10] Crochet, M.J., and Delvaux, V., Numerical simulation of inertial viscoelastic flow with change of type, in: *Nonlinear Evolution Equations That Change Type*, B.L. Keyfitz, M. Shearer Eds, IMA Volumes in Mathematics and its Applications, Vol 27, Springer-Verlag, Berlin, 1991, 47-66.

[11] Dupret, F., and Marchal, J.M., Loss of evolution in the flow of viscoelastic fluids, J. Non-Newtonian Fluid Mech. 20, (1986), 143-171.

[12] El Kissi, N., and Piau, J.M., The different capillary flow regimes of entangled polydimethylsiloxane polymers: macroscopic slip at the wall, hysterisis and cork flow, J. Non-Newtonian Fluid Mech. 37 (1990), 55-94.

[13] Engler, H., On the dynamic shear flow problem for viscoelastic liquids, SIAM J. Math. Anal. 18, (1987), 972-990.

[14] Engquist, B., and Majda, A., Absorbing boundary conditions for the numerical simulation of waves, Mathematics of Computation 139, (1977), 629-651.

[15] Ewards, B.J., and Beris, A.N., Remarks concerning compressible viscoelastic fluid models, J. Non-Newtonian Fluid Mech. 36, (1990), 411-417.

[16] Fraenkel, L.E., On a linear, partly hyperbolic model of viscoelastic flow past a plate, Proc. Roy. Soc. Edinburgh 114 A, (1990), 299-354.

[17] Guillopé, C., and Saut, J.C., Existence results for the flow of viscoelastic fluids with a differential constitutive law, Nonlinear Analysis, Theory, Meth. & Appl. 15, (1990), 849-869.

[18] Guillopé, C., and Saut, J.C., Global existence and one-dimensional nonlinear stability of shearing motions of viscoelastic fluids of Oldroyd type, Math. Model. Numer. Anal. 24 (1990), 369-401.

[19] Guillopé, C., and Saut, J.C., Existence and stability of steady flows of weakly viscoelastic fluids, Proc. Roy. Soc. Edinburgh 119A, (1991), 137-158.

[20] Hakim, A., Analyse mathématique de modèles de fluides viscoélastiques de type White-Metzner, Thèse de Doctorat, Univ. Paris-Sud, 1989, and J. Math. Anal. Appl., to appear.

[21] Hakim, A., Existence of slow steady flows of viscoelastic fluids of White-Metzner type, submitted to publication.

[22] Halpern, L. Artificial boundary conditions for incompletely parabolic perturbations of hyperbolic systems, SIAM J. Math. Anal. 22, (1991), 1256-1283.

[23] Joseph, D.D., *Fluid Dynamics of Viscoelastic Liquids*, Springer-Verlag, Berlin, 1990.

[24] Joseph, D.D., Renardy, M., and Saut, J.C., Hyperbolicity and change of type in the flow of viscoelastic fluids, Arch. Rat. Mech. Anal. 87, (1985), 213-251.

[25] Joseph, D.D., and Saut, J.C., Change of type and loss of evolution in the flow of viscoelastic fluids, J. Non-Newtonian Fluid Mech. 20, (1986), 117-141.

[26] Joseph, D.D., Matta, J., and Chen, K., Delayed die swell, J. Non-Newtonian Fluid Mech. 24, (1987), 31-65.

[27] Joseph, D.D., and Saut, J.C., Short wave instabilities and ill-posed problems, Theoretical and Computational Fluid Dynamics 1, (1990), 191-227.

[28] Keunings, R., in: *Fundamental of Computer Modeling for Polymer Processing*, C.L. Tucker III (Ed.), Carl Hanser Verlag, Munich, 1989.

[29] Kim, J.U., Global smooth solutions for the equations of motion of a nonlinear fluid with fading memory, Arch. Rat. Mech. Anal. 79, (1982), 97-130.

[30] Kwon, T.H., and Shen, S.F., A unified constitutive theory for polymeric liquids I and II, Rheol. Acta 23, (1984), 217-230, and 24, (1985), 175-188.

[31] Laklalech, M., Sur une classe d'équations intégrodifférentielles liées à la mécanique des fluides non newtoniens, Thèse de 3ème cycle, Université Paris VII, 1985.

[32] Larson, R.G., *Constitutive Equations for Polymer Melts and Solutions,* Butterworths, Boston, 1988.

[33] Larson, R.G., Shaqfeh, E.S.G., and Muller, S.J., A purely viscoelastic instability in Taylor-Couette flow, J. Fluid Mech. 218, (1990), 573-600.

[34] Larson, R.G., Muller, S.J. and Shaqfeh, E.S.G., The elastic Taylor-Couette instability for rheologically complex fluids, J. Non-Newtonian Fluid Mech, to appear.

[35] Malkus, D.S., Nohel, J.A., and Plohr, B.J., Analysis of new phenomena in shear flow of non-Newtonian fluids, SIAM J. Appl. Math. 51, (1991), 899-929.

[36] Muller, S.J., Larson, R.G. and Shaqfeh, E.S.G., A purely elastic transition in Taylor-Couette flow, Rheol. Acta 28, (1989), 499-503.

[37] Nohel, J.A., Pego, R.L., and Tzavaras, A.E., Stability of discontinuous steady states in shearing motions of a non-Newtonian fluid, Proc. Roy. Soc. Edinburgh 115A, (1990), 39-59.

[38] Nohel, J.A., and Pego, R.L., Nonlinear stability and asymptotic behaviour of shearing motions of a non-Newtonian fluid, Preprint ETH, Zürich, 1992.

[39] Oldroyd, J.G., On the formulation of rheological equations of state, Proc. Roy. Soc. London A 200, (1950), 523-541.

[40] Pazy, A., *Semigroups of Linear Operators and Applications to Partial Differential Equations,* Applied Mathematical Sciences 44, Springer-Verlag, Berlin, 1983.

[41] Pearson, J.R.A., *Mechanics of Polymer Processings*, Elsevier, London, 1985.

[42] Piau, J.M., El Kissi, N., and Tremblay, B. Influence of upstream instabilities and wall slip on melt fracture and sharkskin phenomena during silicones extension through orifice dies, J. Non-Newtonian Fluid Mech. 34, (1990), 145-180.

[43] Prodi, G., Teoremi di tipo locale per il sistema di Navier-Stokes e stabilità delle soluzioni stazionarie, Rend. Sem. Mat. Univ. Padova 32, (1962), 374-397.

[44] Prüss, J., On the spectrum of C^0 semi-groups, Trans. Amer. Math. Soc. 284, (1984), 847-857.

[45] Renardy, M., Local existence theorems for the first and second initial boundary value problem for a weakly non-Newtonian fluid, Arch. Rat. Mech. Anal. 83, (1983), 229-244.

[46] Renardy, M., Local existence and uniqueness theorem for K-BKZ fluids, Arch. Rat. Mech. Anal. 88, (1985), 83-94.

[47] Renardy, M., Existence of slow steady flows of viscoelastic fluids with differential constitutive equations, Z. Angew. Math. Mech. 65, (1985), 449-451.

[48] Renardy, M., and Renardy Y., Linear stability of plane Couette flow on a upper convected Maxwell fluid. J. Non-Newtonian Fluid Mech. 22, (1986), 23-33.

[49] Renardy, M., Hrusa, W., and Nohel, J., *Mathematical Problems in Viscoelasticity*, Longman Scientific and Technical, Burnt Mill, Harlow, 1987.

[50] Renardy, M., Existence of steady flows of viscoelastic fluids of integral type, Z. Angew. Math. Mech. 68, (1988), 40-44.

[51] Renardy, M., Inflow boundary conditions for steady flows of viscoelastic fluids with differential constitutive laws, Rocky Mount. J. Math. 18, (1989), 445-453.

[52] Renardy, M., and Renardy Y., Stability of shear flows of viscoelastic fluids under perturbations perpendicular to the plane of flow, J. Non-Newtonian Fluid Mech. 32, (1989), 145-155.

[53] Renardy, M., Existence of steady flows of viscoelastic fluids of Jeffreys type with traction boundary conditions, Diff. Integr. Eq 2, (1989), 431-437.

[54] Renardy, M., An alternative approach to inflow boundary conditions for Maxwell fluids in three space dimensions, J. Non-Newtonian Fluid Mech. 36, (1990), 419-425.

[55] Renardy, M., Local existence of solutions of the Dirichlet initial boundary value problem for incompressible hypoelastic materials, SIAM J. Math. Anal. 21, (1990), 1369-1385.
[56] Renardy, M., Short wave instabilities resulting from memory slip, J. Non-Newtonian Fluid Mech. 35, (1990), 73-76.
[57] Renardy, M., A well-posed boundary value problem for supercritical flow of viscoelastic fluids of Maxwell type, in: *Nonlinear Evolution Equations That Change Type*, B.L. Keyfitz, M. Shearer (Ed.), IMA Volumes in Mathematics and its Applications, vol 27, Springer-Verlag, Berlin, 1991, 181-191.
[58] Renardy, M., A rigorous stability proof for plane Couette flow of an upper convected Maxwell fluid at zero Reynolds number, Preprint, Virginia Tech., Blacksburg, 1991.
[59] Renardy, M., and Renardy, Y., Pattern selection in the Bénard problem for a viscoelastic fluid, Z. Angew. Math. Mech. 43, (1992), 154-180.
[60] Rutkevich, I.M., Some general properties of the equations of viscoelastic incompressible fluid mechanics, J. Appl. Math. Mech. (PMM) 33, (1969), 30-39.
[61] Rutkevich, I.M., The propagation of small perturbations in a viscoelastic fluid, J. Appl. Math. Mech (PMM) 34, (1970), 35-50.
[62] Saut, J.C., and Joseph, D.D., Fading memory, Arch. Rat. Mech. Anal. 81, (1983), 53-95.
[63] Schleiniger, G., Calderer, M.C., and Cook, L.P., Embedded hyperbolic regions in a nonlinear model for viscoelastic flow, in: *Current Progress in Hyperbolic Systems : Riemann Problems and Computations,* W.B. Lindquist (Ed.), Contemporary Mathematics 100, American Mathematical Society, Providence, 1990.
[64] Schowalter, W.R., *Mechanics of Non-Newtonian Fluids*, Pergamon Press, New-York, 1978.
[65] Serrin, J., A note in the existence of periodic solutions of the Navier-Stokes equations, Arch. Rat. Mech. Anal. 3, (1959), 120-122.
[66] Shaqfeh, E.S.G., Muller, S.G., and Larson, R.G., The effects of gap width and dilute solution properties on the viscoelastic Taylor-Couette instability, J. Fluid Mech. 235, (1992), 285-317.
[67] Slemrod, M., A hereditary partial differential equation with applications in the theory of simple fluids, Arch. Rat. Mech. Anal. 62, (1976), 303-321.
[68] Slemrod, M., Instability of steady shearing flows in a nonlinear viscoelastic fluid, Arch. Rat. Mech. Anal. 68, (1978), 211-225.
[69] Tajchman, M., Conditions aux limites absorbantes pour un fluide viscoélastique de type differentiel, Thèse de Doctorat, Université Paris-Sud, 1992, and Article in preparation.
[70] Temam, R., *Navier-Stokes Equations*, North-Holland, 3rd ed., Amsterdam, 1984.
[71] Tordella, J.P. Unstable flow of molten polymers, in: F. Eirich (Ed.), *Rheology: Theory and Applications*, Vol 5, Academic Press, New-York, 1969.
[72] Ultman, J.S., and Denn, M.M., Anomalous heat transfer and a wave phenomenon in dilute polymer solutions, Trans. Soc. Rheol. 14, (1970), 307-317.
[73] Valli, A., Periodic and stationary solutions for compressible Navier-Stokes equations via a stability method, Annali Scu. Norm. Sup. Pisa 10, (1983), 607-647.
[74] Zielinska, B.J.A., and Demay, Y., Couette-Taylor instability in viscoelastic fluids, Phys. Rev. A 38, (1988), 897-903.

C. Guillopé and J.C. Saut,
Laboratoire d'Analyse Numérique d'Orsay, Bât. 425,
CNRS and Université Paris-Sud,
91405 ORSAY - FRANCE

Weak Solutions for Thermoconvective Flows of Boussinesq–Stefan Type

JOSÉ-FRANCISCO RODRIGUES

Abstract

In this article we study the existence of weak solutions to the steady-state Boussinesq–Stefan problem consisting of a coupled system between a nonlinear two-phase heat equation with convection and an elliptic variational inequality traducing the non-Newtonian flow in the liquid phase, which may include the solidification or melting of Bingham fluids, pseudo-plastic or dilatant fluids, or also asymptotically Newtonian fluids. Its contents is the following:

1 – Introduction

Fluid flow problems with heat transfer have been a classical topic in the mathematical literature (see, for instance, the books [J], [T2] or [St]), in particular, in the framework proposed by J. Boussinesq in the begining of this century.

On the other hand, heat conduction problems with phase change have been studied extensively in the framework of the Stefan problem (see, for instance, [R4] for references).

In modern technology of materials processing a wide range of situations, such as continuous casting of metals and alloys, laser melting and solidification during welding, glezing, and heat treating of a workpiece, or the crystal growth systems, either by the Czochralski or floating-zone methods for producing silicon and the vertical Bridgman system, lead to high complex problems, where, even in the steady-state, the natural convection in the bulk liquid plays an important role in the shape or control of the melt-solid interface which is a free boundary, i.e., not a *priori* known boundary (see, for instance, [F], [Sz], [CB] or [BS] and the examples sketched in the figures below).

These problems have in common the coupling of the fluid flow and the heat transfer equations and the existence of a solidification front where a jump condition of the heat flux arises, even in the steady-state case when some exterior prescribed convection $\vec{\alpha}$ is imposed, as, for instance, the constant extraction in the continuous casting problem.

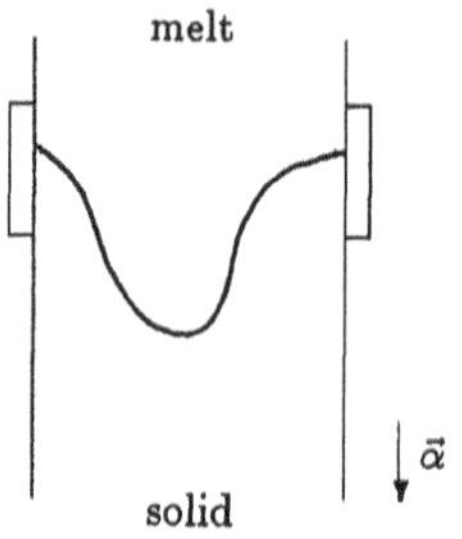

a) Continuous casting with extraction velocity $\vec{\alpha}$

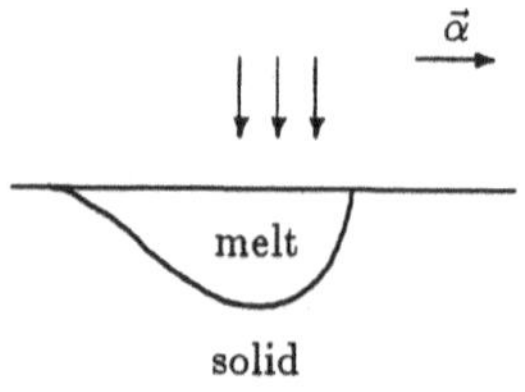

b) Welding with moving heat source

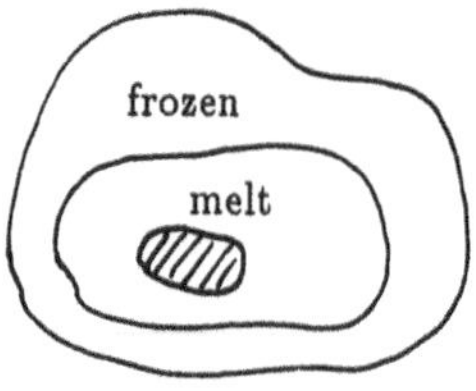

c) Cross section of a partially frozen conducting tube

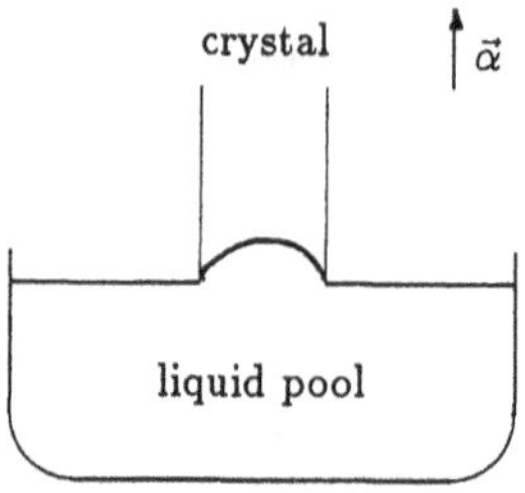

d) Crystal growth system, with pulling velocity $\vec{\alpha}$

From the mathematical point of view, this is a type of Stefan problems with convection given by the Boussinesq approximation. The precise mathematical-physics model is explained in the section 2, where we use a general nonlinear constitutive law for the relation stress–velocity of deformation. This allow us to consider the cases of a Bingham fluid, a pseudo-plastic or dilatant fluids and a large class of asymptotically Newtonian fluids including the models of Prandtl–Eyring, Cross, Williamson and Carreau (see [DL], [C1,2] and their references).

Since classical solutions, in general, cannot be expected, we introduce, in section 3, the variational formulation generalizing the approach of [R1,2,5]. It consists of a nonlinear diffusion–convection problem for the temperature field strongly coupled with a variational inequality for the velocity field, only in the liquid zone determined by the domain where the temperature is below the melting temperature. The existence theorem is stated in section 4 and its proof is based on a suitable approximate–penalized problem, which study is postponed to the section 6. The existence of the approximate solution relies on the Schauder fixed point theorem and the study of two auxiliary problems. The main idea for the proof follows the lines of [R2] and uses an a *priori* L^∞-estimate for the temperature that allows nonlinearities without growth conditions. This requires an extension of the classical weak maximum principle for a class of nonlinear convection–diffusion problems, which is established in section 5.

Finally, in section 7, we discuss a special limit case when the prescribed convection $|\vec{\alpha}| \to 0$, extending the continuous transition result of [R2]. For the limit problem, the Rayleigh–Bénard problem considered in [CDK1], we include the only known result (see [DF]) on the existence and uniqueness of a classical solution for small values of a coupling parameter. For large values of this parameter, multiplicity of solutions is expected as suggested by numerical experiments of [CB] and by the corresponding problem without free boundary, the classical Bénard problem (see [J]).

The use of a temperature dependent penalty term in the velocity field problem is inspired in the earlier works [CDK1] and [CD], that have been extended in [R2], always in the framework of Newtonian fluids. The results of this paper being valid for non-Newtonian fluids are new and have been partially announced in [R4]. The variational formulation we use is particularly adapted to the finite element approximation of the problem, but no mathematical analysis has been considered so far. Although there are some numerical results for special cases in [CB], [VI] and [BS], the error analysis, following the lines of [BL] or [BMP], is an open problem. For the time dependent problem only a bidimensional case has been considered in [CDK2].

2 – The Mathematical–Physics Problem

From the principles of Continuum Mechanics, the general equations governing the flow of an incompressible material with heat transfer in a domain $\Omega \subset \mathbb{R}^n$ (in our physical cases $n = 2$ or $n = 3$) are

(2.1) $$\operatorname{div} \mathbf{v} = \nabla \cdot \mathbf{v} = 0 \qquad \text{(conservation of mass)}$$

(2.2) $$\rho \frac{dv_i}{dt} = \sigma_{ij,j} + f_i \qquad \text{(conservation of momentum)}$$

(2.3) $$\rho \frac{de}{dt} = r - q_{i,i} \qquad \text{(conservation of energy).}$$

Here, using standard notations (see [DL], for instance), $\mathbf{v} = (v_i)$ is the *velocity* vector, $\mathbf{x} = (x_i) \in \overline{\Omega}$ denote Eulerian coordinates, $\rho > 0$ the *density*, σ_{ij} are the components of the (symmetric) *stress tensor*, f_i is a density of forces, e denotes the specific *internal energy* of the medium, q_i are the components of the *heat flux* vector and r is a density of heat supply, including the dissipation effects.

The components of the *velocity of deformation* tensor $\mathbf{D} = (D_{ij})$ are given by

(2.4) $$D_{ij} = \frac{1}{2}(v_{i,j} + v_{j,i}), \qquad \text{where } v_{i,j} = \frac{\partial v_i}{\partial x_j} \,.$$

We use the summation convention (e.g. $v_{i,i} = \nabla \cdot \mathbf{v} = \operatorname{tr} \mathbf{D} = D_{ii}$) and

$$\frac{d}{dt} = \frac{\partial}{\partial t} + v_i \frac{\partial}{\partial x_i} = \frac{\partial}{\partial t} + \mathbf{v} \cdot \nabla$$

denotes the material derivative.

We shall be restricted to steady-state situations and the density, being constant, will be taken $\rho \equiv 1$, so that in (2.2) and (2.3) the operator $\rho \frac{d}{dt}$ will be reduced to the convective operator $\mathbf{v} \cdot \nabla$ only.

We shall consider the material in Ω with two phases separated by an interface Φ, which is *a priori* unknown, and such that the circulating liquid phase Λ is in equilibrium with the solid phase Σ, so that $\Omega = \Lambda \cup \Phi \cup \Sigma$.

Using the constitutive relations between the specific energy and the heat flux with the *absolute temperature* $T = T(x)$, from (2.3), neglecting the heat sources $r \equiv 0$, we obtain

$$(2.5) \qquad v_j\big(e(T)\big)_{,j} = \big(k(T)T_{,i}\big)_{,i} \quad \text{in } \Omega \backslash \Phi = \Lambda \cup \Sigma ,$$

where the thermal conductivity k may be a discontinuous function across Φ. Denoting by T_* the melting temperature at the interface, given by a prescribed constant, we renormalize the temperature by the Kirchhoff transformation

$$\vartheta = K(T) = \int_{T_*}^{T} k(\tau)\, d\tau .$$

Then, the equation (2.5) is transformed at the solid region $\Sigma(\vartheta) = \{\vartheta < 0\}$ and at the liquid bulk $\Lambda(\vartheta) = \{\vartheta > 0\}$ into

$$(2.6) \qquad v_j\big(b(\vartheta)\big)_{,j} = \Delta\vartheta \quad \text{in } \Omega \backslash \Phi = \{\vartheta < 0\} \cup \{\vartheta > 0\} ,$$

where $b = e \circ K^{-1}$ is a known continuous function, and without loss of generality we may assume $b(0) = 0$.

At the interface, we have the melting temperature renormalized to zero and the balance of the heat flux is given by the Stefan condition

$$(2.7) \qquad [\nabla\vartheta]_-^+ \cdot \mathbf{n} = -\lambda\vec{\alpha} \cdot \mathbf{n} \quad \text{on } \Phi = \{\vartheta = 0\} ,$$

where $\lambda > 0$ is essentially the latent heat, $\mathbf{n}$ is a normal vector to Φ and $[\cdot]_-^+$ denotes the jump across Φ, which is supposed to be at a given velocity $\vec{\alpha}$.

The boundary conditions for the renormalized temperature on the fixed boundary $\partial\Omega$, which we suppose decomposed into two parts Γ_D and $\Gamma_N = \partial\Omega \backslash \Gamma_D$, are of mixed type:

$$(2.8) \qquad \vartheta = \vartheta_D \quad \text{on } \Gamma_D \qquad \text{(Dirichlet condition)}$$

$$(2.9) \qquad \frac{\partial\vartheta}{\partial n} = g(x, \vartheta) \quad \text{on } \Gamma_N \qquad \text{(nonlinear Neumann condition)} ,$$

where ϑ_D is a given boundary temperature, which is strictly positive on one part of Γ_D and the strictly negative on its complement. Here $\frac{\partial\vartheta}{\partial n} = \nabla\vartheta \cdot \mathbf{n}$, where $\mathbf{n}$ denotes the unit normal outwards vector on Γ_N.

Since we are going to consider only a steady-state in equilibrium, the velocity field $\mathbf{v} = \mathbf{v}(\mathbf{x})$ in the solid region $\Sigma(\vartheta)$ is assumed equal to the velocity of the melt-solid interface Φ, which is supposed to be the given velocity of the fixed boundary $\partial\Omega$:

$$(2.10) \qquad \mathbf{v} = \vec{\alpha} \quad \text{on } \Sigma(\vartheta) \cup \Phi \cup \partial\Omega ,$$

where we can allow $\vec{\alpha}$ in a certain class of rigid motions.

Consequently, the equation of motion (2.2) is considered only inside the bulk liquid. Introducing the constitutive relation for the incompressible non-Newtonian fluid

$$\sigma_{ij} = -p\,\delta_{ij} + S_{ij} \,, \tag{2.11}$$

where p is the pressure and $\mathbf{S} = (S_{ij})$ is the viscous stress tensor, the steady-state momentum equation (2.2) becames

$$v_j\, v_{i,j} = -p_{,i} + S_{ij,j}(\vartheta, \mathbf{v}) + f_i(\vartheta) \quad \text{in } \Lambda(\vartheta) = \{\vartheta > 0\} \,. \tag{2.12}$$

In the Boussinesq approximation, the variation of the temperature within the fluid, generating driving forces, is taken into account in the nonlinear term

$$f_i = f_i(\vartheta) \,. \tag{2.13}$$

The Boussinesq approximation, neglecting the variability of density due to changes in the temperature, is also relevant to the interface conditions (2.7) and (2.10) since they impose the adherence of the fluid to the solid part and the incorporation of melt into the solid at a rate equal to the growth one.

In the constitutive relation $\mathbf{S} = \mathbf{S}(\vartheta, \mathbf{v})$, we shall consider the temperature dependence only through viscosities coefficients η_k, while the dependence on the velocity $\mathbf{v}$ may be quite general through a potencial convex function $\mathcal{F}$ on the second scalar invariant $D_{II} = \frac{1}{2} D_{ij} D_{ij} = \frac{1}{2} \mathbf{D} : \mathbf{D}$, given in the form

$$\mathcal{F}(\vartheta; \mathbf{v}) = \mathcal{F}(\vartheta; D_{II}(\mathbf{v})) = \sum_{k=1}^{N} \eta_k(\vartheta) F_k(D_{II}(\mathbf{v})) \,. \tag{2.14}$$

In general, we shall assume that, for a fixed temperature $\vartheta = \overline{\vartheta}$, $\mathbf{S}$ is an element of the subdifferential of $\mathcal{F}$ at the point $\mathbf{D}(\mathbf{v})$:

$$\mathbf{S} \in \partial_{\mathbf{D}} \mathcal{F}(\overline{\vartheta}; \mathbf{v}), \qquad \text{at fixed } \overline{\vartheta} \,. \tag{2.15}$$

Recalling that $\partial D_{II} / \partial D_{ij} = D_{ij}$, by the definition of subdifferential, (2.15) is equivalent to the condition

$$\mathcal{F}(\overline{\vartheta}; \mathbf{w}) - \mathcal{F}(\overline{\vartheta}; \mathbf{v}) \geq S_{ij}(D_{ij}(\mathbf{w}) - D_{ij}(\mathbf{v})) \,, \qquad \forall \mathbf{w} \,. \tag{2.16}$$

When all the functions F_k in (2.14) have derivatives F_k', the constitutive relation (2.15) is simply

$$S_{ij} = \Big\{ \sum_{k=1}^{N} \eta_k(\vartheta) F_k'(D_{II}(\mathbf{v})) \Big\} D_{ij}(\mathbf{v}) \,, \tag{2.17}$$

and we can easily recognize some classical examples when we take only one function $F = F_1$ given, for $d = D_{II}(\mathbf{v})$, by:

$$F(d) = 2\,d, \qquad \textit{Newtonian fluid} \,, \tag{2.18}$$

$$F(d) = \frac{2}{p}\, d^{p/2} \,, \qquad \textit{non-Newtonian fluid} \,, \tag{2.19}$$

which is called a *pseudo-plastic* fluid for $1 < p < 2$ and a *dilatant* fluid for $p > 2$; an expression of the form

$$F(d) = 2\int_0^{\sqrt{2d}} \tau\, \mu(\tau)\, d\tau \;, \tag{2.20}$$

for non-Newtonian generalized fluids includes some important *asymptotically Newtonian* models (see [C2], for instance), in which $\mu(s) \underset{s\to\infty}{\longrightarrow} \mu_\infty > 0$, like the

i) *Prandtl–Eyring model*: $\mu(s) = \mu_\infty + a\ln(s\lambda + \sqrt{s^2\lambda^2+1})/s\lambda \quad (a,\lambda > 0)$,

ii) *Cross model*: $\mu(s) = \mu_\infty + a/(1+s^{2/3}) \quad (a > 0)$,

iii) *Williamson model*: $\mu(s) = \mu_\infty + a/(\lambda + s) \quad (a,\lambda > 0)$,

iv) *Carreau model*: $\mu(s) = \mu_\infty + a(1+\lambda s^2)^{(r-2)/2} \quad (a,\lambda > 0,\ 1 < r \le 2)$.

In these four cases, the constitutive law (2.17) corresponds to $F'(d) = 2\mu(s)$, with $s = \sqrt{2d}$, and the function μ verifies the condition

$$\exists P \subset \mathbb{R}_+ :\quad \mu'(s) \ge 0,\ \forall s \in P \ \text{ and } \ s|\mu'(s)| \le \mu(s),\ \forall s \in \mathbb{R}_+ \backslash P \,. \tag{2.21}$$

Another important example of the general law (2.17) is the case of a *Bingham fluid* (see [DL]):

$$S_{ij} = 2\mu D_{ij} + \gamma D_{ij}/D_{II}^{1/2}, \qquad \text{for } D_{II} > 0 \;, \tag{2.22}$$

which corresponds to the combination of the linear law of a Newtonian fluid, with viscosity $\mu = \mu(\vartheta) > 0$, with the plasticity threshold characterized by a parameter $\gamma = \gamma(\vartheta) \ge 0$

$$S_{II}^{1/2} \le \gamma \;\Leftrightarrow\; D_{ij} = 0 \qquad (S_{II} = \tfrac{1}{2} S_{ij} S_{ij}) \;. \tag{2.23}$$

These fluids, corresponding to the limit nondifferentiable case $p = 1$ in (2.19), behave like a rigid body for stresses below the yield limit γ and they move like viscous fluids beyond that limit. They arise in many industrial processes, particularly in chemical and material engineering applications, for instance in rolling and extrusion processes, where the flows may present rigid zones.

In general, we shall consider fluids with the constitutive law (2.15), or rather (2.16), and we notice that the expression (2.14) does not change for translations with rigid motions $\mathcal{R} = \{\mathbf{r} \colon \mathbf{D}(\mathbf{r}) = 0\}$

$$\mathcal{F}(\vartheta, \mathbf{v}) = \mathcal{F}(\vartheta; \mathbf{v} - \mathbf{v}^*) \quad \text{ if } \mathbf{D}(\mathbf{v}^*) \equiv 0 \,.$$

However, for technical reasons, we shall consider only a nonhomogeneous velocity $\vec{\alpha}$ with an additional geometrical restriction:

$$\vec{\alpha} \in \mathcal{R} \;\; (\text{i.e., } \mathbf{D}(\vec{\alpha}) = 0) \quad \text{ and } \quad \vec{\alpha} \cdot \mathbf{n} = 0 \ \text{ on } \Gamma_N \;, \tag{2.24}$$

which is verified in the continuous casting example of [R2,5], where Ω is a cylindrical domain, Γ_N is its lateral boundary and $\vec{\alpha}$ is a constant vector with the direction of the axis of the cylinder.

3 – The Variational Formulation

Our mathematical–physics problem consists of a two-phase Stefan problem with convection (2.6)–(2.7), with mixed boundary conditions (2.8)–(2.9), in a fixed domain Ω, where the unknwons are the temperature ϑ and the free boundary Φ, coupled with the velocity field $\mathbf{v}$, which is known in the solid part Σ up to Φ and in the fixed boundary $\partial\Omega$. For general non-Newtonian fluids, the Boussinesq approximation leads to the equation of motion (2.12) *only in the liquid zone* $\Lambda(\vartheta) = \{\vartheta > 0\}$, which is *a priori* unknown and depends on the temperature, with the constitutive law (2.15).

In order to obtain a weak formulation, we introduce the following functional framework and we recall the usual notations of Sobolev spaces $W^{m,p}(\mathcal{O})$ in an arbitrary open bounded subset $\mathcal{O}$ of $\mathbb{R}^n$ (see [L], [GT] or [R3], for instance). Let $\mathcal{D}(\mathcal{O})$ denote the space of infinite differentiable functions of compact support in $\mathcal{O}$, and

$$\mathcal{V}(\mathcal{O}) = \left\{\mathbf{v} \in [\mathcal{D}(\mathcal{O})]^n \colon \nabla\cdot\mathbf{v} = 0 \text{ in } \mathcal{O}\right\}.$$

For $p > 1$ and $D_{II} = \frac{1}{2}D_{ij}D_{ij}$ we introduce the norm

$$\|\mathbf{v}\|_{\mathbf{V}^p(\mathcal{O})} = \left\{\int_{\mathcal{O}}[D_{II}(\mathbf{v})]^{p/2}\right\}^{1/p} = \|D_{II}^{1/2}(\mathbf{v})\|_{L^p(\mathcal{O})} \tag{3.1}$$

and we define the following Banach spaces

$$\mathbf{V}_p(\mathcal{O}) = \left\{\text{closure of } \mathcal{V} \text{ for } \|\cdot\|_{\mathbf{V}^p(\mathcal{O})}\right\}, \tag{3.2}$$

$$\mathbf{L}^p_\sigma(\mathcal{O}) = \left\{\text{closure of } \mathcal{V} \text{ in } [L^p(\mathcal{O})]^n\right\}, \qquad 1 < p < \infty. \tag{3.3}$$

Let p' be the conjugate exponent of p, $\frac{1}{p} + \frac{1}{p'} = 1$. We remark that if $\mathbf{w} \in \mathbf{L}^p_\sigma(\mathcal{O})$ then

$$\int_{\mathcal{O}} \mathbf{w}\cdot\nabla\varphi = 0, \qquad \forall\varphi\colon \nabla\varphi \in [L^{p'}(\mathcal{O})]^n. \tag{3.4}$$

We shall assume that Ω is a Lipschitz bounded domain. We know that

$$\mathbf{V}_p(\Omega) = \left\{\mathbf{v} \in [W^{1,p}_0(\Omega)]^n \colon \nabla\cdot\mathbf{v} = 0 \text{ in } \Omega\right\}$$

and, by Korn inequality, the norm (3.1) is equivalent in $\mathbf{V}_p(\Omega)$ to the $W^{1,p}_0(\Omega)$-norm, i.e., the L^p-norm of $|\nabla\mathbf{v}|$. Hence, by Poincaré inequality we have

$$\|\mathbf{v}\|_{[L^p(\Omega)]^n} \le C_0\|\mathbf{v}\|_{[W^{1,p}_0(\Omega)]^n} \le C_*\|\mathbf{v}\|_{\mathbf{V}^p(\Omega)}, \qquad \forall\mathbf{v} \in \mathbf{V}^p(\Omega).$$

We also suppose that $\partial\Omega = \Gamma_{\mathrm{D}} \cup \Gamma_{\mathrm{N}}$, with $\mathrm{meas}(\Gamma_{\mathrm{D}}) > 0$, so that, in

$$Z \equiv \left\{\varsigma \in H^1(\Omega)\colon \varsigma = 0 \text{ on } \Gamma_{\mathrm{D}}\right\} \supset H^1_0(\Omega) \equiv W^{1,2}_0(\Omega), \tag{3.5}$$

by Poincaré inequality, the L^2-norm of $|\nabla\varsigma|$ is equivalent to the $H^1(\Omega)$-norm.

We also recall that, by the Sobolev imbedding theorems, we have

$$W^{1,p}(\Omega) \subset L^s(\Omega) \quad \text{for } 1 \le p < n,\ s \le pn/(n-p), \text{ or } \forall s < \infty \text{ if } n = p;$$

$$W^{1,p}(\Omega) \subset C^{0,\lambda}(\overline{\Omega}) \cap L^\infty(\Omega) \quad \text{if } p > n \text{ and } 0 \le \lambda \le 1 - n/p,$$

being these imbeddings compact provided $s < pn/(n-p)$ and $0 \le \lambda < 1 - n/p$. Here $C^{0,\lambda}(\overline{\Omega})$ denotes the space of Hölder continuous functions in $\overline{\Omega}$ and $C^{0,0}(\overline{\Omega}) = C^0(\overline{\Omega})$ of only continuous functions.

If in (2.6) the liquid-solid interface Φ was a smooth boundary, integrating by parts with $\varsigma \in Z$, one could obtain

$$\int_\Omega \nabla\vartheta \cdot \nabla\varsigma + \int_{\Omega\setminus\Phi} [\mathbf{v}\cdot\nabla b(\vartheta)]\,\varsigma = \int_\Phi \varsigma[\nabla\vartheta]_-^+ \cdot \mathbf{n} + \int_{\Gamma_N} \varsigma \frac{\partial\vartheta}{\partial n} . \tag{3.6}$$

Denoting by $\chi_{\{\vartheta>0\}} = \chi_{\Lambda(\vartheta)}$ the characteristic function of the liquid phase, and recalling that, by assumption,

$$\nabla\cdot\vec{\alpha} = \mathbf{D}(\vec{\alpha}) = 0 \ \text{ in } \Omega \qquad \text{and} \qquad \mathbf{n}\cdot\vec{\alpha} = 0 \ \text{ on } \Gamma_N , \tag{3.7}$$

we obtain easily, for arbitrary $\varsigma \in Z$,

$$\int_\Omega \chi_{\Lambda(\vartheta)}(\vec{\alpha}\cdot\nabla\varsigma) = \int_{\Lambda(\vartheta)} \nabla\cdot(\varsigma\,\vec{\alpha}) = \int_\Phi \varsigma\,\vec{\alpha}\cdot\mathbf{n} ,$$

$$\int_{\Omega\setminus\Phi} [\mathbf{v}\cdot\nabla b(\vartheta)]\,\varsigma = \int_\Omega \varsigma\,\nabla\cdot[\mathbf{v}\,b(\vartheta)] = -\int_\Omega b(\vartheta)\,\mathbf{v}\cdot\nabla\varsigma ,$$

since $\mathbf{v} = \vec{\alpha}$ on $\Phi \cup \partial\Omega$. Then, using these relations in (3.6) and recalling the conditions (2.7) and (2.9), we find

$$\int_\Omega \left\{\nabla\vartheta - b(\vartheta)\,\mathbf{v} + \lambda\,\vec{\alpha}\,\chi_{\Lambda(\vartheta)}\right\}\cdot\nabla\varsigma = \int_{\Gamma_N} g(\vartheta)\,\varsigma , \qquad \forall \varsigma \in Z . \tag{3.8}$$

On the other hand, assuming the liquid zone $\Lambda(\vartheta)$ sufficiently smooth and integrating by parts (2.12) with $v_i - w_i$, where $\mathbf{w} = (w_i)$ is a solenoidal vector field, such that $\mathbf{w} = \mathbf{v} = \vec{\alpha}$ on $\partial\Lambda(\vartheta)$, we obtain

$$\int_{\Lambda(\vartheta)} S_{ij}(w_{i,j} - v_{i,j}) + \int_{\Lambda(\vartheta)} v_j\,v_{i,j}(w_i - v_i) = \int_{\Lambda(\vartheta)} f_i(\vartheta)\,(w_i - v_i) . \tag{3.9}$$

By the symmetry of S_{ij} and the constitutive law (2.16) we obtain

$$\int_{\Lambda(\vartheta)} \mathcal{F}(\vartheta,\mathbf{w}) - \int_{\Lambda(\vartheta)} \mathcal{F}(\vartheta,\mathbf{v}) + \int_{\Lambda(\vartheta)} v_j\,v_{i,j}(w_i - v_i) \ge \int_{\Lambda(\vartheta)} f_i(\vartheta)\,(w_i - v_i) , \tag{3.10}$$

which is a variational inequality for $\mathbf{v} = (v_i)$ in $\Lambda(\vartheta)$. Introducing the translated velocity

$$\mathbf{u} = \mathbf{v} - \vec{\alpha} , \tag{3.11}$$

we can write (3.10) for $\mathbf{u}$ with an arbitrary $\Psi = \mathbf{w} - \vec{\alpha} \in \mathbf{V}^p(\Lambda(\vartheta))$ in the form

$$\int_{\Lambda(\vartheta)} \mathcal{F}(\vartheta,\Psi) - \int_{\Lambda(\vartheta)} \mathcal{F}(\vartheta,\mathbf{u}) + \int_{\Lambda(\vartheta)} (u_j+\alpha_j)\,u_{i,j}(\Psi_i - u_i) \ge \int_{\Lambda(\vartheta)} f_i(\vartheta)\,(\Psi_i - u_i) . \tag{3.12}$$

Integrating by parts the convection term, we have formally

$$\int_{\Lambda(\vartheta)} (u_j + \alpha_j)\,u_{i,j}(\Psi_i - u_i) = -\int_{\Lambda(\vartheta)} (u_j + \alpha_j)\,u_i\,\Psi_{i,j} = \int_{\Lambda(\vartheta)} \beta(\mathbf{u},\mathbf{u},\Psi) ,$$

where we set

$$\beta(u,v,w) = -(u_j + \alpha_j)\, v_i\, w_{i,j} \; . \tag{3.13}$$

We observe that this function may be integrated in any open subset $\mathcal{O}$ with $\mathbf{w} \in \mathbf{V}^p(\mathcal{O})$, provided the product $u_j\, v_i \in L^{p'}(\mathcal{O})$.

At this point we have obtained the weak formulation (3.8) for ϑ and (3.12) for $\mathbf{u}$, where the test functions are defined respectively in Ω and in $\Lambda(\vartheta) = \{\vartheta > 0\}$. However, in (3.8) the characteristic function $\chi_{\Lambda(\vartheta)}$, as usually in the weak formulation for the Stefan problem, is replaced by an unknown function χ which is in the Heaviside graph $H(\vartheta)$, allowing a possible mushy region at $\{\vartheta = 0\}$, instead of a smooth interface. On the other hand the necessity that the liquid zone $\Lambda(\vartheta) = \{\vartheta > 0\}$ is at least an open subset of Ω, even without any smoothness requirement, leads us to search for a continuous temperature.

Combining these arguments, we shall consider the following variational formulation of our Boussinesq–Stefan problem:

Definition. The triple $(\vartheta, \chi, \mathbf{u})$ is a weak solution to (2.6)–(2.7) with boundary conditions (2.8)–(2.9) and (2.12)–(2.15) with (2.10), for $\mathbf{v}$ given by (3.11), and $\Lambda(\vartheta) = \{x \in \Omega\colon\ \vartheta(x) > 0\}$, $\Sigma(\vartheta) = \{x \in \Omega\colon\ \vartheta(x) < 0\}$, if

$$\vartheta \in C^0(\Omega) \cap H^1(\Omega)\,, \qquad \vartheta = \vartheta_{\mathrm{D}} \ \text{ on } \Gamma_{\mathrm{D}} \; ; \tag{3.14}$$

$$\chi \in L^\infty(\Omega)\,, \qquad 0 \le \chi_{\Lambda(\vartheta)} \le \chi \le 1 - \chi_{\Sigma(\vartheta)} \le 1 \ \text{ a.e. in } \Omega \; ; \tag{3.15}$$

$$\mathbf{u} \in \mathbf{V}^p(\Omega)\,, \qquad \mathbf{u} = \mathbf{0} \ \text{ a.e. in } \Sigma(\vartheta) \; ; \tag{3.16}$$

$$\int_\Omega \big\{\nabla\vartheta - b(\vartheta)\,(\mathbf{u} + \vec{\alpha}) + \lambda\,\vec{\alpha}\,\chi\big\} \cdot \nabla\varsigma = \int_{\Gamma_{\mathrm{N}}} g(\vartheta)\,\varsigma\,, \qquad \forall \varsigma \in Z \; ; \tag{3.17}$$

$$\int_{\Lambda(\vartheta)} \mathcal{F}(\vartheta, \Psi) - \int_{\Lambda(\vartheta)} \mathcal{F}(\vartheta, \mathbf{u}) + \int_{\Lambda(\vartheta)} \beta(\mathbf{u}, \mathbf{u}, \Psi) \ge \int_{\Lambda(\vartheta)} \mathbf{f}(\vartheta) \cdot (\Psi - \mathbf{u})\,, \qquad \forall \Psi \in \mathbf{V}^p(\Lambda(\vartheta)) \; . \tag{3.18}$$

Remark 3.1. Since, by Sobolev imbedding, $\mathbf{u} \in [L^{pn/(n-p)}(\Omega)]^n$, the convection term $\int_{\Lambda(\vartheta)} \beta(\mathbf{u}, \mathbf{u}, \Psi)$ is well defined provided $p \ge 3n/(n+2)$. □

Remark 3.2. The free boundary Φ is absent from this weak formulation. It can be recovered *a posteriori* as the level set

$$\Phi = \big\{x \in \Omega\colon\ \vartheta(x) = 0\big\} = \partial\Lambda(\vartheta) \cap \partial\Sigma(\vartheta) \; .$$

If the n-dimensional Lebesgue measure of Φ is zero, from (3.15) we see that, in fact, $\chi = \chi_{\Lambda(\vartheta)} = 1 - \chi_{\Sigma(\vartheta)}$. If, in addition, Φ is a Lipschitz surface, from (3.17) we recover the Stefan condition (2.7) and, of course, from (3.16) also the boundary condition $\mathbf{u} = \mathbf{0}$ ($\mathbf{v} = \vec{\alpha}$) on Φ. Consequently, (3.16) can also be regarded as a weak formulation for the free boundary conditions of the velocity. □

Remark 3.3. If, in (2.14) all the F_k are differentiable, then the variational inequality (3.18) is in fact an equality

$$\sum_{k=0}^{N}\int_{\Lambda(\vartheta)}\eta_k(\vartheta)F_k'(D_{II}(\mathbf{u}))D_{ij}(\mathbf{u})D_{ij}(\Psi)+\int_{\Lambda(\vartheta)}\beta(\mathbf{u},\mathbf{u},\Psi)=\int_{\Lambda(\vartheta)}\mathbf{f}(\vartheta)\cdot\Psi,\quad \forall\,\Psi\in\mathbf{V}^p(\Lambda(\vartheta)).$$

In this form, for the special case of a Newtonian fluid, this problem has been considered in [R2] for a special cylindrical geometry, and previously in [CDK1] for the particular case $\mathbf{u}=\mathbf{v}$ ($\vec{\alpha}\equiv 0$) and $\Gamma_{\mathrm{N}}=\emptyset$. □

Remark 3.4. Using the symmetrical Kirchhoff transformation ($\vartheta=\int_T^{T_*}k(\tau)\,d\tau$) the liquid region is represented by $\{\vartheta<0\}$ as in [R2] or [R5], but there exists no essential change in the weak formulation of the Boussinesq–Stefan problem. □

Remark 3.5. As in [R2] we can also consider a more general nonlinear boundary condition on Γ_{N}, namely by setting

$$\text{(2.9}'\text{)}\qquad \frac{\partial\vartheta}{\partial n}\in G(\vartheta)\quad \text{on } \Gamma_{\mathrm{N}}\ ,$$

where G is a multivalued maximal monotone graph, including the cases of boundary climatization or an ambiguous cooling (see [DL]). This corresponds in the weak formulation to introduce a solution as a set $(\vartheta,\chi,\mathbf{u},g)$, with $g\in L^1(\Gamma_{\mathrm{N}})$, such that, $g\in G(\vartheta)$ a.e. in Γ_{N}, substituting $g(\vartheta)$ in the right hand side of (3.17). □

4 – Existence of Weak Solutions

In order to show the existence of at least one weak solution, we shall assume the following set of assumptions on the functions and for the data:

$$\text{(4.1)}\qquad \eta_k\in C^0(\mathbb{R}),\quad F_k\in C^0(\mathbb{R}_+),\quad \eta_k,F_k\geq 0,\quad F_k(0)=0\qquad (\forall\,k=1,...,N)\ ;$$

$$\text{(4.2)}\qquad \text{each } F_k \text{ is convex and } F_k(d)\leq C(d^{p/2}+1),\quad \text{for some } C>0\ ;$$

$$\text{(4.3)}\qquad \exists\,k^*,\ C_*>0,\ \eta_*>0\colon \begin{cases} F_{k^*}(d)\geq C_*\,d^{p/2}, & \forall\,d>0;\ F_{k^*} \text{ is strictly convex},\\ \eta_{k^*}(t)\geq\eta_* & \forall\,t\in\mathbb{R}\ ;\end{cases}$$

$$\text{(4.4)}\qquad b\colon \mathbb{R}\to\mathbb{R},\quad \mathbf{f}\colon \mathbb{R}_+\to\mathbb{R}^n \ \text{ are continuous functions,}\quad \mathbf{f}(0)=\mathbf{0}\ ;$$

$$\text{(4.5)}\qquad \begin{cases} g=g(x,t)\colon \Gamma_{\mathrm{N}}\times\mathbb{R}\to\mathbb{R} \ \text{ is monotone decreasing and continuous in } t\\ \text{for fixed } x\in\Gamma_{\mathrm{N}},\ g(\cdot,t)\in L^q(\Gamma_{\mathrm{N}}) \text{ for fixed } \tau, \text{ with } q>2 \text{ if } n=3,\ q=2=n\ ;\end{cases}$$

$$\text{(4.6)}\qquad \exists\,M>0\colon g(x,t)t\leq 0 \ \text{ for } |t|\geq M,\quad \text{a.e. } x\in\Gamma_{\mathrm{N}}\ ;$$

$$\text{(4.7)}\qquad \vartheta_{\mathrm{D}}\in H^{1/2}(\partial\Omega) \ \text{ and }\ -M\leq\vartheta_{\mathrm{D}}(x)\leq M\quad \text{a.e. } x\in\partial\Omega\ .$$

Theorem 4.1. *Under the assumptions (4.1)–(4.7) and (2.24), with $p > 3/2$ if $n = 2$ and $p > 9/5$ if $n = 3$, there exists a weak solution $(\vartheta, \chi, \mathbf{u})$ to the Boussinesq–Stefan problem in the sense of definition (3.14)–(3.18), such that, the temperature $\vartheta \in C^{0,\alpha}(\Omega)$ satisfies*

$$|\vartheta(x)| \leq M\,, \quad \forall x \in \Omega\,. \tag{4.8}$$

Remark 4.1. We observe that we do not require any growth assumption on the nonlinearities η_k, b, $\mathbf{f}$ or g in t, since the conditions (4.6) and (4.7) yield bounded temperatures ϑ. □

Remark 4.2. The conditions (4.1)–(4.2) imply that, for each $\tau \in L^\infty(\mathcal{O})$, the functional

$$J_{\mathcal{O}}(\tau;\mathbf{v}) = \int_{\mathcal{O}} \mathcal{F}(\tau;\mathbf{v}) = \sum_{k=0}^{N} \int_{\mathcal{O}} \eta_k(\tau) F_k(D_{II}(\mathbf{v})) \tag{4.9}$$

is a convex, continuous and nonnegative function defined over all $\mathbf{V}^p(\mathcal{O})$. In addition, (4.3) implies that $J_{\mathcal{O}}$ is strictly convex and coercive in the following sense:

$$J_{\mathcal{O}}(\tau;\mathbf{v}) \geq \nu_* \|\mathbf{v}\|^p_{\mathbf{V}^p(\mathcal{O})}\,, \quad \forall v \in \mathbf{V}^p(\mathcal{O}) \text{ with } \nu_* = C_* \eta_* > 0\,. \;\Box \tag{4.10}$$

Remark 4.3. It is clear that the assumptions (4.1)–(4.3) include all the examples given in section 2, in particular, the case of the Bingham fluid (2.22) corresponding to

$$\mathcal{F}(\vartheta;\mathbf{v}) = 2\mu(\vartheta) D_{II}(\mathbf{v}) + \gamma(\vartheta) D_{II}^{1/2}(\mathbf{v})\,.$$

On the other hand, it is easy to see that all the given four examples of asymptotically Newtonian fluids, satisfy $0 < m \leq \mu(s) \leq M$ and correspond, through (2.20), to the Hilbertian case $p = 2$, as well as the Newtonian fluid (2.18). □

Proof of Theorem 4.1. The method of proof follows the lines of [R2] and consists in passing to the limit in an appropriate penalization of the solid zone with a regularization of some data.

For each $\varepsilon > 0$, let χ_ε be defined by

$$\chi_\varepsilon(t) = \begin{cases} 1 & \text{for } t \leq -\varepsilon \\ 1 - \dfrac{t}{\varepsilon} & \text{for } -\varepsilon \leq t \leq -\varepsilon/2 \\ 0 & \text{for } t \geq -\varepsilon/2 \end{cases} \tag{4.11}$$

and consider a family of functions $\tilde{\vartheta}_\varepsilon \in C^{0,1}(\overline{\Omega})$, uniformly bounded in $H^1(\Omega)$ and whose trace $\tilde{\vartheta}_\varepsilon \to \vartheta_D$ in $H^{1/2}(\Gamma_D)$, as $\varepsilon \to 0$, and also a family of $b_\varepsilon \in C^{0,1}(\mathbb{R})$ with $b_\varepsilon \to b$ uniformly on compact subsets as $\varepsilon \to 0$.

We shall prove in Section 6 that, for each $\varepsilon > 0$, there exists at least one solution $(\vartheta_\varepsilon, \mathbf{u}_\varepsilon)$ to the following approximate problem:

Problem ($\mathbf{P}_\varepsilon$): *Find* $\vartheta_\varepsilon \in C^0(\overline{\Omega}) \cap H^1(\Omega)$, $\vartheta_\varepsilon|_{\Gamma_D} = \tilde{\vartheta}_\varepsilon$, $\mathbf{u}_\varepsilon \in \mathbf{V}^p(\Omega)$, *such that,*

$$\int_\Omega \Big\{ \nabla\vartheta_\varepsilon - b_\varepsilon(\vartheta_\varepsilon)\,(\mathbf{u}_\varepsilon + \vec{\alpha}) + \lambda\vec{\alpha}[1 - \chi_\varepsilon(\vartheta_\varepsilon)] \Big\} \cdot \nabla\varsigma = \int_{\Gamma_N} g(\vartheta_\varepsilon)\varsigma\,, \quad \forall \varsigma \in Z\,, \tag{4.12}$$

$$\int_\Omega \mathcal{F}(\vartheta_\varepsilon, \mathbf{v}) - \int_\Omega \mathcal{F}(\vartheta_\varepsilon, \mathbf{u}_\varepsilon) + \int_\Omega \beta(\mathbf{u}_\varepsilon, \mathbf{u}_\varepsilon, \mathbf{v}) + \frac{1}{\varepsilon}\int_\Omega \chi_\varepsilon(\vartheta_\varepsilon)\,\mathbf{u}_\varepsilon \cdot (\mathbf{v} - \mathbf{u}_\varepsilon) \geq \tag{4.13}$$

$$\geq \int_\Omega \mathbf{f}(\vartheta_\varepsilon) \cdot (\mathbf{v} - \mathbf{u}_\varepsilon)\,, \quad \forall \mathbf{v} \in \mathbf{V}^p(\Omega)\,.$$

Moreover, we shall prove the following *a priori* estimates, independently of ε, $0<\varepsilon<1$:

(4.14) $\|\vartheta_\varepsilon\|_{H^1(\Omega)} \leq C\,, \quad -M \leq \vartheta_\varepsilon(x) \leq M, \ \forall x \in \overline{\Omega}\,;$

(4.15) $\|\vartheta_\varepsilon\|_{C^{0,\gamma}(K)} \leq C(K)$, for any compact subset $K \subset \Omega$ and some $0<\gamma<1$;

(4.16) $\|\mathbf{u}_\varepsilon\|_{\mathbf{V}^p(\Omega)} \leq C$.

From these estimates, by compactness, there exists a subsequence, still indexed by ε, such that, for $\varepsilon \to 0$:

(4.17) $\vartheta_\varepsilon \to \vartheta$ in $H^1(\Omega)$-weak, uniformly in any compact subset $K \subset \Omega$,
and their trace in $L^r(\partial\Omega)$, $\forall r < \infty$, and pointwise a.e. on $\partial\Omega$;

(4.18) $1 - \chi_\varepsilon(\vartheta_\varepsilon) \rightharpoonup \chi$ in $L^\infty(\Omega)$-weak* ;

(4.19) $\mathbf{u}_\varepsilon \to \mathbf{u}$ in $\mathbf{V}^p(\Omega)$-weak and in $[L^2(\Omega)]^n$-strong $(n = 2, 3)$;

for some limit functions $\vartheta \in H^1(\Omega)$, $\chi \in L^\infty(\Omega)$ and $\mathbf{u} \in \mathbf{V}^p(\Omega)$.

Due to (4.15), we have $\vartheta \in C^{0,\gamma}(\Omega)$ and we can define the liquid zone and the solid zone, respectively, by

$$\Lambda(\vartheta) = \big\{x \in \Omega\colon\ \vartheta(x) > 0\big\} \quad \text{and} \quad \Sigma(\vartheta) = \big\{x \in \Omega\colon\ \vartheta(x) < 0\big\}\,,$$

which are open subsets of Ω.

Since $0 \leq 1 - \chi_\varepsilon(\vartheta_\varepsilon) \leq 1$, in the limit we also obtain $0 \leq \chi \leq 1$ a.e. in Ω. For any compact subset $K^+ \subset \Lambda(\vartheta)$, we have $\min_K \vartheta > 0$ and, by the uniform convergence (4.17), we have $1 - \chi_\varepsilon(\vartheta_\varepsilon) = 1$ in K^+ for all ε sufficiently small. Hence $\chi \geq \chi_{\Lambda(\vartheta)}$. Analogously, for any compact subset $K^- \subset \Sigma(\vartheta)$, we find $1 - \chi_\varepsilon(\vartheta_\varepsilon) = 0$ in K^- for ε small, $\varepsilon \leq \varepsilon_0(K^-)$, and then $\chi \leq 1 - \chi_{\Sigma(\vartheta)}$, which shows that χ satisfies (3.15). Also letting $\mathbf{v} = \mathbf{0}$ in (4.13), we obtain

$$\frac{1}{\varepsilon}\int_{K^-} |\mathbf{u}_\varepsilon|^2 \leq \frac{1}{\varepsilon}\int_\Omega \chi_\varepsilon(\vartheta_\varepsilon)\,|\mathbf{u}_\varepsilon|^2 \leq \max_{|\tau|\leq M} |\mathbf{f}(\tau)| \int_\Omega |\mathbf{u}_\varepsilon| \leq C\,,$$

where $C > 0$ is independent of $\varepsilon \leq \varepsilon_0(K^-)$, by (4.17). Therefore we obtain $\mathbf{u} = \mathbf{0}$ a.e. in K^- and, since K^- is arbitrary in $\Sigma(\vartheta)$, we conclude (3.16).

Since $b_\varepsilon(\vartheta_\varepsilon) \to b(\vartheta)$ pointwise and is uniformly bounded, from (4.19) we deduce $b_\varepsilon(\vartheta_\varepsilon)\mathbf{u}_\varepsilon \to b(\vartheta)\mathbf{u}$ in $[L^2(\Omega)]^n$. Since, by (4.17) and (4.5), $g(\vartheta_\varepsilon) \to g(\vartheta)$ in $L^q(\Gamma_N)$, we can pass easily to limit $\varepsilon \to 0$ in (4.12), obtaining (3.17).

In order to pass to the limit in (4.13), we take an arbitrary $\Psi \in [\mathcal{D}(\Lambda(\vartheta))]^n$, such that $\nabla \cdot \Psi = 0$, and extend Ψ by $\mathbf{0}$ in Ω. Since $\chi_\varepsilon(\vartheta_\varepsilon) = 0$ in $\operatorname{supp}\Psi \subset \Lambda(\vartheta)$, for all $\varepsilon \leq \varepsilon_\Psi$, we obtain from (4.13)

$$\int_{\Lambda(\vartheta)} \mathcal{F}(\vartheta_\varepsilon, \Psi) + \int_{\Lambda(\vartheta)} \beta(\mathbf{u}_\varepsilon, \mathbf{u}_\varepsilon, \Psi) \geq \int_\Omega \mathcal{F}(\vartheta_\varepsilon, \mathbf{u}_\varepsilon) + \frac{1}{\varepsilon}\int_\Omega \chi_\varepsilon(\vartheta_\varepsilon)\, |\mathbf{u}_\varepsilon|^2 + \int_\Omega \mathbf{f}(\vartheta_\varepsilon)\cdot(\Psi - \mathbf{u}_\varepsilon)$$

$$\geq \int_{\Lambda(\vartheta)} \mathcal{F}(\vartheta_\varepsilon, \mathbf{u}_\varepsilon) + \int_{\Lambda(\vartheta)} \mathbf{f}(\vartheta_\varepsilon)\cdot\Psi - \int_\Omega \mathbf{f}(\vartheta_\varepsilon)\cdot\mathbf{u}_\varepsilon\,, \qquad \forall \varepsilon \leq \varepsilon_\Psi\,. \tag{4.20}$$

Recalling the assumptions and the convergences (4.17) and (4.19), we easily get, as $\varepsilon \to 0$,

$$\int_{\Lambda(\vartheta)} \mathcal{F}(\vartheta_\varepsilon, \Psi) \to \int_{\Lambda(\vartheta)} \mathcal{F}(\vartheta, \Psi)$$

$$\int_{\Lambda(\vartheta)} \beta(\mathbf{u}_\varepsilon, \mathbf{u}_\varepsilon, \Psi) \to \int_{\Lambda(\vartheta)} \beta(\mathbf{u}, \mathbf{u}, \Psi) \qquad \text{(recall (3.13))}$$

$$\int_{\Lambda(\vartheta)} \mathbf{f}(\vartheta_\varepsilon)\cdot\Psi \to \int_{\Lambda(\vartheta)} \mathbf{f}(\vartheta)\cdot\Psi$$

$$\int_\Omega \mathbf{f}(\vartheta_\varepsilon)\cdot\mathbf{u}_\varepsilon \to \int_\Omega \mathbf{f}(\vartheta)\cdot\mathbf{u} = \int_{\Lambda(\vartheta)} \mathbf{f}(\vartheta)\cdot\mathbf{u} \qquad \text{(by (4.4) and (3.16))}.$$

On the other hand, we pass to the limit in the remainder term

$$\liminf_{\varepsilon\to0} \int_{\Lambda(\vartheta)} \mathcal{F}(\vartheta_\varepsilon, \mathbf{u}_\varepsilon) = \liminf_{\varepsilon\to0} \Big\{ \int_{\Lambda(\vartheta)} \mathcal{F}(\vartheta, \mathbf{u}_\varepsilon) + \sum_{k=1}^N \int_{\Lambda(\vartheta)} [\eta_k(\vartheta_\varepsilon) - \eta_k(\vartheta)]\mathbf{F}_k(D_{II}(\mathbf{u}_\varepsilon)) \Big\}$$

$$\geq \int_{\Lambda(\vartheta)} \mathcal{F}(\vartheta, \mathbf{u})\,, \tag{4.21}$$

by using the lower semi-continuity of the energy and the property

$$\int_{\Lambda(\vartheta)} \Big\{ [\eta_k(\vartheta_\varepsilon) - \eta_k(\vartheta)]\mathbf{F}_k(D_{II}(\mathbf{u}_\varepsilon)) \Big\} \to 0 \qquad \text{as } \varepsilon \to 0\,,$$

which holds by Lebesgue theorem, the estimates (4.14) and (4.16) and the pointwise convergence $\eta_k(\vartheta_\varepsilon) \to \eta_k(\vartheta)$.

Therefore we obtain (3.18) by letting $\varepsilon \to 0$ in (4.20), first for smooth Ψ and, by density, also for all $\Psi \in \mathbf{V}^p(\Lambda(\vartheta))$. Consequently we have shown that the triple $(\vartheta, \chi, \mathbf{u})$ solves the problem (3.14)–(3.18), which ends the proof of the Theorem 4.1. ∎

Remark 4.4. In the case of Remark 3.5, we need a further regularization in (4.12): replace g by the Yosida regularization g_ε of the maximal monotone graph $G_x(\cdot) = G(x, \cdot)$, $x \in \Gamma_N$, which must satisfy the following assumptions generalizing (4.5) and (4.6): $[-M, M] \subset \{t \in \mathbb{R} : G(x,t) \neq \emptyset\}$, $G_x(M) \subset]-\infty, 0]$, $G_x(-M) \subset [0, +\infty[$ and the function $x \mapsto g^0(x,t)$ (i.e., the smallest number in absolute value of $G_x(t)$) is in $L^q(\Gamma_N)$, for all $|t| \leq M$, with $q = 2 = n$ and $q > 2$ if $n = 3$ (see [R2], for details). □

5 – A Nonlinear Convection–Diffusion Problem

In the study of the approximate problem (4.12)–(4.13), it appears a special case of the following nonlinear convection–diffusion problem:

Problem (Θ): Find $\vartheta \in C^0(\overline{\Omega}) \cap H^1(\Omega)$, such that, $\vartheta = \tilde{\vartheta}$ on Γ_D and

$$\int_\Omega \{\nabla\vartheta + W(x,\vartheta)\} \cdot \nabla\varsigma = \int_{\Gamma_N} g(x,\vartheta)\,\varsigma\,, \quad \forall \varsigma \in Z\,. \tag{5.1}$$

Here, we assume the same conditions (4.5) and (4.6) on g and, for the Dirichlet data, we replace (4.7) by

$$\tilde{\vartheta} \in C^{0,1}(\overline{\Omega}) \quad \text{and} \quad -M \le \tilde{\vartheta}(x) \le M, \ \ \forall x \in \overline{\Omega}\,. \tag{5.2}$$

The convective term $W : \Omega \times \mathbb{R} \to \mathbb{R}^n$ is a Carathéodory function, such that, for some λ, $1/2 \le \lambda \le 1$, and $k \in L^2(\Omega)$:

$$|W(x,t) - W(x,s)| \le k(x)\,|t-s|^\lambda, \quad \text{a.e. } x \in \Omega, \ \ \forall t,s \in [-M,M]\,; \tag{5.3}$$

$$|W(x,t)| \le \omega_0(x), \quad \text{a.e. } x\in\Omega, \ \ \forall t\in[-M,M], \ \ \text{for } \omega_0\in L^p(\Omega), \ \ p>n\ge 2\,; \tag{5.4}$$

$$\int_\Omega W(x,t)\cdot\nabla\varsigma \ge 0 \quad \forall\varsigma \in Z\colon \varsigma \ge 0 \text{ in } \Omega; \ \forall t \in [-M,M]\,. \tag{5.5}$$

Remark 5.1. If, for each $t \in [-M,M]$, $W(\cdot,t) \in [H^1(\Omega)]^n$, the assumption (5.5) is equivalent to the double condition

$$\nabla_x \cdot W \le 0 \ \text{ a.e. in } \Omega \quad \text{and} \quad W\cdot\mathbf{n} \ge 0 \ \text{ on } \Gamma_N\,. \tag{5.6}$$

These conditions are sufficient in order to use the weak maximum principle in (5.1), which corresponds to the weak formulation for the mixed problem

$$\begin{cases} \nabla\cdot[\nabla\vartheta + W(x,\vartheta)] = 0 & \text{in } \Omega \\ \vartheta = \tilde{\vartheta} \ \text{ on } \Gamma_D \quad \text{and} & \partial\vartheta/\partial n = g(x,\vartheta) \ \text{ on } \Gamma_N. \ \square \end{cases}$$

Proposition 5.1. *Under the preceding assumptions, namely (4.5)–(4.6) and (5.2)–(5.5) there exists a unique solution to Problem (Θ), which satisfies $\vartheta \in C^{0,\gamma}(\overline{\Omega})$, for some γ, $0 \le \gamma < 1$, and*

$$\|\vartheta\|_{L^\infty(\Omega)} \le M\,. \tag{5.7}$$

Proof: The uniqueness follows as a special case of [CM]. The existence result is based on the a *priori* estimate (5.7), an estimate due to Stampacchia [S] and a fixed point argument already used in [R2] for a particular case.

In order to obtain the a *priori* bound $\vartheta \le M$, take $\varsigma = (\vartheta - M)^+$ in (5.1). We have

$$\int_\Omega |\nabla(\vartheta - M)^+|^2 = -\int_\Omega W(x,\vartheta)\cdot\nabla(\vartheta-M)^+ + \int_{\Gamma_N} g(x,\vartheta)\,(\vartheta-M)^+ \le 0\,,$$

by the assumptions (5.5) and (4.6), respectively. Indeed, if we assume W as in Remark 5.1 and we define

$$\overline{W}_M(x,t) = \begin{cases} \int_M^t W(x,s)\,ds & \text{if } t > M \\ 0 & \text{if } t \le M, \end{cases}$$

it is clear that $\overline{W}_M$ also satisfies (5.6). Hence, by integrating by parts, we obtain

$$\int_\Omega W(x,\vartheta(x))\cdot\nabla(\vartheta - M)^+\,dx = \int_\Omega \nabla_x\cdot\overline{W}_M(x,\vartheta(x))\,dx - \int_{\{\vartheta>M\}}\Big(\int_M^{\vartheta(x)}\nabla_x\cdot W(x,t)\,dt\Big)\,dx$$

$$\ge \int_\Omega \nabla_x\cdot\overline{W}_M = \int_{\Gamma_N}\mathbf{n}\cdot\overline{W}_M \ge 0\,. \tag{5.8}$$

When $W(x,t)$ is not smooth in x, we can argue with its regularization $W_\delta(x,t)$ (for instance, by convolution with a mollifier in x, for fixed t, $W_\delta(x,t) = (W(\cdot,t)*\rho_\delta)(x) \underset{\delta\to 0}{\longrightarrow} W(x,t)$), and then passing to the limit in the corresponding inequality (5.8).

Similarly, we obtain $\vartheta \ge -M$, by taking $\varsigma = (\vartheta + M)^-$.

Now we complete the proof by applying the Schauder fixed point theorem (as in [R2]) to the compact mapping

$$\mathcal{S}: B_K \to B_K \quad \text{with } B_K = \left\{\tau\in C^0(\overline{\Omega}): \|\tau\|_{C^0(\overline{\Omega})}\le K\right\}$$

defined by the unique solution $\sigma = \mathcal{S}(\tau)$ of the mixed linear problem: find $\sigma\in H^1(\Omega)$, such that $\sigma = \tilde{\vartheta}$ on Γ_D and

$$\int_\Omega \nabla\sigma\cdot\nabla\varsigma = \int_{\Gamma_N} g^M(\tau)\,\varsigma - \int_\Omega W^M(\tau)\cdot\nabla\varsigma\,, \qquad \forall\varsigma\in Z\,. \tag{5.9}$$

Here g^M and W^M denote the truncated functions of g and W, respectively, i.e., $W^M(x,t) = W(x,\min(M,\max(-M,t)))$, and the constant $K > M$ is given by Stampacchia estimate [S] for the solution of (5.9), for some Hölder exponent γ, $0<\gamma<1$ (with $g_0(x) = \max(|g(x,M)|,|g(x,-M)|)$, recall (4.5)):

$$\|\sigma\|_{C^{0,\gamma}(\overline{\Omega})} \le C\left(\|\omega_0\|_{L^p(\Omega)} + \|g_0\|_{L^q(\Gamma_N)} + \|\tilde{\vartheta}\|_{C^{0,1}(\overline{\Omega})}\right) \equiv K\,. \tag{5.10}$$

Since the fixed point $\vartheta_* = \mathcal{S}(\vartheta_*)$ also verifies the estimate (5.7), it is clear that $\vartheta_* = \vartheta$ is the solution of the Problem (Θ). ∎

Remark 5.2. The Theorem 2 is a generalization of Theorem 2 of [R2] and the existence part of it also holds if we relax the assumption (5.3) to continuity only and if we relax the monotonicity of g in (4.5) to the assumption of $\exists\, g_0\in L^q(\Gamma_N)$, $q > n-1$: $|g(x,t)|\le g_0(x)$, a.e. $x\in\Gamma_N$, $\forall t\in[-M,M]$. □

In order to apply this result to the problem (4.12) we need also a continuous dependence result which is contained in the next proposition.

Proposition 5.2. *For arbitrary Hölder continuous real functions $a,\nu\in C^{0,1/2}(\mathbb{R})$, for any $\vec{\alpha}$ satisfying (2.24) and for any $\mathbf{w}\in\mathbf{L}^s_\sigma(\Omega)$, $s > n \ge 2$, there exists a unique solution $\vartheta = \vartheta(\mathbf{w})$ to (5.1), corresponding to*

$$W(x,t) = a(t)\,\vec{\alpha}(x) + \nu(t)\,\mathbf{w}(x)\,. \tag{5.11}$$

Moreover, if $\mathbf{w}_\delta \rightharpoonup \mathbf{w}_0$ *in* $\mathbf{L}^s_\sigma(\Omega)$*-weak, then the corresponding solutions* $\vartheta_\delta = \vartheta(\mathbf{w}_\delta) \to \vartheta_0 = \vartheta(\mathbf{w}_0)$ *in* $H^1(\Omega)$*-weak and in* $C^{0,\lambda}(\overline{\Omega})$*-strong for any* $0 \leq \lambda < \gamma < 1$.

Proof: The first part is a corollary of Theorem 5.1, since its assumptions are immediately satisfied (recall that $\mathbf{w}$ and $\vec{\alpha}$ satisfying (3.4) and (2.24), respectively, imply the condition (5.5) for W).

To prove the continuous dependence result, let $\mathbf{w}_\delta \rightharpoonup \mathbf{w}_0$ in $\mathbf{L}^s_\sigma(\Omega)$-weak. From the estimate (5.7) and (5.10) for ϑ_δ, we have

$$\|\vartheta_\delta\|_{C^{0,\gamma}(\overline{\Omega})} \leq K \quad \text{(independently of } \delta).$$

Letting $\varsigma = \vartheta_\delta - \vartheta_0$ in (5.1) for ϑ_δ, we easily find

$$\|\vartheta_\delta\|_{H^1(\Omega)} \leq C \quad \text{(independently of } \delta),$$

and, for a subsequence still indexed by δ, we have

$$\vartheta_\delta \to \vartheta \quad \text{in } H^1(\Omega)\text{-weak and in } C^{0,\lambda}(\overline{\Omega}), \quad 0 < \lambda < \gamma \, .$$

Passing to the limit in (5.1) for $\mathbf{w}_\delta$ we find that ϑ solves (5.1) corresponding to $\mathbf{w}_0$ and by uniqueness $\vartheta = \vartheta_0$ and the whole sequence converges as stated. ∎

Remark 5.3. In the Proposition 5.2, if also $\mathbf{w}_\delta \to \mathbf{w}_0$ strongly in $[L^2(\Omega)]^n$ then also $\vartheta_\delta \to \vartheta_0$ in $H^1(\Omega)$-strong. To obtain this strong convergence it is enough to take $\varsigma_\delta = \vartheta_\delta - \vartheta_0$ in (5.1) for ϑ_δ and for ϑ_0. Using $\nabla\varsigma_\delta \rightharpoonup 0$ in $[L^2(\Omega)]^n$-weak, we conclude from

$$\int_\Omega |\nabla\varsigma_\delta|^2 = \int_{\Gamma_N} [g(\vartheta_\delta) - g(\vartheta_0)]\,\varsigma_\delta - \\ - \int_\Omega \Big\{[a(\vartheta_\delta) - a(\vartheta_0)]\vec{\alpha} + [\nu(\vartheta_\delta) - \nu(\vartheta_0)]\mathbf{w}_0 + \nu(\vartheta_\delta)\,(\mathbf{w}_\delta - \mathbf{w}_0)\Big\} \cdot \nabla\varsigma_\delta \to 0. \ ∎$$

6 – Existence of Approximate Solutions

We begin by studying an auxiliary class of elliptic variational inequalities. For $\tau \in L^\infty(\Omega)$ and $\mathbf{w} \in \mathbf{L}^r_\sigma(\Omega)$, for some $r \geq 2$, find $\mathbf{u} \in \mathbf{V}^p(\Omega)$, such that,

$$\int_\Omega \mathcal{F}(\tau,\mathbf{v}) - \int_\Omega \mathcal{F}(\tau,\mathbf{u}) + \int_\Omega \beta(\mathbf{w},\mathbf{u},\mathbf{v}) + \int_\Omega [a(\tau)\mathbf{u} - \mathbf{f}(\tau)]\cdot(\mathbf{v}-\mathbf{u}) \geq 0, \quad \forall\, \mathbf{v} \in \mathbf{V}^p(\Omega). \tag{6.1}$$

Here $a \in C^0(\mathbb{R}^n)$ may be any nonnegative continuous function, $\mathcal{F}$ is defined in (2.14) with the assumptions (4.1)–(4.3). Recalling (3.13), we observe that, if $1/r + 1/s + 1/p = 1$, by Hölder inequality, we have

$$\left|\int_\Omega \beta(\mathbf{w},\mathbf{u},\mathbf{v})\right| \leq C\,\|\vec{\alpha} + \mathbf{w}\|_{L^r(\Omega)}\,\|\mathbf{u}\|_{L^s(\Omega)}\,\|\mathbf{v}\|_{\mathbf{V}^p(\Omega)} \, . \tag{6.2}$$

Since by Sobolev inequality $\mathbf{V}^p(\Omega) \subset \mathbf{L}^{np/(n-p)}_\sigma(\Omega)$, for every $r \geq np/(np+p-2n)$, we have

$$\int_\Omega \beta(\mathbf{w},\mathbf{u},\mathbf{v}) = -\int_\Omega \beta(\mathbf{w},\mathbf{v},\mathbf{u}), \quad \forall\, \mathbf{w} \in \mathbf{L}^r_0(\Omega), \ \forall\, \mathbf{u},\mathbf{v} \in \mathbf{V}^p(\Omega) \, . \tag{6.3}$$

Proposition 6.1. *There exists a unique solution* $\mathbf{u} \in \mathbf{V}^p(\Omega)$ *to the variational inequality (6.1), with* $p > 2n/(n+2)$ *and* $r \geq \max(np/(np+p-2n), 2)$. *Moreover if*

$$(6.4) \qquad \tau_\delta \to \tau_0 \ \text{ in } L^\infty(\Omega) \quad \text{and} \quad \mathbf{w}_\delta \rightharpoonup \mathbf{w}_0 \ \text{ in } \mathbf{L}^r_\sigma(\Omega)\text{-weak},$$

then the respective solutions $\mathbf{u}_\delta = \mathbf{u}(\tau_\delta, \mathbf{w}_\delta)$ *and* $\mathbf{u}_0 = \mathbf{u}(\tau_0, \mathbf{w}_0)$ *satisfy*

$$(6.5) \qquad \mathbf{u}_\delta \to \mathbf{u}_0 \quad \text{in } \mathbf{V}^p(\Omega)\text{-weak and in } \mathbf{L}^s_\sigma(\Omega)\text{-strong},$$

for any $s < np/(n-p)$ *if* $p < n$ *or any* $s < +\infty$ *if* $p \geq n$.

Proof: i) *Existence.* Defining the operator $A\colon V \to V'$ by

$$(6.6) \qquad \langle A\mathbf{u}, \mathbf{v}\rangle = \int_\Omega \beta(\mathbf{w}, \mathbf{u}, \mathbf{v}) + \int_\Omega [a(\tau)\mathbf{u} - \mathbf{f}(\tau)] \cdot \mathbf{v}, \quad \forall \mathbf{u}, \mathbf{v} \in V \equiv \mathbf{V}^p(\Omega)\ ,$$

we easily see that A is continuous and monotone, i.e.

$$\langle A\mathbf{u} - A\mathbf{v}, \mathbf{u} - \mathbf{v}\rangle = \int_\Omega a(\tau)\,|\mathbf{u} - \mathbf{v}|^2 \geq 0, \qquad \text{since } a \geq 0\ .$$

Setting $J(\mathbf{v}) = \int_\Omega \mathcal{F}(\tau, \mathbf{v}) : V \to [0, +\infty[$, from Remark 4.2, we observe that J is a strictly convex functional satisfying (4.10), i.e., if $\|\mathbf{v}\| = \|\mathbf{v}\|_{\mathbf{V}^p(\Omega)}$, we have

$$(6.7) \qquad J(\mathbf{v}) \geq \nu_*\, \|\mathbf{v}\|^p, \qquad \forall v \in V \equiv \mathbf{V}^p(\Omega)\ .$$

Since, we have $\|\mathbf{v}\|_{L^p(\Omega)} \leq C_0\|\mathbf{v}\|$, $\forall \mathbf{v} \in V$, from (6.6) we obtain

$$\langle A\mathbf{v}, \mathbf{v}\rangle \geq -C\|\mathbf{v}\|_{L^1(\Omega)} \geq -C_*\|\mathbf{v}\|\ .$$

Then, from (6.7), we obtain the following coerciveness property

$$\big\{\langle A\mathbf{v}, \mathbf{v}\rangle + J(\mathbf{v})\big\}/\|\mathbf{v}\| \to +\infty \qquad \text{as } \|\mathbf{v}\| \to \infty\ ,$$

and by a classical theorem of [L, pag. 251], there exists a solution of the variational inequality

$$(6.8) \qquad \mathbf{u} \in V\colon J(\mathbf{v}) - J(\mathbf{u}) + \langle A\mathbf{u}, \mathbf{v} - \mathbf{u}\rangle \geq 0, \qquad \forall \mathbf{v} \in V\ ,$$

which is another way of writing (6.1).

ii) *Uniqueness.* If $\mathbf{u}$ and $\hat{\mathbf{u}}$ are two different solutions to (6.8), taking $\mathbf{v} = \hat{\mathbf{u}}$ in (6.8) and $\mathbf{v} = \mathbf{u}$ in the corresponding inequality for $\hat{\mathbf{u}}$, by the monotonicity of A, we obtain

$$\langle A\mathbf{u} - A\hat{\mathbf{u}}, \mathbf{u} - \hat{\mathbf{u}}\rangle = 0\ .$$

Setting $\mathbf{v} = \frac{1}{2}(\mathbf{u} + \hat{\mathbf{u}})$ in (6.8) for $\mathbf{u}$ and for $\hat{\mathbf{u}}$, by addition, we have

$$2\,J\Big(\frac{\mathbf{u} + \hat{\mathbf{u}}}{2}\Big) = 2\,J\Big(\frac{\mathbf{u} + \hat{\mathbf{u}}}{2}\Big) - \frac{1}{2}\langle A\mathbf{u} - A\hat{\mathbf{u}}, \mathbf{u} - \hat{\mathbf{u}}\rangle \geq J(\mathbf{u}) + J(\hat{\mathbf{u}})\ ,$$

which is a contradiction with the strict convexity of J:

$$J\left(\frac{\mathbf{u}+\hat{\mathbf{u}}}{2}\right) < \frac{1}{2}J(\mathbf{u}) + \frac{1}{2}J(\hat{\mathbf{u}}) .$$

iii) *Continuous Dependence.* By setting $\mathbf{v} = \mathbf{0}$ in (6.1), we obtain

$$\nu_* \|\mathbf{u}_\delta\|^p \leq \int_\Omega \mathcal{F}(\tau_\delta, \mathbf{u}_\delta) \leq \int_\Omega \mathbf{f}(\tau_\delta) \cdot \mathbf{u}_\delta \leq C\|\mathbf{u}_\delta\|_{L^1} \leq C'\|\mathbf{u}_\delta\|$$

and, consequently, we have the a *priori* estimate

$$\|\mathbf{u}_\delta\| \leq C_* \quad \text{(independently of } \delta). \tag{6.9}$$

Then, for a subsequence and for some $\mathbf{u} \in \mathbf{V}^p(\Omega)$, we have

$$\mathbf{u}_\delta \to \mathbf{u} \quad \text{in } \mathbf{V}^p(\Omega)\text{-weak and } \mathbf{L}^s_\sigma(\Omega)\text{-strong}, \quad \forall s < np/(n-p) .$$

In order to pass to the limit in

$$\int_\Omega \mathcal{F}(\tau_\delta, \mathbf{v}) - \int_\Omega \mathcal{F}(\tau_\delta, \mathbf{u}_\delta) + \int_\Omega \beta(\mathbf{w}_\delta, \mathbf{u}_\delta, \mathbf{v}) + \int_\Omega [a(\tau_\delta)\mathbf{u}_\delta - \mathbf{f}(\tau_\delta)] \cdot (\mathbf{v} - \mathbf{u}_\delta) \geq 0 \tag{6.10}$$

we use

$$\liminf \int_\Omega \mathcal{F}(\tau_\delta, \mathbf{u}_\delta) \geq \int_\Omega \mathcal{F}(\tau_0, \mathbf{u}) \quad \text{(as in (4.21))},$$

and we observe that, for any $\mathbf{v} \in \mathcal{V}(\Omega) = \{\mathbf{v} \in [\mathcal{D}(\Omega)]^n \colon \nabla \cdot \mathbf{v} = 0 \text{ in } \Omega\}$

$$\int_\Omega \mathcal{F}(\tau_\delta, \mathbf{v}) \to \int_\Omega \mathcal{F}(\tau_0, \mathbf{v})$$

and

$$\int_\Omega \beta(\mathbf{w}_\delta, \mathbf{u}_\delta, \mathbf{v}) \to \int_\Omega \beta(\mathbf{w}_0, \mathbf{u}, \mathbf{v}) ,$$

since $r \geq 2$ and $\mathbf{u}_\delta \to \mathbf{u}$ strongly, in particular in $\mathbf{L}^2_\sigma(\Omega)$.

The last term of (6.10) passes easily to the limit, and we obtain that $\mathbf{u}$ solves

$$\int_\Omega \mathcal{F}(\tau_0, \mathbf{v}) - \int_\Omega \mathcal{F}(\tau_0, \mathbf{u}) + \int_\Omega \beta(\mathbf{w}_0, \mathbf{u}, \mathbf{v}) + \int_\Omega [a(\tau_0)\mathbf{u} - \mathbf{f}(\tau_0)] \cdot (\mathbf{v} - \mathbf{u}) \geq 0, \quad \forall \mathbf{v} \in \mathbf{V}(\Omega),$$

and, by density of $\mathcal{V}(\Omega)$ in $\mathbf{V}^p(\Omega)$ and by uniqueness of the solution to (6.1), we conclude $\mathbf{u} = \mathbf{u}_0$ and (6.5) holds. ∎

Remark 6.1. The a *priori* estimate (6.9) shows that, in fact, the solution $\mathbf{u}$ of (6.1) is independent of the choice of $\mathbf{w}$ in $\mathbf{L}^r_\sigma(\Omega)$ and depends only on τ through $\|\mathbf{f}(\tau)\|_{L^\infty(\Omega)}$. □

Remark 6.2. With the notations of the proof of Proposition 6.1 and recalling that the definition of the subdifferential of J at the point $\mathbf{u} \in V$ is given by the set

$$\partial J(u) = \left\{\xi \in V' \colon J(\mathbf{v}) - J(\mathbf{u}) \geq \langle \xi, \mathbf{v} - \mathbf{u} \rangle, \ \forall \mathbf{v} \in V\right\} ,$$

from (6.8) we can rewrite the variational inequality in the form

$$-A(\mathbf{u}) \in \partial J(\mathbf{u}). \ \square$$

Theorem 6.2. *Assume (4.1)–(4.6), (2.24), $p > 3/2$ if $n = 2$ or $p > 9/5$ if $n = 3$, $\tilde{\vartheta}_\varepsilon$ satisfies (5.2), χ_ε is defined by (4.11) and b_ε is Lipschitz continuous. Then for each $\varepsilon > 0$, there exists at least a solution $(\vartheta_\varepsilon, \mathbf{u}_\varepsilon) \in [H^1(\Omega) \cap C^{0,\gamma}(\overline{\Omega})] \times \mathbf{V}^p(\Omega)$, for some $0 < \gamma < 1$, to the approximate problem (P_ε). Moreover it satisfies the estimates (4.14)–(4.16).*

Proof: The existence result relies on the Schauder fixed point theorem in the following convex, closed, bounded subset of $\mathbf{L}^r_\sigma(\Omega)$, given by

$$\mathcal{B}_R = \left\{\mathbf{v} \in \mathbf{L}^r_\sigma(\Omega) \colon \|\mathbf{v}\|_{\mathbf{L}^r} \leq R\right\} \quad \text{with } r=6 \text{ if } n=2 \text{ or } r=9/2 \text{ if } n=3 , \tag{6.11}$$

where $R > 0$ is a constant to be defined by an *a priori* estimate. We define a nonlinear mapping $\mathcal{T} \colon \mathbf{L}^r_\sigma(\Omega) \to \mathcal{B}_R$, which will be shown to be continuous and compact.

i) *Definition of $\mathcal{T}$:* for any $\mathbf{w} \in \mathbf{L}^r_\sigma(\Omega)$ we define $\tau_\varepsilon = \tau_\varepsilon(\mathbf{w})$ as the unique solution of the Problem (Θ), given by the Proposition 5.2 corresponding to the choice

$$W(x,t) = \lambda\, \vec{\alpha}[1 - \chi_\varepsilon(t)] - b_\varepsilon(t)[\mathbf{w} + \vec{\alpha}] ;$$

we know that $\tau_\varepsilon \in C^{0,\gamma}(\overline{\Omega})$, for some $0 < \gamma < 1$, and $|\tau_\varepsilon| \leq M$ in $\overline{\Omega}$; with this τ_ε and with the same $\mathbf{w}$, we define $\mathbf{w}_\varepsilon = \mathbf{u}(\tau_\varepsilon, \mathbf{w})$ as being the unique solution of the variational inequality (6.1) given by the Proposition 6.1, corresponding to the choice of $a(t) = \frac{1}{\varepsilon}\chi_{\varepsilon(t)}$; we know that $\mathbf{w}_\varepsilon \in \mathbf{V}^p(\Omega) \subset \mathbf{L}^s_\sigma(\Omega)$, with $s = 2p/(2-p) > 6$ $(p > 3/2)$ for $n = 2$ and $s = 3p/(3-p) > 9/2$ $(p > 9/5)$ for $n = 3$; we set $\mathbf{w}_\varepsilon = \mathcal{T}\mathbf{w}$.

ii) *$\mathcal{T}(\mathcal{B}_R) \subset \mathcal{B}_R$:* as observed in Remark 6.1, by taking $\mathbf{v} = \mathbf{0}$ in (6.1), recalling the Sobolev imbedding $\mathbf{V}^p(\Omega) \subset \mathbf{L}^r_\sigma(\Omega)$, from the estimate (6.9) we obtain

$$\|\mathbf{w}_\varepsilon\|_{\mathbf{L}^r(\Omega)} \leq C_{r,p} \|\mathbf{w}_\varepsilon\|_{\mathbf{V}^p(\Omega)} \leq C_* C_{r,p} \equiv R , \tag{6.12}$$

which determines the choice of R, independently of $\mathbf{w} \in \mathbf{L}^r_\sigma(\Omega)$ and of $\varepsilon > 0$; hence if $\mathbf{w} \in \mathcal{B}_R$, $\mathbf{w}_\varepsilon = \mathcal{T}(\mathbf{w}) \in \mathcal{B}_R$.

iii) *Continuity and compactness of $\mathcal{T}$:* take any sequence $\mathbf{w}_\delta \rightharpoonup \mathbf{w}_0$ in $\mathbf{L}^r_\sigma(\Omega)$-weak; from the Proposition 5.2, we have $\tau_\delta(\mathbf{w}_\delta) \to \tau_0(\mathbf{w}_0)$ uniformly in $\overline{\Omega}$, and the Proposition 6.1 implies then $\mathcal{T}(\mathbf{w}_\delta) \to \mathcal{T}(\mathbf{w}_0)$ strongly in $\mathbf{L}^r_\sigma(\Omega)$, by the compactness of the imbedding $\mathbf{V}^p(\Omega) \subset \mathbf{L}^r_\sigma(\Omega)$ due to the choice of r (see (6.11)) and the assumption $p > 3/2$ if $n = 2$ or $p > 9/5$ if $n = 3$.

Therefore we obtain a solution to the approximate problem (P_ε) with the fixed point

$$\mathbf{u}_\varepsilon = \mathcal{T}(\mathbf{u}_\varepsilon) \quad \text{and} \quad \vartheta_\varepsilon = \tau(\mathbf{u}_\varepsilon) .$$

Finally the *a priori* estimates (4.14), (4.15) and (4.16) follow immediately from (5.13) and (5.7), from (5.12) and from (6.12), respectively. ∎

7 – Remarks on the Two-Phase Rayleigh–Bénard Problem

It is obvious that all the preceding results hold for the particular case $\vec{\alpha} = \mathbf{0}$. This problem with Dirichlet boundary condition for the temperature on the whole $\partial\Omega$, was considered in [CB] for a Newtonian fluid, from a numerical point of view, under the name of two-phase Rayleigh–Bénard problem. We shall show that this problem can be obtained as the limit for $\alpha = |\vec{\alpha}| \to 0$ of the Boussinesq–Stefan problem, under the following assumption on the nonlinear flux condition on Γ_N:

$$g(x,t) = a(x)\, h(\vartheta_D(x) - t), \qquad \text{for } t \in \mathbb{R} \text{ and } x \in \Gamma_N , \tag{7.1}$$

where ϑ_D is given by (4.7), a is a bounded function satisfying

$$0 < \frac{1}{k\,\alpha} \le a(x) \le \frac{k}{\alpha}, \qquad \text{a.e. } x \in \Gamma_N , \tag{7.2}$$

for constants $k \ge 1$ and $\alpha > 0$, and h is a monotone increasing function, such that, for some $q \ge 1$,

$$h \in C^0(\mathbb{R}) \quad \text{and} \quad h(t)\,t \ge |t|^q, \qquad \forall t \in \mathbb{R} . \tag{7.3}$$

For instance, we can take $h(t) = t|t|^{q-2}$, as an example.

Following [CDK1], we introduce the definition of weak solution of the Rayleigh–Bénard problem.

Definition. We say that $(\vartheta_0, \mathbf{u}_0)$ is a weak solution of (2.6)–(2.7) with $\vec{\alpha} = \mathbf{0}$ and with Dirichlet boundary condition, and of (2.12)–(2.15), with homogeneous conditions, if

$$\vartheta_0 \in C^0(\Omega) \cap H^1(\Omega), \qquad \vartheta = \vartheta_D \text{ on } \partial\Omega , \tag{7.4}$$

$$\mathbf{u}_0 \in \mathbf{V}^p(\Omega), \qquad \mathbf{u}_0 = \mathbf{0} \text{ a.e. in } \Sigma(\vartheta_0) = \{x \in \Omega\colon \vartheta_0(x) < 0\} , \tag{7.5}$$

$$\int_\Omega \{\nabla\vartheta_0 - b(\vartheta_0)\mathbf{u}_0\} \cdot \nabla\xi = 0, \qquad \forall\, \xi \in H_0^1(\Omega) , \tag{7.6}$$

$$\int_{\Lambda(\vartheta_0)} \mathcal{F}(\vartheta_0, \Psi) - \int_{\Lambda(\vartheta_0)} \mathcal{F}(\vartheta_0, \mathbf{u}_0) + \int_{\Lambda(\vartheta_0)} \beta_0(\mathbf{u}_0, \mathbf{u}_0, \Psi) \ge \int_{\Lambda(\vartheta_0)} \mathbf{f}(\vartheta_0) \cdot (\Psi - \mathbf{u}_0) , \quad \forall\, \Psi \in \mathbf{V}^p(\Lambda(\vartheta_0)), \tag{7.7}$$

where $\Lambda(\vartheta_0) = \{x \in \Omega\colon \vartheta_0(x) > 0\}$ and β_0 is given by (3.13) with $\vec{\alpha} = \mathbf{0}$.

The next theorem, giving a weak continuous dependence property, extends the Theorem 4 of [R2] and gives also an existence result for solutions to the problem (7.4)–(7.7).

Theorem 7.1. *Under the assumptions of the Theorem 4.1 and (7.1)–(7.3), when $\alpha = |\vec{\alpha}| \to 0$ there exists solutions $(\vartheta_\alpha, \mathbf{u}_\alpha)$ to (3.14)–(3.18) and a subsequence, still denoted by $\alpha \to 0$, such that*

$$\vartheta_\alpha \to \vartheta_0 \quad \text{in } H^1(\Omega)\text{-strong and in } C^{0,\gamma}(Q) , \tag{7.8}$$

$$\mathbf{u}_\alpha \to \mathbf{u}_0 \quad \text{in } \mathbf{V}^p(\Omega)\text{-weak and in } \mathbf{L}^r_\sigma(\Omega)\text{-strong} , \tag{7.9}$$

for any compact subset $Q \subset \Omega$, for some $0 < \gamma < 1$, for $p > 3/2$ and $6 \le r < 2p/(2-p)$ if $n = 2 > p$, or $p > 9/5$ and $9/2 \le r < 3p/(3-p)$ if $n = 3 > p$, or any $r < +\infty$ if $p > n = 2, 3$, where $(\vartheta_0, \mathbf{u}_0)$ is a solution to (7.4)–(7.7).

Proof: From the Theorem 4.1, we have the a *priori* estimates

$$\|\vartheta_\alpha\|_{L^\infty(\Omega)} \le M \quad \text{and} \quad \|\mathbf{u}_\alpha\|_{\mathbf{V}^p(\Omega)} \le C \ , \tag{7.10}$$

with M and $C > 0$ independent of $\alpha \le 1$. Then, letting $\varsigma = \vartheta_\alpha - \Theta$ in (3.17), where $\Theta \in H^1(\Omega)$ is the harmonic function in Ω such that $\Theta = \vartheta_D$ on $\partial\Omega$, we obtain, using (7.2) and (7.3):

$$\frac{1}{2}\int_\Omega |\nabla(\vartheta_\alpha - \Theta)|^2 + \frac{1}{k\alpha}\int_{\Gamma_N} |\vartheta_\alpha - \vartheta_D|^q \le C', \quad \text{for any } 0 < \alpha \le 1 \ . \tag{7.11}$$

Then, we also have, for some $0 < \gamma' < 1$ and any compact subset $Q \subset \Omega$:

$$\|\vartheta_\alpha\|_{H^1(\Omega)} + \|\vartheta_\alpha\|_{C^{0,\gamma'}(Q)} \le C'' \quad (\text{independently of } 0 < \alpha \le 1),$$

and we can select a subsequence satisfying (7.8) and (7.9).

The limit ϑ_0 satisfies (7.4) since, from (7.11) we have

$$\int_{\Gamma_N} |\vartheta_0 - \vartheta_D|^q \le \liminf_{\alpha\to 0} \int_{\Gamma_N} |\vartheta_\alpha - \vartheta_D|^q = 0 \ ,$$

and also solves the equation (7.6), because we can pass easily to the limit $\alpha \to 0$ in (3.17) for any $\varsigma \in H_0^1(\Omega) \subset Z$.

The strong convergence $\vartheta_\alpha \to \vartheta_0$ in $H^1(\Omega)$, follows as in [R2], by letting $\varsigma = \vartheta_\alpha - \vartheta_0$ in (3.17), which yields (with $R_\alpha \to 0$ as $\alpha \to 0$)

$$\begin{aligned}\int_\Omega |\nabla\vartheta_0|^2 &\le \liminf_{\alpha\to 0}\int_\Omega |\nabla\vartheta_\alpha|^2 \le \limsup_{\alpha\to 0}\int_\Omega |\nabla\vartheta_\alpha|^2 \\ &\le \lim_{\alpha\to 0}\Big\{\int_\Omega \nabla\vartheta_\alpha\cdot\nabla\vartheta_0 + R_\alpha\Big\} = \int_\Omega |\nabla\vartheta_0| \ .\end{aligned}$$

On the other hand, since $\vartheta_\alpha \to \vartheta_0$ uniformly in compact subsets K of Ω, if $K \subset \Sigma(\vartheta_0)$ then $K \subset \Sigma(\vartheta_\alpha)$ for all α sufficiently small. Since $\mathbf{u}_\alpha = \mathbf{0}$ a.e. in $\Sigma(\vartheta_\alpha)$, we have also $\mathbf{u}_0 = 0$ a.e. in K, and since $K \subset \Sigma(\vartheta_0)$ is arbitrary, (7.5) holds.

To pass to the limit $\alpha \to 0$ in (3.18), we take first an arbitrary $\Psi \in \mathcal{V}(\Lambda(\vartheta_0))$. Since $\operatorname{supp}\Psi \subset \Lambda_0 \equiv \Lambda(\vartheta_0)$, by the uniform convergence, we have $\operatorname{supp}\Psi \subset \Lambda_\alpha \equiv \Lambda(\vartheta_\alpha)$ for all $\alpha \le \alpha_\Psi$ sufficiently small. Hence

$$\begin{aligned}&\lim_\alpha \int_{\Lambda_\alpha} \mathcal{F}(\vartheta_\alpha, \Psi) = \lim_\alpha \int_{\Lambda_0} \mathcal{F}(\vartheta_\alpha, \Psi) = \int_{\Lambda_0} \mathcal{F}(\vartheta_0, \Psi) \ , \\ &\lim_\alpha \int_{\Lambda_\alpha} \beta_\alpha(\mathbf{u}_\alpha, \mathbf{u}_\alpha, \Psi) = \lim_\alpha \int_{\Lambda_0} \beta_\alpha(\mathbf{u}_\alpha, \mathbf{u}_\alpha, \Psi) = \int_{\Lambda_0} \beta_0(\mathbf{u}_0, \mathbf{u}_0, \Psi) \ , \\ &\lim_\alpha \int_{\Lambda_\alpha} \mathbf{f}(\vartheta_\alpha)\cdot\Psi = \lim_\alpha \int_{\Lambda_0} \mathbf{f}(\vartheta_\alpha)\cdot\Psi = \int_{\Lambda_0} \mathbf{f}(\vartheta_0)\cdot\Psi \ .\end{aligned}$$

Recalling that $\mathbf{f}(0) = \mathbf{0}$, using (3.16) and (7.5), we have

$$\int_{\Lambda_\alpha} \mathbf{f}(\vartheta_\alpha)\cdot\mathbf{u}_\alpha = \int_\Omega \mathbf{f}(\vartheta_\alpha)\cdot\mathbf{u}_\alpha \;\to\; \int_\Omega \mathbf{f}(\vartheta_0)\cdot\mathbf{u}_0 = \int_{\Lambda_0} \mathbf{f}(\vartheta_0)\cdot\mathbf{u}_0, \qquad \text{as } \alpha\to 0.$$

Using the lower semi-continuity of the energy, as in (4.21), we have

$$\liminf_{\alpha\to 0}\int_K \mathcal{F}(\vartheta_\alpha,\mathbf{u}_\alpha)\geq \int_K \mathcal{F}(\vartheta_0,u_0)$$

for any measurable subsets K of Ω. Choosing, in particular, any compact subset $K\subset\Lambda_0$, for all $\alpha\leq\alpha_K$ sufficiently small, we have $K\subset\Lambda_\alpha$ and we conclude

$$\liminf_{\alpha\to 0}\int_{\Lambda_\alpha}\mathcal{F}(\vartheta_\alpha,\mathbf{u}_\alpha)\geq \liminf_{\alpha\to 0}\int_K \mathcal{F}(\vartheta_\alpha,\mathbf{u}_\alpha)\geq \int_K \mathcal{F}(\vartheta_0,\mathbf{u}_0)\ .$$

Therefore, $\mathbf{u}_0$ solves

$$\int_{\Lambda_0}\mathcal{F}(\vartheta_0,\Psi)+\int_{\Lambda_0}\beta_0(\mathbf{u}_0,\mathbf{u}_0,\Psi)\geq\int_{\Lambda_0}\mathbf{f}(\vartheta_0)(\Psi-\mathbf{u}_0)+\int_K\mathcal{F}(\vartheta_0,\mathbf{u}_0)$$

for any $\Psi\in\mathcal{V}(\Lambda_0)$ and for any compact subset K of the open set $\Lambda_0=\Lambda(\vartheta_0)$. Then, taking an increasing sequence of compact subsets $K_n\subset\Lambda_0$ such that $\bigcup_n K_n=\Lambda_0$, we conclude that $\mathbf{u}_0$ verifies (7.7) for every $\Psi\in\mathcal{V}(\Lambda_0)$ and, by density, also for any $\Psi\in\mathbf{V}^p(\Lambda_0)$. ∎

In general, we don't know any regularity result on the free boundary $\Phi_\alpha=\{\vartheta_\alpha=0\}$ associated with the weak solution ϑ_α of the Boussinesq–Stefan problem (case $\alpha>0$) or of the Rayleigh–Bénard problem (case $\alpha=0$).

An interesting, but very special, problem have been studied by DiBenedetto and Friedman in [DF], corresponding to the later case for an incompressible homogeneous fluid which motion, in a two dimensional annular domain, is governed by the Navier–Stokes equations. For completeness let us describe their results.

From now on, let us suppose that the "ice–water" domain $\Omega\subset\mathbb{R}^2$ is an annulus, which smooth boundary $\partial\Omega=\Gamma_+\cup\Gamma_-$ (of class $C^{3,\lambda}$) corresponds to the contact with an outer frozen zone and with a inner heating core. The equations (2.6), (2.12), with the constitutive law (2.18) and with $\vec{\alpha}\equiv 0$, lead to find the temperature $\vartheta\colon\Omega\to\mathbb{R}$ and the velocity $\mathbf{v}\colon\Lambda=\{\vartheta>0\}\to\mathbb{R}^2$, and the pressure $p\colon\Lambda\to\mathbb{R}$, such that,

$$-\Delta\vartheta+b\mathbf{v}\cdot\nabla\vartheta=0\ \text{ in }\Omega,\qquad\text{with }\ \vartheta|_{\Gamma_+}=1,\ \ \vartheta|_{\Gamma_-}=-1\ ,\tag{7.12}$$

$$-\mu\Delta\mathbf{v}+(\mathbf{v}\cdot\nabla)\mathbf{v}+\nabla p=b\vartheta\mathbf{j}\ \text{ and }\ \nabla\cdot\mathbf{v}=0\quad\text{in }\Lambda\ ,\tag{7.13}$$

$$\mathbf{v}=\mathbf{0}\quad\text{in }\ \Omega\backslash\Lambda\cup\Gamma_+\ .\tag{7.14}$$

Here, $\mu>0$ represents the constant viscosity, $b>0$ a dimensionless constant related to the Prandtl and Rayleigh numbers (see [CB]) and $\mathbf{j}=(0,1)$. We observe that, since $\vec{\alpha}=\mathbf{0}$, from the Stefan condition (2.7), the jump across the free boundary $\partial\Lambda\cap\Omega$ disapears and (2.7) is only a transmission condition. The main result, which proof may be found in [DF] may be summarized in the following theorem.

Theorem 7.2 [DF]. i) *For any $b>0$ there exists a solution to the problem (7.12)–(7.13), in the sense that $\vartheta\in W^{2,p}(\Omega)\cap C^{1,\gamma}(\overline{\Omega})$, $\forall 1\leq p<\infty$, $\forall 0\leq\gamma<1$, $\mathbf{v}\in\mathbf{V}^2(\Lambda)$, $p\in L^2(\Lambda)$, $\Lambda=\{\vartheta>0\}$ and $\Phi=\{\vartheta=0\}=\partial\Lambda\cap\Omega$ is a $C^{1,\delta}$-curve in Ω, $\forall 0<\delta<1$.*

ii) *If, in addition,* Γ_+ *and* Γ_- *are star shaped with respect to the origin, for a sufficiently small* $b > 0$, *the solution* $(\vartheta, \mathbf{v}, \Phi)$ *is unique and classical, in the sense that, for every* $0 < \gamma < 1$

$$\vartheta \in C^{2,\gamma}(\overline{\Omega}) \cap C^{\infty}(\Omega \backslash \Phi), \quad \mathbf{v} \in C^{1,\gamma}(\overline{\Lambda}) \cap C^{\infty}(\Lambda), \quad p \in C^{0,\gamma}(\overline{\Lambda}) \cap C^{\infty}(\Lambda)$$

and the interface $\Phi = \{\vartheta = 0\}$ *is given, in polar coordinates, by a* $C^{2,\gamma}$*-graph.* ∎

Remark 7.1. It is clear that the solutions given in this theorem are also weak solutions in the sense of (7.4)–(7.7), being now the variational inequality (7.7) equivalent to the equation in $\Lambda = \Lambda(\vartheta) = \{\vartheta > 0\}$:

$$\mu \int_{\Lambda} \nabla \mathbf{v} \cdot \nabla \Psi - \int_{\Lambda} \mathbf{v} \cdot (\mathbf{v} \cdot \nabla) \Psi = b \int_{\Lambda} \vartheta \mathbf{j} \cdot \Psi, \qquad \forall \Psi \in \mathbf{V}^2(\Lambda)\ . \ \square$$

Remark 7.2. Actually in [DF], the method used to obtain the results of Theorem 7.2 is based on the search of the stream function $\varphi : \Omega \to \mathbb{R}$, linked to the velocity by $\mathbf{v} = (\varphi_{,2}, -\varphi_{,1}) = \nabla^* \varphi$, which requires to replace the equation for $\mathbf{v}$ by the fourth order equation

$$\Delta^2 \varphi - \nabla^* \varphi \cdot \nabla(\Delta \varphi) = b\, \vartheta_{,1} \quad \text{in } \Lambda = \{\vartheta > 0\}\ ,$$

with appropriate boundary conditions and the constraint $\varphi = 0$ on $\Omega \backslash \Lambda$. In [DF] no results are stated about the pressure, but they follow by the general theory of Navier–Stokes equations (see [T1]). □

REFERENCES

[BMP] C. Bernardi, B. Métivet, B. Pernaud-Thomas – *Couplage des équations de Navier–Stokes et de la chaleur: le modéle et son approximation par éléments finis*, Prépub. Lab. Anal. Numer., Univ. Paris VI, 1991.

[BL] J. Boland, W. Layton – *Error analysis for finite element methods for steady natural convection problems*, Numer. Funct. Anal. and Optimiz., **11** (1990), 449–483.

[BS] B. Basu, J. Srinivasan – *Numerical study of steady-state laser melting problem*, Int. J. Heat Mass Transfer, **31** (1988), 2331–2338.

[CD] J.R. Cannon, E. Dibenedetto – *The Steady State Stefan Problem with Convection, with Mixed Temperature and non-linear Heat Flux Boundary Conditions*, in "Free Boundary Problems", Ed. by E. MAGENES, Istit. Naz. Alta Matematica, Roma, 1980, Vol. I, 231–265.

[CDK1] J.R. Cannon, E. Dibenedetto, G.H. Knightly – *The Steady State Stefan Problem with Convection*, Arch. Rational Mech. Anal., **73** (1980), 79–97.

[CDK2] J.R. Cannon, E. DiBenedetto and G.H. Knightly – *The Bidimensional Stefan Problem with Convection: the Time Dependent Case*, Comm. P.D.E., **8** (1983), 1549–1604.

[C1] D. Cioranescu – *Sur une Class de Fluides Non-Newtoniens*, Appl. Math. Optimiz., **3** (1977), 263–282.

[C2] D. Cioranescu – *Quelques Exemples de Fluides Newtoniens Généralisés* (in this volume).

[CB] *C.J. Chang, R.A. Brown* – *Natural Convection in Steady Solidification: Finite Element Analysis of a Two-Phase Rayleigh-Bénard Problem*, J. Comput. Phys., **53** (1984), 1–27.

[CM] *M. Chipot, G. Michaille* – *Uniqueness results and monotonicity properties for strongly nonlinear elliptic variational inequalities*, Ann. Scuola Norm. Sup. Pisa, **16** (1989), 137–166.

[DF] *E. DiBenedetto, A. Friedman* – *Conduction-convection problems with change of phase*, J. Diff. Eqs., **62** (1986), 129–185.

[DL] *G. Duvaut, J.L. Lions* – *Les inéquations en Méchanique et en Physique*, Dunod, Paris, 1972.

[F] *M.C. Flemings* – *Solidification Processing*, McGraw-Hill, New York, 1974.

[GT] *D. Gilbarg, N.S. Trudinger* – *Elliptic Partial Differential Equations of Second Order*, Springer, New York, 1977.

[J] *D.D. Joseph* – *Stability of Fluid Motions I-II*, Springer-Verlag, Berlin, 1976.

[La] *O.A. Ladyzhenskaya* – *The Mathematical Theory of Viscous Incompressible Flow*, Gordon and Breach, New York, 1969.

[L] *J.L. Lions* – *Quelques Méthodes de Résolution des Problèmes aux Limites Non-Linéaires*, Dunod–Gauthier–Villars, Paris, 1969.

[R1] *J.F. Rodrigues* – *Aspects of the Variational Approach to a Continuous Casting Problem*, Research Notes Math., Pitman, **120** (1985), 72–83.

[R2] *J.F. Rodrigues* – *A steady-state Boussinesq–Stefan problem with continuous extraction*, Annali di Mat. pura ed appl. (iv) **144** (1986), 203–218.

[R3] *J.F. Rodrigues* – *Obstacle Problems in Mathematical Physics*, North-Holland, Amsterdam, 1987.

[R4] *J.F. Rodrigues* – *The Stefan problem revisited*, in Int. Ser. Num. Math. (Birkhäuser) **88** (1989), 129–190.

[R5] *J.F. Rodrigues* – *A Steady-state solidification problem for a NonNewtonian fluid*, in Pitman Res. Notes Math. Ser., **186** (1990), 699–702.

[S] *G. Stampacchia* – *Problemi al contorno ellittici con dati discontinui dotati di soluzioni holderiane*, Ann. Mat. Pura Appl., **51** (1960), 1–32.

[St] *B. Straughan* – *The energy method, stability, and nonlinear convection*, Springer-Verlag, Berlin, New York, 1992.

[Sz] *J. Szekely* – *On Some Free and Moving Boundary Problems in Materials Processing*, in Pitman Res. Notes in Math., **185** (1990), 222–242.

[T1] *R. Temam* – *Navier–Stokes Equations*, North-Holland, Amsterdam, 1977.

[T2] *R. Temam* – *Infinite Dimensional Dynamic Systems in Mechanics and Physics*, Springer, New York, 1988.

[VI] *P.N. Vabishchevich* and *O.P. Iliev* – *Numerical solution of conjugate heat and mass transfer problems with phase changes taken into account*, Diff. Eqs., **23**, nº 7 (1988), 746–750.

José-Francisco Rodrigues,
CMAF/University of Lisbon,
Av. Prof. Gama Pinto, 2
1699 LISBOA Codex – PORTUGAL

Boundary and Initial-Boundary Value Problems for the Navier-Stokes Equations in Domains with Noncompact Boundaries

V.A. SOLONNIKOV

This paper is written on the basis of lectures given by the author at a mathematical school on the Navier-Stokes equations in Lisbon in September 1991, and it is concerned with questions of the solvability of stationary and evolution problems of a viscous incompressible flow in domains with noncompact boundaries. A special attention has been given to these problems in recent 15 years when it became evident that they should be formulated in a slightly different way, compared to problems in bounded or exterior domains: in addition to standard initial and boundary conditions there should be prescribed such quantities as fluxes of the velocity vector field through some apertures or differences of limiting values of the pressure at infinity. Another characteristic feature of these problems consists in the fact that very often their solutions have unbounded energy integral which means that a basic *a priori* estimate given by the energy inequality may become deficient. In recent years there was elaborated a special techniques of integral estimates (so called "techniques of the Saint-Venant's principle") that makes it possible to construct and to investigate such solutions. Finally, some problems with noncompact free boundaries were considered.

This paper consists of four sections. In §1 there are given formulations of boundary and initial-boundary value problems for domains with several "exits into infinity" containing infinite cones, and there are presented basic results concerning the solvability of these problems in classes of vector fields with finite energy integrals (i.e. with a finite Dirichlet integral in a stationary case). A special attention is given to evolution problems. In §2 the existence theorem is presented for the boundary value problem of a steady viscous flow in the domain with "exits into infinity" having a form of cylinders (or modified cylinders). In §3 evolution problems are investigated in classes of vector fields with infinite energy integrals. Finally, §4 is devoted to boundary and initial-boundary value problems for the Navier-Stokes equations in bounded two-dimensional domains with discontinuous boundary data. These problems are also investigated in a class of vector fields with unbounded energy by means of estimates of Saint-Venant's type. In comparison with §§1–3, the results of §4 are relatively new.

In all cases we restrict ourselves with the construction of generalized solutions in the possibly largest class of functions, and we leave aside problems of regularity of solutions. Because of this reason we do not present here the theory of free boundary problems with noncompact free boundaries. They are usually solved in classes of sufficiently regular functions, and in these problems there arise additional difficulties due to the presence of an unknown boundary and to more complicated boundary conditions. Methods of investigation of free boundary problems in noncompact domains are presented in papers [6, 30, 35, 38, 43, 47] and others.

The author brings his cordial gratitude to the organizers of the summer school on the

Navier-Stokes equations in Lisbon, in the first line to Professor J. F. Rodrigues, and also to all the lecturers and participants of the school.

1 – Formulation and existence theorems for boundary and initial-boundary value problems (the case of finite energy)

The motion of a viscous incompressible fluid is governed by the system of the Navier-Stokes equations for the vector field of velocities $\mathbf{v}(x,t) = (v_1, \dots, v_n)$, $n = 2,3$ and a scalar pressure $p(x,t)$. This system has a form

$$\mathbf{v}_t + (\mathbf{v} \cdot \nabla)\,\mathbf{v} - \nu\,\Delta \mathbf{v} + \frac{1}{\rho}\,\nabla p = \mathbf{f}\,, \qquad \operatorname{div} \mathbf{v} = 0\,, \tag{1.1}$$

where ν (viscosity) and ρ (density) are constant coefficients, $\nabla = (\frac{\partial}{\partial x_1}, \dots, \frac{\partial}{\partial x_n})$, $\nabla p = \operatorname{grad} p$, $(\mathbf{v} \cdot \nabla)\mathbf{v} = \sum_{k=1}^{n} v_k \frac{\partial \mathbf{v}}{\partial x_k}$, $\mathbf{f}$ is a given vector field of external forces. In what follows we set $\rho = 1$.

Initial-boundary value problem for the system (1.1) consists in finding its solution satisfying the initial condition

$$\mathbf{v}(x,0) = \mathbf{v}_0(x) \tag{1.2}$$

and some conditions at the boundary S of the domain $\Omega \subset \mathbb{R}^n$ filled with a liquid. If this boundary is formed by the walls of the container of the liquid, then there should be formulated so called adherence boundary condition

$$\mathbf{v}|_{x \in S} = 0\,. \tag{1.3}$$

The boundary value problems for the stationary Navier-Stokes equations has a form

$$-\nu\,\Delta \mathbf{v} + (\mathbf{v} \cdot \nabla)\,\mathbf{v} + \nabla p = \mathbf{f}, \quad \operatorname{div} \mathbf{v} = 0, \quad (x \in \Omega \subset \mathbb{R}^n)\,, \qquad \mathbf{v}|_S = 0\,. \tag{1.4}$$

For bounded domains Ω, these formulations are complete. For unbounded domains, conditions at infinity should be also prescribed. In the case of stationary problem in an "exterior" domain $\Omega = \mathbb{R} \backslash K$ (K is a compact) a typical condition at infinity is $\mathbf{v}|_{|x|=\infty} = \mathbf{v}^{\infty} = \text{const}$.

Fundamental results concerning the solvability of problems (1.1)–(1.3) and (1.4) were obtained by the late fifties. In particular, existence theorem was proved for problem (1.4) in an arbitrary bounded domain and also for the exterior problem with arbitrary $\mathbf{v}^{\infty}$ (see [23, 24]). This was done by the method of functional analysis on a very simple analytical basis provided by an estimate of the Dirichlet integral of $\mathbf{v}$. One of the central ideas of the proof was the exclusion of pressure and the replacement of the system (1.4) by an integral identity

$$\nu \int_{\Omega} \nabla \mathbf{v} : \nabla \vec{\eta}\; dx - \sum_{k=1}^{n} \int_{\Omega} v_k \mathbf{v} \cdot \frac{\partial \vec{\eta}}{\partial x_k}\; dx = \int_{\Omega} \mathbf{f} \cdot \vec{\eta}\; dx \tag{1.5}$$

obtained after multiplication of $-\nu\Delta\mathbf{v} + (\mathbf{v} \cdot \nabla)\mathbf{v} + \nabla p = \mathbf{f}$ by an arbitrary vector field $\vec{\eta} \in \dot{J}^{\infty}(\Omega)$ ($\dot{J}^{\infty}(\Omega)$ is the set of divergence free vector fields from $C_0^{\infty}(\Omega)$) and integration over Ω. The term containing pressure vanishes:

$$\int_\Omega \nabla p \cdot \vec{\eta}\, dx = -\int_\Omega p \operatorname{div} \vec{\eta}\, dx = 0 .$$

Vector field $\mathbf{v}$ vanishing at the boundary and satisfying (1.5) is called a generalized (weak) solution of problem (1.4). It is sought in the space $\mathcal{H}(\Omega)$ which is a closure of $\dot{J}^\infty(\Omega)$ in the norm of the Dirichlet integral $\int_\Omega |\nabla \mathbf{v}|^2\, dx$ (we have used here the notations

$$\nabla \mathbf{v} = \left(\frac{\partial v_i}{\partial x_k}\right)_{i,k=1,\dots,n} \quad \text{and} \quad \nabla \mathbf{v} : \nabla \vec{\eta} = \sum_{i,k} \frac{\partial v_i}{\partial x_k}\frac{\partial \eta_i}{\partial x_k} \ ).$$

For a linearized problem, i.e. when the second integral in (1.5) is absent, the existence of a unique solution is a simple consequence of the Riesz representation theorem. The proof of the solvability of a nonlinear problem is based on the Leray-Schauder principle (see [24]).

This scheme applies also in the case of unbounded Ω (at least when $\mathbf{v} \to 0$ as $|x| \to \infty$) but, as it was made clear by J. Heywood [11], the fact that $\mathbf{v} \in \mathcal{H}(\Omega)$ may not only imply $\operatorname{div} \mathbf{v} = 0$ and $\mathbf{v}|_S = 0$, but also have some other physically significant consequences.

Consider a domain $\Omega_H \subset \mathbb{R}^3$ whose boundary S is a plane $x_1 = 0$ with one or several apertures Σ, for example, let $\Sigma = \{x_1 = 0,\ |x_1|^2 + |x_2|^2 < 1\}$; in the two-dimensional case set $\Omega_H = \mathbb{R}^2 \backslash S$, $S = S_+ \cup S_-$, $S_+ = \{x_1 = 0,\ x_2 \geq 1\}$, $S_- = \{x_1 = 0,\ x_2 \leq -1\}$.

For arbitrary $\mathbf{u} \in \dot{J}^\infty(\Omega_H)$ (and consequently for $\mathbf{u} \in \mathcal{H}(\Omega_H)$, there holds the relation

$$\int_\Sigma \mathbf{u} \cdot \mathbf{n}\, dS = \int_\Sigma u_1 dx_1 dx_2 = 0 .$$

For $\Omega \subset \mathbb{R}^n$, $n = 2, 3$, define $\mathcal{D}(\Omega)$ as the closure of the set of all vector fields $\mathbf{u} \in C_0^\infty(\Omega)$ in the norm of the Dirichlet integral, and set $\hat{\mathcal{H}}(\Omega) = \{\mathbf{u} \in \mathcal{D}(\Omega),\ \operatorname{div} \mathbf{u} = 0\}$. Clearly, $\hat{\mathcal{H}}(\Omega) \supseteq \mathcal{H}(\Omega)$. It is shown in [11] that the spaces $\hat{\mathcal{H}}(\Omega)$ and $\mathcal{H}(\Omega)$ coincide for arbitrary bounded or exterior domain with a smooth boundary but $\dim \hat{\mathcal{H}}(\Omega_H)/\mathcal{H}(\Omega_H) = 1$.

Example 1: Vector field $\mathbf{A} \in \hat{\mathcal{H}}(\Omega_H)/\mathcal{H}(\Omega_H)$. In the case $n = 2$, set $\mathbf{A} = (\frac{\partial \varsigma}{\partial x_2}, -\frac{\partial \varsigma}{\partial x_1})$ where $\varsigma \in C^\infty(\mathbb{R}^2)$ is a function satisfying the conditions:

$$\varsigma|_{S_+} = 1\,, \quad \varsigma|_{S_-} = 0\,, \quad \nabla\varsigma|_{S_+} = 0\,, \quad \nabla\varsigma|_{S_-} = 0$$

and inequalities $0 \leq \varsigma \leq 1$, $|D^j \varsigma(x)| \leq \frac{C(|j|)}{|x|^{|j|}}$, $|j| = 1, 2$. It follows that $\mathbf{A}_{x_i}$ decay at the infinity as $|x|^{-2}$ and $\mathbf{A} \in \mathcal{D}(\Omega_H)$. At the same time

$$\int_\Sigma A_1 dx_2 = \varsigma(0,1) - \varsigma(0,-1) = 1 .$$

For $n = 3$, let $\mathbf{A} = \operatorname{rot} \psi \mathbf{H}$ where

$$\mathbf{H} = \frac{1}{2\pi}\left(0, -\frac{x_3}{x_2^2 + x_3^2}, \frac{x_2}{x_2^2 + x_3^2}\right) \tag{1.6}$$

and $\psi(x)$ is an infinitely differentiable function which is equal to 1 in a neighborhood of S and to zero for small values of $x_2^2 + x_3^2$. If we suppose that

$$|D^j\psi(x)| \leq \frac{C(|j|)}{|x|^{|j|}}, \qquad |j| = 1, 2 ,$$

then $\mathbf{A}$, $\mathbf{A}_{x_i} \in L_2(\Omega)$. Finally, in virtue of the Stokes formula

$$\int_\Sigma A_1 dx_2 dx_3 = \int \psi \mathbf{H} \cdot dl = 1 .$$

J. Heywood concluded that the formulation (1.4) of a stationary problem in Ω_H should be completed by the prescription of a total flux through the aperture Σ:

$$\int_\Sigma \mathbf{v} \cdot \mathbf{n}\, dS \equiv \int_\Sigma v_1 dx_1 dx_2 = Q \tag{1.7}$$

(in the above formulation it is understood implicitly that $Q = 0$). A generalized solution of problem (1.4), (1.6) can be found in the form $\mathbf{v} = \mathbf{u} + Q\mathbf{A}$ with $\mathbf{u} \in \mathcal{H}(\Omega_0)$ satisfying the integral identity

$$\nu \int_{\Omega_H} \nabla \mathbf{u} : \nabla \vec{\eta}\, dx - \sum_{k=1}^n \int_{\Omega_H} (u_k Q\mathbf{A} + QA_k \mathbf{u}) \cdot \frac{\partial \vec{\eta}}{\partial x_k}\, dx - \sum_{k=1}^n \int_{\Omega_H} u_k \mathbf{u} \cdot \frac{\partial \vec{\eta}}{\partial x_k}\, dx =$$
$$= \int_{\Omega_H} \left(\mathbf{f} \cdot \vec{\eta}\, dx - \nu Q \nabla \mathbf{a} : \nabla \vec{\eta} + Q^2 \sum_{k=1}^n A_k \mathbf{A} \cdot \frac{\partial \vec{\eta}}{\partial x_k} \right) dx \quad \forall \vec{\eta} \in \mathcal{H}(\Omega_H) .$$

Since $\mathbf{A} \in L_4(\Omega_H)$, the existence of $\mathbf{u}$ is easily proved at least for small Q and $\mathbf{f}$; a special choice of $\mathbf{A}$ make it possible to establish the solvability of the problem for arbitrary data (see [26]).

In the paper [25] of O.A. Ladyzhenskaya and the author it is shown that $\hat{\mathcal{H}}(\Omega) \neq \mathcal{H}(\Omega)$ when the domain Ω has several "exits to infinity", i.e. $\Omega^{(R)} = \{x \in \Omega \colon |x| > R\}$, $R \gg 1$, is a union of a finite number of connected unbounded components. Clearly, Ω_H has two "exits".

Let us present some results of the paper [25]. Denote by $W_2^l(\Omega)$ the usual S.L. Sobolev space with norm

$$\|u\|^2_{W_2^l(\Omega)} = \sum_{|j| \leq l} \int_\Omega |D^j u(x)|^2 dx$$

and let $\overset{\circ}{W}{}_2^l(\Omega)$ be a closure of $C_0^\infty(\Omega)$ in this norm. For arbitrary bounded domain $\omega \subset \mathbb{R}^n$, $n = 2, 3$, consider an auxiliary problem of finding a vector field $\mathbf{u} \in W_2^1(\Omega)$ such that

$$\operatorname{div} \mathbf{u} = r, \quad \mathbf{u}|_{\partial\omega} = 0 \tag{1.8}$$

with a given $r \in L_2(\Omega)$ satisfying a necessary condition

$$\int_\Omega r dx = 0 . \tag{1.9}$$

Theorem 1.1. *Let ω possess the following property:*

$$\omega = \omega^{(\delta)} \cup \left\{ \bigcup_{i=1}^{N} \bigcup_{z \in \Gamma_i} K_i(z) \right\} , \tag{1.10}$$

where $\omega^{(\delta)} = \{x \in \omega : \operatorname{dist}(x, \partial\omega) > \delta\}$, $\delta > 0$, Γ_i is a closed set contained in $\bar{\omega}$, $K_i(z) = \{x \in \mathbb{R}^n : x = z + y,\ y \in K_i\}$, K_i is a finite circular cone of a height h_i with a vertex at the origin, and the basement $\sigma_i(z)$ of each cone $K_i(z)$ is contained in $\omega^{(\delta)}$. Then problem (1.8) with arbitrary $r \in L_2(\omega)$ satisfying (1.9) has a solution $u = \mathcal{M} r \in \overset{\circ}{W}{}_2^l(\omega)$, and

$$\|\nabla \mathbf{u}\|_{L_2(\omega)} \le C_1 \|r\|_{L_2(\omega)} \tag{1.11}$$

with a constant C_1 independent of r. The operator $\mathcal{M}$ is linear.

Remark. Clearly, problem (1.8) is underdetermined, and $\mathcal{M}$ is defined not in a unique way. Condition (1.10) is fulfilled for domains with lipschitzean boundaries (in this case $\Gamma_i \subset \partial\omega$), and also for domains with reentrant sharp corners and edges, such as Ω_H. Concerning this problem see [4, 7, 37, 41].

Theorem 1.2. *For any bounded or exterior domain satisfying condition (1.10), $\hat{\mathcal{H}}(\Omega) = \mathcal{H}(\Omega)$.*

Theorem 1.3. *Suppose that the domain $\Omega \subset \mathbb{R}^n$, $n = 2, 3$, possesses the following properties:*

1) *$\Omega = \Omega_{R_0} \cup G_1 \cup \ldots \cup G_m$ where $\Omega_{R_0} = \{x \in \Omega : |x| \le R_0\}$, $R_0 > 0$, and G_i are connected unbounded domains.*
2) *Each G_i contains an infinite cone C_i and $\hat{\mathcal{H}}(G_i) = \mathcal{H}(G_i)$.*
3) *For a certain $R_1 > R_0$, the domain Ω_{R_1} is connected and $\hat{\mathcal{H}}(\Omega_{R_1}) = \mathcal{H}(\Omega_{R_0})$.*
4) *All the domains $\omega_i = G_i \cap (\Omega_{R_1} \setminus \Omega_{R_0})$ satisfy the hypothesis of Theorem 1.1.*

Then $\dim \hat{\mathcal{H}}(\Omega)/\mathcal{H}(\Omega) = m - 1$.

Conditions $\hat{\mathcal{H}}(G_i) = \mathcal{H}(G_i)$, $i = 1, \ldots, m$ are fulfilled if all the domains $\omega_{iR} = (\Omega_{2R} \setminus \Omega_R) \cap G_i$ satisfy the hypothesis of Theorem 1.1 and the estimates (1.11) for all these domains hold with a common constant C_1.

More general domains are considered in [18, 19, 39].

Let us present one of the methods of construction of the basis in $\hat{\mathcal{H}}(\Omega)/\mathcal{H}(\Omega)$ for domains with several "exits to infinity" (see [26, 39, 41]).

Example 2: Vector field with a non-zero flux through the cross-section Σ_k and Σ_j of "exits" G_k and G_j and with zero net flux through other Σ_i.

Let $b_{k_j} \subset \Omega$ be an infinite contour composed of two half-axes (for instance, axes of symmetry of C_j and C_k) connected by a finite smooth curve.

In the case $n = 2$, b_{k_j} decomposes S into two parts S'_{k_j} and S''_{k_j} situated on different sides of b_{k_j}, and we may set

$$\mathbf{A}_{k_j} = \left(\frac{\partial \varsigma_{k_j}}{\partial x_2}, -\frac{\partial \varsigma_{k_j}}{\partial x_1} \right) ,$$

where $\varsigma_{k_j} \in C^\infty(\mathbb{R}^2)$ is equal to 1 in a neighborhood of S'_{k_j} and to zero in a neighborhood of S''_{k_j}. Clearly, ς_{k_j} may be chosen in such a way that

$$|D^q \varsigma_{k_j}| \leq C|x|^{-q}, \qquad |q| \leq 1 , \tag{1.12}$$

which implies $\frac{\partial}{\partial x_m} \mathbf{A}_{k_j} \in L_2(\Omega)$.

In the three-dimensional case, we denote by ς_{k_j} an infinitely differentiable function which is equal to 1 in a neighborhood of the whole boundary S and to zero in a neighborhood of b_{k_j}, and we set

$$\mathbf{A}_{k_j} = \operatorname{rot} \varsigma_{k_j} \mathbf{H}_{k_j} , \qquad \mathbf{H}_{k_j} = \frac{1}{4\pi} \int_{b_{k_j}} \frac{\mathbf{x} - \mathbf{y}}{|x - y|^3} \times d\mathbf{l}_y$$

(the vector field (1.6) corresponds to the case when b_{k_j} is the x_1-axis). In virtue of the Stokes formula,

$$\int_{\Sigma_i} \mathbf{A}_{k_j} \cdot \mathbf{n} \, dS = \int_{\partial \Sigma_i} \mathbf{H}_{k_j} \cdot d\mathbf{l}$$

and the right-hand side equals ± 1 for $i = j$ or $i = k$; otherwise it vanishes. If we choose ς_{k_j} satisfying (1.12), then it may be shown (see [39]) that $\mathbf{A}_{k_j}, \nabla \mathbf{A}_{k_j} \in L_2(\Omega)$.

The basis in $\hat{\mathcal{H}}(\Omega)/\mathcal{H}(\Omega)$ in both cases ($n = 2$ and $n = 3$) is formed by vector fields $\mathbf{A}_{1m}, \ldots, \mathbf{A}_{m-1,m}$.

If Ω satisfies the hypotheses of Theorem 1.3, problem (1.4) should be completed by conditions

$$\int_{\Sigma_j} \mathbf{v} \cdot \mathbf{n} \, dS = Q_j , \qquad j = 1, \ldots, m-1 , \tag{1.13}$$

from which we can easily compute $\int_{\Sigma_m} \mathbf{v} \cdot \mathbf{n} \, dS$. The solvability of problem (1.4), (1.13) is proved exactly in the same way as in the case $\Omega = \Omega_H$.

Additional conditions may be written in a different form. It may be shown that the pressure corresponding to a weak solution $\mathbf{v}$ of the problem (1.4), (1.13), tends to finite limits p_j as $x \to \infty$, $x \in G_j$. Instead of (1.13), one can prescribe differences $p_j - p_m$:

$$p_j - p_m = \beta_j , \qquad j = 1, \ldots, m-1 \tag{1.14}$$

(for $\Omega = \Omega_H$, this was also proposed by J. Heywood in [11], see also [34, 40]). Following [40], we define a weak solution of problem (1.4), (1.14) as a vector field $\mathbf{v} \in \hat{\mathcal{H}}(\Omega)$ satisfying the integral identity

$$\nu \int_\Omega \nabla \mathbf{v} : \nabla \vec{\eta} \, dx - \int_\Omega \sum_{k=1}^{3} v_k \mathbf{v} \cdot \frac{\partial \vec{\eta}}{\partial x_k} \, dx + \sum_{j=1}^{m-1} \beta_j \int_{\Sigma_j} \vec{\eta} \cdot \mathbf{n} \, dS = \int_\Omega \mathbf{f} \cdot \vec{\eta} \, dx$$

for arbitrary $\vec{\eta} = \sum_{k=1}^{m-1} \lambda_k \mathbf{A}_{km}(x) + \vec{\eta}'$, $\lambda_k \in \mathbb{R}$, $\vec{\eta}' \in \dot{J}^\infty(\Omega)$. This set is dense in $\hat{\mathcal{H}}(\Omega)$.

The following theorem is proved in [10].

Theorem 1.4. *For arbitrary* $\mathbf{f} \in \hat{\mathcal{H}}^*(\Omega)$ *($\hat{\mathcal{H}}^*(\Omega)$ is a dual space to $\hat{\mathcal{H}}(\Omega)$) and for arbitrary β_j problem (1.4), (1.14) has at least one weak solution, and to this solution there corresponds a unique, up to a constant, pressure* $p \in L_{2,\mathrm{loc}}(\Omega)$ *which tends to a* *constant limit* p_j *as* $|x| \to \infty$, $x \in G_j$. *These limits satisfy the relations (1.14), and* $p - p_j \in L_2(G_j)$.

We see that existence theorems for problem (1.4), (1.13) and (1.4), (1.14) are of the same type as for problem (1.4) in bounded or exterior domains. The asymptotic behavior of the solution at infinity is studied only for the problem of J. Heywood with $\Omega = \Omega_H$, by W. Borchers and K. Pileckas [8] and Huakang Chang [12] whose results may be considered as analogues of well-known results of R. Finn [9] and K. I. Babenko [5] for exterior domains. They permit to establish the uniqueness of weak solutions of the above problems with small data.

We now turn to evolution problems. Generalized solutions of evolution problems usually belong to the space $W_2^1(\Omega)$ for almost all $t \in (0,T)$, hence basic functional spaces in a nonstationary case are $\hat{\mathring{J}}{}_2^1(\Omega)$ which is defined as the closure of $\dot{J}^\infty(\Omega)$ in $W_2^1(\Omega)$-norm and $\mathring{J}_2^1(\Omega) = \{\mathbf{u} \in \mathring{W}_2^1(\Omega) : \operatorname{div} \mathbf{u} = 0\}$.

Theorem 1.5. *If the domain* $\Omega \in \mathbb{R}^3$ *satisfies the hypotheses of Theorem 1.3, then* $\dim \hat{\mathring{J}}{}_2^1(\Omega)/\mathring{J}_2^1(\Omega) = m-1$, *and the vector fields* $(\mathbf{A}_{1m}, ..., \mathbf{A}_{(m-1)m})$ *from the Example 2 form a basis in* $\hat{\mathring{J}}{}_2^1(\Omega)/\mathring{J}_2^1(\Omega)$. *For a domain* $\Omega \subset \mathbb{R}^2$ *satisfying the same conditions,* $\hat{\mathring{J}}{}_2^1(\Omega) = \mathring{J}_2^1(\Omega)$.

Let us make short comments concerning the last statement of the theorem. It may be shown that in an "exit" $G_j \subset \mathbb{R}^2$ containing an infinite cone C_j there exist no divergence free vector fields $\mathbf{v} \in W_2^1(G_j)$ vanishing on S and having non-zero net flux through the cross-section of G_j. Let $\Sigma_j = \{x \in G_j : |x - \bar{x}| = r\}$ where $\bar{x}$ is the vertex of C_j. We have:

$$Q_j^2 \le \Big(\int_{\Sigma_j} |\mathbf{v}|\, dS\Big)^2 \le 2\pi r \int_{\Sigma_j} |\mathbf{v}|^2\, dS\ ,$$

hence,

$$Q_j^2 \ln\frac{r_2}{r_1} \le 2\pi \int_{r_1}^{r_2} dr \int_{\Sigma_j} |\mathbf{v}|^2\, dS < \infty\,, \qquad \forall r_2 > r_1\ ,$$

which implies $Q_j = 0$.

It follows that additional conditions in evolution problem (considered in classes of vector fields with a finite energy) should be prescribed only in the case $n = 3$. They have a form

$$\int_{\Sigma_j} \mathbf{v} \cdot \mathbf{n}\, dS = Q_j(t)\,, \qquad j = 1, ..., m-1\ , \tag{1.15}$$

or

$$p_j(t) - p_m(t) = \beta_j(t)\,, \qquad j = 1, ..., m-1\ . \tag{1.16}$$

A generalized (weak) solution of problem (1.1)–(1.3), (1.15) can be defined as a divergence free vector field $\mathbf{v} \in L_2(0,T; \hat{\mathring{J}}{}_2^1(\Omega))$ satisfying conditions (1.15) and an integral identity

$$\int_0^T dt \int_\Omega \left(-\mathbf{v}\cdot\vec{\eta}_t + \nu\nabla\mathbf{v} : \nabla\vec{\eta} - \sum_{k=1}^{3} v_k\mathbf{v}\cdot\frac{\partial\vec{\eta}}{\partial x_k}\right) dx =$$

$$= \int_0^T \int_\Omega \mathbf{f}\cdot\vec{\eta}\; dxdt + \int_\Omega \mathbf{v}_0(x)\cdot\vec{\eta}(x,0)\; dx\;,$$

with arbitrary $\vec{\eta} \in L_2(0,T;\overset{\circ}{J}{}_2^1(\Omega))$ having the derivative $\vec{\eta}_t \in L_2(Q_T)$ $(Q_T = \Omega\times(0,T))$ and vanishing for $t = T$.

By a weak solution of problem (1.1)–(1.3), (1.16) we mean a vector field $\mathbf{v} \in L_2(0,T;\hat{\overset{\circ}{J}}{}_2^1(\Omega))$ satisfying an integral identity

$$\int_0^T dt \int_\Omega \left(-\mathbf{v}\cdot\vec{\eta}_t + \nu\nabla\mathbf{v} : \nabla\vec{\eta} - \sum_{k=1}^{3} v_k\mathbf{v}\cdot\frac{\partial\vec{\eta}}{\partial x_k}\right) dx =$$

$$= \int_0^T \int_\Omega \mathbf{f}\cdot\vec{\eta}\; dxdt + \int_\Omega \mathbf{v}_0(x)\cdot\vec{\eta}(x,0)\; dx - \sum_{j=1}^{m-1}\int_0^T \beta_j(t)dt\int_{\Sigma_j}\vec{\eta}\cdot\mathbf{n}\,dS\;, \qquad (1.17)$$

with arbitrary $\vec{\eta} \in L_2(0,T;\hat{\overset{\circ}{J}}{}_2^1(\Omega))$ such that $\vec{\eta}_t \in L_2(Q_T)$ and $\vec{\eta}(x,T) = 0$.

The solvability of both problems can be established by the method elaborated for (1.1)–(1.3) in bounded domains (see [24]).

Denote by $L_{q,r}(Q_T)$ the space $L_r(0,T;L_q(\Omega))$ with the norm

$$\|\mathbf{f}\|_{L_{q,r}(Q_T)} = \left(\int_0^T \|\mathbf{f}\|^r_{L_q(\Omega)}\, dt\right)^{\frac{1}{r}}.$$

Theorem 1.6. *Suppose that* $\mathbf{f} \in L_{2,1}(Q_T)$, $Q_j \in C(0,T)$ $(T<\infty)$, $\mathbf{w}_0 \equiv \mathbf{v}_0 - \sum_{j=1}^m Q_j(0)\mathbf{A}_{jm} \in \overset{\circ}{J}(\Omega)$ *($\overset{\circ}{J}$ is the closure of $\dot{J}^\infty$ in L_2-norm). Then problem (1.1)–(1.3), (1.15) has a weak solution* $\mathbf{v} = \mathbf{u} + \sum_{k=1}^m Q_j(t)\,\mathbf{A}_{jm}$ *where* $\mathbf{u} \in L_2(0,T;\overset{\circ}{J}{}_2^1(\Omega))$. *If, in addition,* $\mathbf{f} \in L_2(Q_T)$, $\mathbf{f}_t \in L_{2,1}(Q_T)$, $Q_{jt} \in L_2(0,T)$, $\mathbf{w}_0 \in W_2^2(\Omega)\cap \overset{\circ}{J}{}_2^1(\Omega)$, *then in the interval* $t \in (0,T_1)$ *with a certain* $T_1 < T$ *a weak solution has a derivative* $\mathbf{v}_t \in L_2(0,T_1;W_2^1(\Omega))$, *and in this class it is unique. If the norms of* $\mathbf{f}$, Q_j, $\mathbf{w}_0$ *in the above spaces are small enough, then the solution possessing the time derivative exists in whole interval* $(0,T)$, *and the case* $T=\infty$ *is not excluded.*

Theorem 1.7. *Let* $\mathbf{f} \in L_{2,1}(Q_T)$, $(T<\infty)$. $\beta_j \in L_1(0,T)$ *and* $\mathbf{v}_0 \in \hat{\overset{\circ}{J}}(\Omega)$ *where* $\hat{\overset{\circ}{J}}(\Omega)$ *is the closure of* $\hat{\overset{\circ}{J}}{}_2^1(\Omega)$ *in L_2-norm. Then problem (1.1)–(1.3), (1.16) has a weak solution* $\mathbf{v} \in L_2(0,T;\hat{\overset{\circ}{J}}{}_2^1(\Omega))$. *If* $\mathbf{f} \in L_2(Q_T)\cap L_{2,1}(Q_T)$, $\mathbf{f}_t \in L_{2,1}(Q_T)$, $\beta_j \in W_2^1(0,T)$, $\mathbf{v}_0 \in W_2^2(\Omega)\cap\hat{\overset{\circ}{J}}{}_2^1(\Omega)$, *then in the interval* $(0,T_1)$, $T_1<T$, *there exists a unique weak solution possessing the derivative* $\mathbf{v}_t \in L_2(0,T_1;W_2^1(\Omega))$, *and* $T_1 = T \le \infty$, *provided that the norms of the data* $\mathbf{f}$, β_j, $\mathbf{v}_0$ *are small.*

In the preceding papers [11, 25, 26, 40] the existence theorem for the problem (1.1)-(1.3), (1.15) is announced without extended proof. Theorem 1.7 has not been even formulated.

Proof of Theorem 1.7: Let $\mathbf{a}_1, \mathbf{a}_2, \ldots$ be an orthonormal basis in $\hat{\overset{\circ}{J}}(\Omega)$ such that $\mathbf{a}_1, \ldots, \mathbf{a}_{m-1} \in \hat{\overset{\circ}{J}}{}_2^1(\Omega)/\overset{\circ}{J}{}_2^1(\Omega)$, and $\mathbf{a}_j, j \geq m+1$, are smooth elements of $\overset{\circ}{J}{}_2^1(\Omega)$ with a compact support. We apply Galerkin's method and find an approximate solution of (1.1)-(1.3), (1.16) in the form $\mathbf{v}^{(n)}(x,t) = \sum_{k=1}^n C_{kn}\mathbf{a}_n(x)$ where C_{kn} are found from the system of differential equations $(k = 1, 2, \ldots)$

$$\frac{dC_{kn}(t)}{dt} = \frac{d}{dt}\int_\Omega \mathbf{v}^{(n)} \cdot \mathbf{a}_k\, dx = \tag{1.18}$$

$$= -\nu \int_\Omega \nabla \mathbf{v}^{(n)} : \nabla \mathbf{a}_k\, dx + \sum_{j=1}^3 \int_\Omega v_j^{(n)} \mathbf{v}^{(n)} \cdot \frac{\partial \mathbf{a}_k}{\partial x_j}\, dx + \int_\Omega \mathbf{f} \cdot \mathbf{a}_k\, dx - \sum_{j=1}^m \int_0^T \beta_j(t) \int_{\Sigma_j} \mathbf{a}_k \cdot \mathbf{n}\, dS$$

and initial conditions $C_{kn}(0) = \int_\Omega \mathbf{v}_0 \cdot \mathbf{a}_k\, dx$. It is easy to see that

$$\int_\Omega \mathbf{v}_t^{(n)} \cdot \mathbf{v}^{(n)}\, dx + \nu \int_\Omega |\nabla \mathbf{v}^{(n)}|^2 dx = \int_\Omega \mathbf{f} \cdot \mathbf{v}^{(n)}\, dx - \sum_j \int_0^T \beta_j(t) \int_{\Sigma_j} \mathbf{v}^{(n)} \cdot \mathbf{n}\, dS\ ,$$

which yields "the energy inequality"

$$\max_{t \leq T} \|\mathbf{v}^{(n)}(\cdot, t)\|^2_{L_2(\Omega)} + \nu \int_0^T dt \int_\Omega |\nabla \mathbf{v}^{(n)}|^2 dx \leq C_1(T) F_0(T)$$

with

$$F_0(T) = \|\mathbf{v}_0\|^2_{L_2(\Omega)} + \int_0^T \sum_j |\beta_j(t)|^2 dt + \Big(\int_0^T \|\mathbf{f}\|_{L_2(\Omega)} dt\Big)^2 .$$

Now, when we integrate (1.18) with respect to time from t to $t + \Delta t$ and multiply by $C_{kn}(t + \Delta t) - C_{kn}(t)$, we easily obtain the estimate

$$\|\Delta_t \mathbf{v}^{(n)}\| \leq \nu^{1/2} \Big\| \int_t^{t+\Delta t} |\nabla \mathbf{v}^{(n)}| d\tau \Big\|^{\frac{1}{2}} \|\Delta_t \nabla \mathbf{v}^{(n)}\|^{\frac{1}{2}} + \Big\| \int_t^{t+\Delta t} |\mathbf{v}^{(n)}|^2 d\tau \Big\|^{\frac{1}{2}} \|\Delta_t \nabla \mathbf{v}^{(n)}\|^{\frac{1}{2}} \tag{1.19}$$

$$+ \Big\| \int_t^{t+\Delta t} |\mathbf{f}| d\tau \Big\|^{\frac{1}{2}} \|\Delta_t \mathbf{v}^{(n)}\|^{\frac{1}{2}} + \sum_j |\Sigma_j|^{\frac{1}{4}} \Big(\int_t^{t+\Delta t} |\beta_j(\tau)| d\tau \Big)^{\frac{1}{2}} \|\Delta_t \mathbf{v}^{(n)}\|^{\frac{1}{2}}_{L_2(\Sigma_j)},$$

where $\Delta_t \mathbf{v}^{(n)} = \mathbf{v}^{(n)}(x, t + \Delta t) - \mathbf{v}^{(n)}(x,t)$ and $\|\cdot\| = \|\cdot\|_{L_2(\Omega)}$.
The integration of (1.19) with respect to $t \in (0, T - \Delta t)$ yields (see [24], ch. VI)

$$\|\Delta_t \mathbf{v}^{(n)}\|_{L_{2,1}(Q_{T-\Delta t})} \leq C_2(T)\sqrt{\Delta t}$$

which means that $\mathbf{v}^{(n)}$ are uniformly continuous with respect to t in the norm $L_{2,1}(Q_{T-\Delta t})$. If follows that there exists a subsequence $\mathbf{v}^{(n_j)}$ that is convergent weakly in $L_2(0,T; W_2^1(\Omega))$ to a certain $\mathbf{v} \in L_2(0,T; W_2^1(\Omega))$ and strongly in $L_2(Q'_T)$ in arbitrary $Q'_T = \Omega' \times (0,T)$

where Ω' is a bounded subdomain of Ω. This makes it possible to pass to the limit in the integral identity

$$\int_0^T dt \int_\Omega \left(-\mathbf{v}^{(n_j)} \cdot \vec{\eta}_{Mt} + \nu \nabla \mathbf{v}^{(n_j)} : \nabla \vec{\eta}_M - \sum_{k=1}^{3} v_k^{(n_j)} \mathbf{v}^{(n_i)} \cdot \frac{\partial \vec{\eta}_M}{\partial x_k} \right) dx =$$

$$= \int_0^T \int_\Omega \mathbf{f} \cdot \vec{\eta}_M \, dx dt + \int_\Omega \mathbf{v}_0(x) \cdot \vec{\eta}_M(x,0)\, dx - \sum_{j=1}^{m} \int_0^T \beta_j(t) dt \int_{\Sigma_j} \vec{\eta}_M \cdot \mathbf{n}\, dS, \qquad (1.20)$$

where $\vec{\eta}_M = \sum_{k=1}^{M} d_k(t)\mathbf{a}_k(x)$, $d_k(t)$ are arbitrary smooth functions vanishing for $t = T$, and $M \leq n_i$ (this identity is an evident consequence of (1.17)). The passage to the limit in linear terms is trivial. Consider the difference

$$\int_0^T dt \int_\Omega [v_k^{(n_i)} \mathbf{v}^{(n_j)} - v_k \mathbf{v}] \cdot \frac{\partial \vec{\eta}_M}{\partial x_k} dx = \int_0^T dt \int_{\Omega_R} [v_k^{(n_i)} \mathbf{v}^{(n_j)} - v_k \mathbf{v}] \cdot \frac{\partial \vec{\eta}_m}{\partial x_k} dx$$

$$+ \int_0^T dt \int_{\Omega \setminus \Omega_R} [v_k^{(n_i)} \mathbf{v}^{(n_j)} - v_k \mathbf{v}] \cdot \frac{\partial \vec{\eta}_m}{\partial x_k} dx + \int_0^T dt \int_\Omega [v_k^{(n_i)} \mathbf{v}^{(n_j)} - v_k \mathbf{v}] \cdot \frac{\partial \vec{\eta}'_m}{\partial x_k} dx , \qquad (1.21)$$

where $\vec{\eta}_m = \sum_{k=1}^{m} d_k(t)\, \mathbf{a}_k(x)$, and $\vec{\eta}'_m = \sum_{k=m+1}^{M} d_k(t)\, \mathbf{a}_k(x)$ is a vector field with a compact support. The second integral in the right-hand side of (1.19) can be made arbitrarily small by taking large R, since $|\mathbf{a}_{kx_j}| \leq \frac{C_3}{|x|^2}$ for $k \leq m$. Other integrals are small when n_i is large enough, so for a fixed $\vec{\eta}_M$ the left-hand side in (1.21) tends to zero as $n_j \to \infty$. Clearly, the limiting vector field $\mathbf{v}$ will satisfy (1.17) with $\vec{\eta} = \vec{\eta}_M$ for arbitrary fixed $\vec{\eta}_M$ and consequently for arbitrary $\vec{\eta} \in L_2(0,T; \overset{\circ}{\hat{J}}{}_2^1(\Omega))$ such that $\vec{\eta}_t \in L_2(Q_T)$, $\vec{\eta}(x,T) = 0$.

Let us prove the second part of the theorem. We require additionally that $\{\mathbf{a}_j\}$ be a basis in $W_2^2(\Omega) \cap \overset{\circ}{\hat{J}}{}_2^1(\Omega)$ and that $\sum_{k=0}^{m} C_{kn}(0)\mathbf{a}_k \to \mathbf{v}_0$ in $W_2^2(\Omega)$ as $n \to \infty$, and we obtain an additional estimate for $\mathbf{v}_t^{(n)}$. When we differentiate (1.18) with respect to t, multiply it by $\frac{dC_{kn}}{dt}$ and make the summation with respect to k, we easily obtain

$$\frac{1}{2}\frac{d}{dt}\|\mathbf{v}_t^{(n)}\|^2 + \nu \int_\Omega |\nabla \mathbf{v}_t^{(n)}|^2\, dx - \sum_k \int_\Omega (v_{kt}^{(n)} \mathbf{v}^{(n)} + v_k^{(n)} \mathbf{v}_t^{(n)}) \cdot \mathbf{v}_{tx_k}^{(n)}\, dx =$$

$$= \int_\Omega \mathbf{f}_t \cdot \mathbf{v}_t^{(n)}\, dx - \sum_{j=1}^{m} \beta'_j(t) \int_{\Sigma_j} \mathbf{v}_t^{(n)} \cdot \mathbf{n}\, dS . \qquad (1.22)$$

We estimate the integral with nonlinear terms making use of the well known multiplicative inequality

$$\|\mathbf{u}\|_{L_3(\Omega)} \leq C_4 \|\nabla \mathbf{u}\|^{\frac{1}{2}} \|\mathbf{u}\|^{\frac{1}{2}}, \quad \forall \mathbf{u} \in \overset{\circ}{W}{}_2^1(\Omega) ,$$

and of S. L. Sobolev's imbedding theorem

$$\|\mathbf{u}\|_{L_6(\Omega)} \leq C_5 \|\nabla \mathbf{u}\|, \quad \forall \mathbf{u} \in \overset{\circ}{W}{}_2^1(\Omega) .$$

This leads to the inequality

$$\Big|\sum_k \int_\Omega (v_{kt}^{(n)}\mathbf{v}^{(n)} + v_k^{(n)}\mathbf{v}_t^{(n)}) \cdot \mathbf{v}_{tx_k}^{(n)} dx\Big| \le 2\|\nabla \mathbf{v}_t^{(n)}\|\|\mathbf{v}^{(n)}\|_{L_6(\Omega)}\|\mathbf{v}_t^{(n)}\|_{L_3(\Omega)}$$
$$\le 2C_4C_5\|\mathbf{v}_t^{(n)}\|^{\frac{1}{2}}\|\nabla\mathbf{v}^{(n)}\|\|\nabla\mathbf{v}_t^{(n)}\|^{\frac{3}{2}} .$$

Making use also of another imbedding theorem

$$\Big|\int_{\Sigma_j} \mathbf{v}_t \cdot \mathbf{n}\, dS\Big| \le |\Sigma_j|^{\frac{1}{2}}\|\mathbf{v}_t^{(n)}\|_{L_2(\Sigma_j)} \le C_6\|\mathbf{v}_t^{(n)}\|_{W_2^1(\Omega)}$$

we easily deduce from (1.22) the estimate

$$\frac{1}{2}\sup_{\tau\le t}\|\mathbf{v}_t^{(n)}(\cdot,\tau)\|^2 + \nu\int_0^t \|\nabla \mathbf{v}_\tau^{(n)}\|^2 d\tau \le F_1(t) + 4C_4C_5\int_0^t \|\mathbf{v}_\tau^{(n)}\|^{\frac{1}{2}}\|\nabla\mathbf{v}^{(n)}\|\|\nabla\mathbf{v}_\tau^{(n)}\|^{\frac{3}{2}} d\tau$$

with

$$F_1(t) = 4\Big\{\Big(\int_0^t \|\mathbf{f}_\tau\| d\tau\Big)^2 + \Big(\int_0^t |b| d\tau\Big)^2 + \int_0^t |b|^2 d\tau + \frac{1}{2}\|\mathbf{v}_t^{(n)}(\cdot,0)\|^2\Big\}$$

and $|b| = \sum_j |\beta_j|^2$. In addition, we have

$$\frac{1}{2}\|\nabla\mathbf{v}^{(n)}\|^2 \le \frac{1}{2}\|\nabla\mathbf{v}^{(n)}(\cdot,0)\|^2 + \int_0^t \|\nabla\mathbf{v}^{(n)}\|\|\nabla\mathbf{v}_\tau^{(n)}\|\, d\tau$$
$$\le F_2(t) + \frac{1}{2\nu}\int_0^t \|\nabla\mathbf{v}_\tau^{(n)}\|^2 d\tau ,$$

where

$$F_2(t) = \frac{1}{2}\|\nabla\mathbf{v}^{(n)}(\cdot,0)\|^2 + C_1(t)F_0(t) .$$

Hence,

$$\frac{1}{2}\sup_{\tau\le t}\|\mathbf{v}_t^{(n)}\|^2 + \frac{1}{2}\sup_{\tau\le t}\|\nabla\mathbf{v}^{(n)}\|^2 + \nu\int_0^t \|\nabla\mathbf{v}_\tau^{(n)}\|^2 d\tau \le$$
$$\le F_3(t) + C_6\int_0^t \|\mathbf{v}_\tau^{(n)}\|^{\frac{1}{2}}\|\nabla\mathbf{v}^{(n)}\|\|\nabla\mathbf{v}_\tau^{(n)}\|^{\frac{3}{2}} d\tau , \qquad (1.23)$$

where

$$F_3(t) = F_1(t)\Big(1 + \frac{1}{2\nu^2}\Big) + F_2(t), \qquad C_6 = 4C_4C_5\Big(1 + \frac{1}{2\nu^2}\Big) .$$

Let us estimate the left-hand side of (1.23) on a certain finite time interval. Applying the Young's inequality we arrive at

$$\sup_{\tau\le t}\|\mathbf{v}_t^{(n)}\|^2 + \sup_{\tau\le t}\|\nabla\mathbf{v}^{(n)}\|^2 \le 4F_3(t) + C_7\int_0^t \|\mathbf{v}_\tau^{(n)}\|^2\|\nabla\mathbf{v}^{(n)}\|^4\, d\tau .$$

Consequently, $Z(t) = \int_0^t(\|\mathbf{v}_\tau^{(n)}\|^6 + \|\nabla\mathbf{v}^{(n)}\|^6)\, d\tau$ satisfies the inequality

$$Z'(t) \le (4F_3(t) + C_7 Z(t))^3$$

which leads to

$$t \geq \int_0^{Z(t)} \frac{dz}{(4F_3(t)+C_7z)^3} = \frac{1}{2C_7}\Big(\frac{1}{(4F_3)^2} - \frac{1}{(4F_3+Z)^2}\Big)$$

or

$$Z(t) \leq \frac{2C_7t(4F_3(t))^3}{\sqrt{1-2C_7t(4F_3)^2}(1+\sqrt{1-2C_7t(4F_3)^2})} . \tag{1.24}$$

This inequality proves the boundedness of the left-hand side of (1.23) on the time interval $(0,T)$ with $T_1 < \frac{1}{32C_7F_3^2(T)}$.

Let us show that the expression $F_3(T)$ containing the norms $\|\mathbf{v}_t^{(n)}(\cdot,0)\|^2$ and $\|\nabla\mathbf{v}^{(n)}(\cdot,0)\|^2$ is uniformly bounded. It follows from (1.18) that

$$\begin{aligned}\|\mathbf{v}_t^{(n)}(\cdot,0)\|^2 &= \nu\int_\Omega \Delta\mathbf{v}^{(n)}\cdot\mathbf{v}_t^{(n)}(x,0)\,dx - \sum_{i=1}^3\int_\Omega v_i^{(n)}(x,0)\frac{\partial\mathbf{v}^{(n)}}{\partial x_i}\cdot\mathbf{v}_t^{(n)}(x,0)\,dx\\ &+\int_\Omega \mathbf{f}(x,0)\cdot\mathbf{v}_t^{(n)}(x,0)\,dx - \sum_{k,j=1}^m \beta_j(0)\int_{\Sigma_j}\mathbf{a}_k(x)\cdot\mathbf{n}\,dS\frac{dC_k(0)}{dt}\end{aligned}$$

and as a consequence

$$\begin{aligned}\|\mathbf{v}_t^{(n)}(\cdot,0)\| &\leq \nu\|\Delta\mathbf{v}^{(n)}(\cdot,0)\| + \|\mathbf{v}^{(n)}(\cdot,0)\|_{L_4(\Omega)}\|\nabla\mathbf{v}^{(n)}(\cdot,0)\|_{L_4(\Omega)}\\ &+\|\mathbf{f}(\cdot,0)\|_{L_2(\Omega)} + \sum_{j=1}^m|\beta_j(0)|\Big(\sum_{k=1}^m\|\mathbf{a}_k\|^2_{L_2(\Sigma_j)}\Big)^{\frac{1}{2}} .\end{aligned}$$

The right-hand side is uniformly bounded, since $\mathbf{v}^{(n)}(\cdot,0)\to\mathbf{v}_0$ in $W_2^2(\Omega)$ as $n\to\infty$. (1.24) gives a uniform estimate of the left-hand side of (1.23) in a time interval $(0,T_1)$ independent of n. If we take in (1.20) $d_k(t)=0$ for $t\geq T_1$, we can easily pass to the limit in all the terms and establish the existence of a weak solution with all properties indicated in the formulation of the theorem.

The uniqueness of the solution is proved by the following arguments. The difference of any two solutions $\mathbf{w}=\mathbf{v}'-\mathbf{v}''$ satisfies the integral identity:

$$\int_0^t d\tau\int_\Omega(\mathbf{w}_\tau\cdot\vec{\eta}+\nu\nabla\mathbf{w}:\nabla\vec{\eta})\,dx = \sum_k\int_0^t d\tau\int_\Omega(\mathbf{v}'_k\mathbf{w}+w_k\mathbf{v}'')\frac{\partial\vec{\eta}}{\partial x_k}\,dx$$

and homogeneous initial conditions. If we take here $\vec{\eta}=\mathbf{w}$ and make use of the boundedness of $\sup_{t\leq\tau}(\|\mathbf{v}'\|_{W_2^1(\Omega)}+\|\mathbf{v}''\|_{W_2^1(\Omega)})$, we easily arrive at

$$\frac{1}{2}\|\mathbf{w}(\cdot,t)\|^2 + \nu\int_0^t\|\nabla\mathbf{w}\|^2\,d\tau \leq C_8\int_0^t\|\mathbf{w}\|^{\frac{1}{2}}\|\mathbf{w}\|^{\frac{3}{2}}_{W_2^1(\Omega)}\,d\tau .$$

This inequality implies $\|\mathbf{w}(\cdot,t)\|^2\leq C_9\int_0^t\|\mathbf{w}\|^2\,d\tau$ and consequently $\mathbf{w}=0$.

It remains to establish the existence of a solution in the time interval $(0,T)$ when the data are small. We have:

$$\begin{aligned}&\int_0^t\|\mathbf{v}_t^{(n)}\|^{\frac{1}{2}}\|\nabla\mathbf{v}^{(n)}\|\|\nabla\mathbf{v}_t^{(n)}\|^{\frac{3}{2}}d\tau\leq\\ &\leq\sup_{\tau\leq t}\|\mathbf{v}_\tau^{(n)}\|^{\frac{1}{2}}\|\nabla\mathbf{v}^{(n)}\|^{\frac{1}{2}}\Big(\int_0^t\|\nabla\mathbf{v}^{(n)}\|^2d\tau\Big)^{\frac{1}{4}}\Big(\int_0^t\|\nabla\mathbf{v}_t^{(n)}\|^2d\tau\Big)^{\frac{3}{4}}\end{aligned}$$

So it follows from (1.23) that

$$S_n(t) \le F_3(t) + C_7 F_0^{\frac14}(t) S_n^{\frac54}(t) \ ,$$

where $S_n(t)$ is the left-hand side of (1.23) and $C_7 = C_6 \frac{2^{\frac14} c_1^{\frac14}}{\nu}$. The function $\varphi(s) = F_3 + C_7 F_0^{\frac14} s^{\frac54}$ has positive first and second derivatives for $s \in \mathbb{R}_+$, and $\varphi'(s) = 1$ for $s = s_0 = \frac{1}{F_0}\left(\frac{4}{5C_7}\right)^4$. Suppose that $F_3(t) \le \frac15 s_0(t)$ i.e.

$$5F_0(t)F_3(t) \le \left(\frac{4}{5C_7}\right)^4 \tag{1.25}$$

Then $\varphi(s_0) = F_3(t) + \frac45 s_0 \le s_0$, and the graphs of the functions $\varphi(s)$ and s have at least one point of intersection (see fig.1)

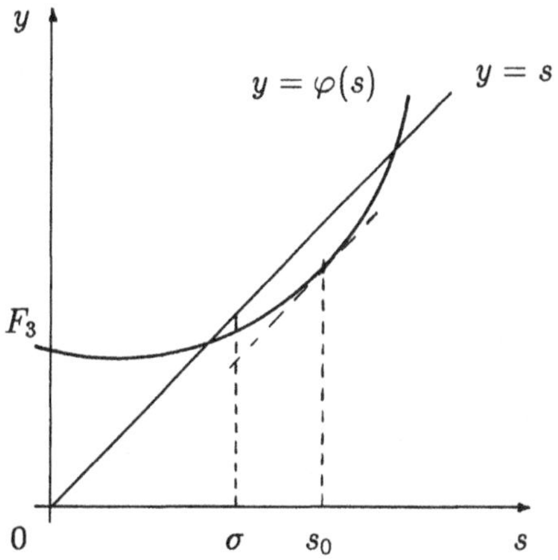

Fig. 1

Let δ be a minimal solution of the equation $s = \varphi(s)$; clearly, $\delta \ge F_3(t)$. For arbitrary $\tau \le t$ we have

$$\begin{aligned} S_n(\tau) &\le F_3(\tau) + C_7 F_0^{\frac14}(\tau) S_n^{\frac54}(\tau) \\ &\le F_3(t) + C_7 F_0^{\frac14}(t) S_n^{\frac54}(\tau) = \varphi(S_n(\tau)) \ ; \end{aligned} \tag{1.26}$$

moreover, it follows from (1.23) that

$$S_n(0) = \frac12 (\|\mathbf{v}_t^{(n)}(\cdot,0)\|^2 + \|\nabla \mathbf{v}^{(n)}(\cdot,0)\|^2) \le F_3(0) \le F_3(t) \le \delta \ .$$

Since $S_n(\tau)$ is a continuous function of τ and (1.26)) holds for all $\tau \in [0,t]$, we may conclude that $S_n(\tau) \le \delta$ for all $\tau \in [0,t]$. Hence, $S_n \le F_3 + C_7 F_0^{\frac14} s_0^{\frac14} S_n(t) \le F_3 + \frac45 S_n(t)$ and $S_n(t) \le 5F_3(t)$, $t \in (0,T)$, which gives a uniform estimate of $S_n(t)$ under the smallness condition (1.25).

The proof of Theorem 1.7 is completed.

Theorem 1.6 is proved by similar arguments, if we introduce a new unknown vector field $\mathbf{u} = \mathbf{v} - \sum_{j=1}^{m-1} Q_j(t)\, \mathbf{A}_{jm} \in L_2(0,T; \overset{\circ}{J}{}_2^1(\Omega))$.

2 – Solvability of stationary problems in the class of vector fields with an infinite Dirichlet integral

We have already seen on the example of problem (1.1)–(1.3), (1.15) in two-dimensional case that the condition of the boundedness of the energy integral may be too restrictive. Similar examples can be constructed for stationary problems.

Example 3: Consider the domain $G \subset \mathbb{R}^n$, $n = 2, 3$, defined by the inequality

$$|x'| \equiv \sqrt{x_2^2 + \ldots + x_n^2} < g(x_1)\,, \qquad x_1 > 0\,. \tag{2.1}$$

Let $\Sigma_t = \{x \in G : x_1 = t \geq 0\}$, $S = \{x \in \mathbb{R}^2 : |x'| = g(x_1), x_1 \geq 0\}$; suppose that $g(t) \geq g_0 > 0$. For arbitrary divergence free $\mathbf{v}(x)$ vanishing on S and satisfying the condition $\int_{\Sigma_t} v_1 dx' = Q$ we have

$$Q^2 \leq C_1 |g(t)|^{n-1} \int_{\Sigma_t} v_1^2 dx' \leq C_2 |g(t)|^{n+1} \int_{\Sigma_t} |\nabla' v_1|^2 dx'\,,$$

hence

$$Q^2 \int_{t_1}^{t_2} \frac{dt}{g^{n-1}(t)} \leq C_1 \int_{t_1}^{t_2} dt \int_{\Sigma_t} v_1^2 dx'\,, \qquad Q^2 \int_{t_1}^{t_2} \frac{dt}{g^{n+1}(t)} \leq C_2 \int_{t_1}^{t_2} dt \int_{\Sigma_t} |\nabla' v_1|^2 dx'\,.$$

These inequalities show that conditions

$$\int_{t_1}^{\infty} g^{-n+1}(t) dt < \infty \quad \text{and} \quad \int_{t_1}^{\infty} g^{-n-1}(t) dt < \infty$$

are necessary for the existence of a solenoidal $\mathbf{v}$ such that $\mathbf{v} \in L_2(G)$ or $\nabla \mathbf{v} \in L_2(G)$, respectively, vanishing on S and having non-zero flux Q. (It is reasonable to call the domain G "large" in this case). The functions $g(x_1) = g_0 x_1 + g_1$ and $g(x_1) = g_0$ correspond to a conical and cylindrical Ω, respectively.

For a cylinder $\Omega = \mathbb{R} \times \Sigma$ a solution of problem (1.4), (1.13) with $\mathbf{f} = 0$ is well known, it is the Poiseuille flow.

Example 4: Poiseuille flow in a cylinder $\Omega = \mathbb{R} \times \Sigma$.

The velocity vector field corresponding to the Poiseuille flow is given by the formula $\mathbf{v}(x) = (v(x'), \ldots, 0)$ where $v(x')$ is a solution of the Dirichlet problem

$$-\nu \Delta v(x') + k = 0 \quad (x' \in \Sigma)\,, \qquad v|_{\partial\Sigma} = 0\,.$$

The corresponding pressure is $p = kx_1 + p_1$, $p_1 = \text{const}$. The constant k is uniquely determined by the total flux $Q = \int_\Sigma v_1\, dx'$ or by pressure drop per unit length along x_1-axis.

As a characteristic example we consider problem (1.4), (1.13) in the domain $\Omega \subset \mathbb{R}^n$, $n = 2, 3$ with the following properties:

1) Ω has two "exits to infinity", i.e. $\Omega = \Omega_0 \cup G_1 \cup G_2$ where Ω_0 is a bounded domain and

$$G_i = \left\{ y^{(i)} \in \Sigma(y_1^{(i)}),\; y_1^{(i)} > 0 \right\}, \qquad i = 1, 2\,, \tag{2.2}$$

$\{y_1^{(1)},\dots,y_n^{(1)}\}$ and $\{y_1^{(2)},\dots,y_n^{(2)}\}$ are two different Cartesian coordinate systems, $y^{(i)\prime}=y_2^{(i)}$ for $n=2$ and $y^{(i)\prime}=(y_2^{(i)},y_3^{(i)})$ for $n=3$. The boundary $S=\partial\Omega$ is lipzschitzean.

2) $\Sigma(y_1^{(i)})$ is a bounded domain which is contained in a circle $|y^{(i)\prime}|\le r_0$ and contains a circle $|y^{(i)\prime}|\le r_1$.

3) For every domain $\omega_i(t)=\{y^{(i)}\in\Sigma(y_1^{(i)}),\ t<y_1^{(i)}<t+1\}$, problem (1.8) is solvable and constants C_1 in (1.11) may be taken independent of t.

4) In Ω there exists a divergence free vector field $\mathbf{A}(x)$ which is uniformly bounded together with its derivatives and satisfies the conditions

$$\mathbf{A}|_S=0,\qquad \int_\Sigma \mathbf{A}\cdot\mathbf{n}\,dS=1$$

for arbitrary cross-section Σ of Ω (for instance, for $\Sigma=\Sigma(y_1^{(i)})$).

The most important restriction on Ω is 3). The assumption that $m=2$ and G_i may be given by (2.2) is only technical. In fact, the "exits" G_i may be rather twisted, important is the fact that they may be cut into bounded pieces for which the assumption 3) holds. The vector field $\mathbf{A}$ may be constructed as explained in Example 2, i.e.

$$\mathbf{A}=\operatorname{rot}\varsigma\mathbf{H},\quad \mathbf{H}=\frac{1}{4\pi}\int_l\frac{\mathbf{x}-\mathbf{y}}{|x-y|^3}\times d\mathbf{l}_y\quad\text{in the three dimensional case.}$$

The contour l may be composed of two half-axes ($y^{(i)\prime}=0,\ y_1^{(i)}\ge 0$) connected by a smooth finite curve contained in Ω_0. The function should vanish in the neighborhood of l and be equal to 1 in the neighborhood of S. If we let ς be equal to the so called Hopf cut-off function, then $\mathbf{A}$ will satisfy the inequality

$$|\mathbf{A}(x)|\le\frac{\epsilon}{\operatorname{dist}(x,\Gamma)}\,,\tag{2.3}$$

where ϵ is an arbitrary fixed small number (see the details in [26, 39, 41]).

Consider in Ω a stationary problem

$$-\nu\Delta\mathbf{v}+(\mathbf{v}\cdot\nabla)\mathbf{v}+\nabla p=\mathbf{f},\qquad \operatorname{div}\mathbf{v}=0\qquad(x\in\Omega\subset\mathbb{R}^n)\,,\tag{2.4}$$

$$\mathbf{v}|_S=0,\qquad\int_\Sigma\mathbf{v}\cdot\mathbf{n}\,dS=Q\,.\tag{2.5}$$

By a weak solution of this problem we shall mean a divergence free $\mathbf{v}\in W_2^1(\Omega')$ for arbitrary bounded $\Omega'\subset\Omega$ satisfying (2.5) and the integral identity

$$\nu\int_\Omega\nabla\mathbf{v}:\nabla\vec{\eta}\,dx-\sum_k\int_\Omega v_k\mathbf{v}\cdot\frac{\partial\vec{\eta}}{\partial x_k}\,dx=\int_\Omega\mathbf{f}\cdot\vec{\eta}\,dx$$

for arbitrary $\vec{\eta}\in\mathcal{H}(\Omega)$ with a compact support.

Let $\Omega_t=\Omega_0\cup G_{1t}\cup G_{2t}$, $G_{it}=\{x\in G_i\colon 0\le y_1^{(i)}\le t\}$. Concerning $\mathbf{f}$ we shall suppose that this vector field has a finite norm

$$|||\mathbf{f}|||^2=\sup_{t>0}q^{-1}(t)\|\mathbf{f}\|^2_{\mathcal{H}^*(\Omega_t)}\,,\tag{2.6}$$

where $\mathcal{H}^*(\Omega_t)$ is a dual space to $\mathcal{H}(\Omega_t)$ and $q(t)$ is a positive increasing continuously differentiable function satisfying the relation $\lim t^{-3}q(t) \to 0$ (as $t \to \infty$) and inequality

$$q(t) + \beta t \geq \mu_1 q'(t) + \mu_2 q'^{\frac{3}{2}}(t), \quad t \geq t_0 > 0, \tag{2.7}$$

with constants β, μ_1, μ_2 defined later (see (2.20)) (for instance, (2.7) is satisfied by $q(t) = t^h$, $h < 3$).

The following theorems are proved in [27] (in the case $\mathbf{f} = 0$).

Theorem 2.1. *Suppose that Ω satisfies the above conditions 1)-4) and $\mathbf{A}(x)$ satisfies (2.3). For arbitrary $\mathbf{f}$ with a finite norm (2.6) and arbitrary $Q \in \mathbb{R}$ problem (2.4), (2.5) has at least one solution such that*

$$\|\nabla \mathbf{v}\|^2_{L_2(\Omega_t)} \leq C_3(q(t) + t + b), \quad b > 1. \tag{2.8}$$

Theorem 2.2. *If $\mathbf{v}_1$ and $\mathbf{v}_2$ are two weak solutions of (2.4), (2.5) possessing the properties (2.8) and*

$$\|\nabla \mathbf{v}_i\|^2_{L_2(\Omega_0)} \leq \delta, \quad \|\nabla \mathbf{v}_i\|^2_{L_2(\omega_l(t))} \leq \delta, \tag{2.9}$$

with a small $\delta > 0$, then $\mathbf{v}_1 = \mathbf{v}_2$.

Remark. Inequalities (2.9) may be established for arbitrary weak solution obtained in Theorem 2.1 in the case of small Q and $\mathbf{f} = 0$ (see [27]).

Theorem 2.3. *Suppose that $\mathbf{f} = 0$ and that the domain $G_i \backslash G_{it_0}$ is cylindrical, i.e. $\Sigma(y_1^{(i)})$ is independent of $y_1^{(i)}$ for $y_1^{(i)} > t_0$, and let Q be small. Then a weak solution of problem (2.4), (2.5) tends to the Poiseuille flow as $y_1^{(i)} \to \infty$ with an exponential rate.*

Problem (2.4), (2.5) is closely connected with the so called J. Leray's problem in which it is required to find a solution of the Navier-Stokes equations in Ω with $\mathbf{f} = 0$ that vanishes on S and tends to corresponding Poiseuille flows as $y_1^{(i)} \to \infty$, $i = 1, 2$ (it should be assumed, of course, that the both exits G_1 and G_2 are cylindrical). For small Q the solvability of this problem was proved by C. I. Amick [1,2] and C. I. Amick and L. E. Fraenkel [3]. In the latter article, there was investigated also the problem (2.4), (2.5) in domains whose "exits" expand at infinity, in particular, in "large" domains (in the terminology of Example 3 and for these domains the solvability of (2.4), (2.5) was established without smallness assumptions. In comparison with the J. Leray's problem, the formulation (2.4), (2.5) is less restrictive. We study problem (2.4), (2.5), making use of a special techniques of integral estimates proposed by R. Toupin [48] and I. Knowles [21] for the purpose of justification of the Saint-Venant's principle in the elasticity theory. After the papers [48, 21] there appeared a series of papers of different authors (see for instance [22, 31–33]) where this techniques was applied to elliptic and parabolic equations in domains of different shape, mainly for the purpose of establishing uniqueness theorems. O. A. Oleinik and G. I. Iosif'ian [32] use integral estimates "of Saint-Venant's type" also in the proof of the existence of solutions of Dirichlet problem for second order elliptic equations with unbounded Dirichlet integrals.

As far as the Navier-Stokes equations are concerned, we refer to the papers of C. Horgan and L. Wheeler [13, 14] and G. I. Iosif'ian [16, 17], C. Horgan and L. Wheeler establish the result presented above as theorem 2.3. G. I. Iosif'ian proves uniqueness theorem for solutions of the Stokes and Navier-Stokes equation in a class of vector fields with unbounded Dirichlet integrals. In all these papers the techniques of integral estimates "of Saint-Venant's type" is intensively used.

Let us describe this techniques and prove Theorems 2.1–2.3.

The following simple lemma is crucial in obtaining integral estimates of solution of (2.4), (2.5) in domains Ω_t.

Lemma 2.1. *Let $y(t)$ and $\varphi(t)$ be continuously differentiable positive increasing functions given in the interval (t_1, t_2) and satisfying the inequalities*

$$y(t) \le \psi(y'(t)) + (1-\delta)\varphi(t) \quad \text{and} \quad \delta\varphi(t) \ge \psi(\varphi'(t)), \qquad t \in (t_1, t_2) . \tag{2.10}$$

where $\delta \in (0,1)$ and $\psi(p)$ is a positive increasing function of a positive argument p. If

$$y(t_2) \le \varphi(t_2) , \tag{2.11}$$

then $y(t) \le \varphi(t)$ for all $t \in (t_1, t_2)$.

Proof: If $y(t_0) > \varphi(t_0)$ for a certain $t_0 \in (t_1, t_2)$, then $\delta y(t_0) < \psi(y'(t_0))$ and

$$\psi(y'(t_0)) > \delta y'(t_0) > \delta\varphi(t_0) \ge \psi(\varphi'(t_0)) ,$$

hence, $y'(t_0) > \varphi'(t_0)$ and $y(t) > \varphi(t)$ for all $t > t_0$ which contradicts (2.11).

Lemma 2.2. *Let a positive continuously differentiable increasing function $y(t)$ satisfy the inequality*

$$y(t) \le a_1 y'(t) + a_2 y'^{\frac{3}{2}}(t) \qquad (a_1, a_2 > 0, \forall t > 0) . \tag{2.12}$$

Then $y(t) \ge a_3 t^3$ for large t with $a_3 > 0$.

Proof: Suppose that $y(t_0) \equiv y_0 > 0$. Then $y'(t_0) \ge \alpha_0 > 0$, and $y(t) \ge y_0$, $y'(t) \ge \alpha_0$ for all $t \ge t_0$. Hence $y'(t) \to \infty$ as $t \to \infty$ and $y'(t) \ge 1$ for $t \ge \tau$. Inequality (2.12) implies $y(t) \le (a_1 + a_2)\, y'^{\frac{3}{2}}(t)$ or $y'(t)/y^{\frac{2}{3}}(t) \ge (a_1 + a_2)^{-\frac{2}{3}}$ for $t \ge \tau$. Integrating this inequality we prove the lemma.

Let us proceed to the proof of main results of this section.

Proof of the Theorem 2.1: A new unknown vector field $\mathbf{u} = \mathbf{v} - \mathbf{a}$, $\mathbf{a} = Q\mathbf{A}$, should satisfy the conditions

$$\mathbf{u}|_S = 0 , \qquad \int_\Sigma \mathbf{u} \cdot \mathbf{n}\, dS = 0$$

and the integral identity

$$\nu \int_\Omega \left[\nabla \mathbf{u} : \nabla \vec{\eta} - \sum_{k=1}^{n} (a_k \mathbf{u} + u_k \mathbf{a} + u_k \mathbf{u}) \cdot \frac{\partial \vec{\eta}}{\partial x_k} \right] dx =$$

$$= \int_\Omega \left(\mathbf{f} \cdot \vec{\eta} - \nu \nabla \mathbf{a} : \nabla \vec{\eta} + \sum_{k=1}^{n} a_k \mathbf{a} \cdot \frac{\partial \vec{\eta}}{\partial x_k} \right) dx \tag{2.13}$$

for arbitrary $\vec{\eta} \in \mathcal{H}(\Omega)$ with a compact support. We shall find $\mathbf{u}$ as a limit of solutions $\mathbf{u}^{(N)}$ of auxiliary problems in Ω_N defined as elements of $\mathcal{H}(\Omega_N)$ satisfying the integral identity

$$\nu \int_{\Omega_N} \left[\nabla \mathbf{u}^{(N)} : \nabla \vec{\eta} - \sum_{k=1}^{n} (a_k \mathbf{u}^{(N)} + u_k^{(N)} \mathbf{a} + u_k^{(N)} \mathbf{u}^{(N)}) \cdot \frac{\partial \vec{\eta}}{\partial x_k} \right] dx =$$

$$= \int_{\Omega_N} \left[\mathbf{f} \cdot \vec{\eta} - \nu \nabla \mathbf{a} : \nabla \vec{\eta} + \sum_{k=1}^{n} a_k \mathbf{a} \cdot \frac{\partial \vec{\eta}}{\partial x_k} \right] dx \equiv \mathcal{L}(\vec{\eta}) \;, \qquad (2.14)$$

with arbitrary $\vec{\eta} \in \mathcal{H}(\Omega_N)$. The solvability of these problems follows from results of O. A. Ladyzhenskaya [24]. If we substitute $\vec{\eta} = \mathbf{u}^{(N)}$ into (2.14) and make use of (2.3) with an appropriate ϵ, we obtain the energy inequality

$$\begin{aligned} \|\nabla \mathbf{u}^{(N)}\|^2_{L_2(\Omega_N)} &\le C_4(\|\mathbf{f}\|^2_{\mathcal{H}^*(\Omega_N)} + \|\nabla \mathbf{a}\|^2_{L_2(\Omega_N)} + \|\mathbf{a}\|^4_{L_4(\Omega_N)}) \\ &\le C_4 |||\mathbf{f}|||^2 q(N) + C_5 |Q|^2 (1 + Q^2) N \;. \end{aligned} \qquad (2.15)$$

The objective of following arguments is the estimate of the integral $\int_{\Omega_t} |\nabla \mathbf{u}^{(N)}|^2 \, dx$ with arbitrary $t < N - 1$. To this end, we substitute into (2.14) the test function $\vec{\eta} = \mathbf{U}_t^{(N)}$ where $\mathbf{U}_t^{(N)}(x)$ is defined as follows: $\mathbf{U}_t^{(N)}(x) = \mathbf{u}^{(N)}(x)$ for $x \in \Omega_t$, $\mathbf{U}_t^{(N)} = 0$ for $x \in \Omega \backslash \Omega_{t+1}$, and in $\omega_i(t)$, $\mathbf{U}_t^{(N)}$ is defined as a solution of the problem

$$\nabla \cdot \mathbf{U}_t^{(N)} = 0 \;\; (x \in \omega_i(t)) \,, \quad \mathbf{U}_t^{(N)}|_{\Sigma_i(t)} = \mathbf{u}^{(N)}|_{\Sigma_i(t)} \,, \quad \mathbf{U}_t^{(N)}|_{\partial\omega_i(t)\backslash\Sigma_i(t)} = 0 \;. \qquad (2.16)$$

In its turn, the solution of this problem may be found in the form

$$\mathbf{U}_t^{(N)} = \lambda_t(x) \mathbf{u}^{(N)}(x) + \mathbf{V}_t^{(N)}(x) \;,$$

where λ_t is a smooth function of $y_1^{(i)}$ which is equal to 1 for $y_1^{(i)} \le t + 1/8$ and to zero for $y_1^{(i)} \ge t + 7/8$ and $\mathbf{V}_t^{(N)}(x)$ satisfies the relations

$$\nabla \cdot \mathbf{V}_t^{(N)} = -\nabla \cdot \mathbf{u}^{(N)} \lambda_t = -\mathbf{u}^{(N)} \cdot \nabla \lambda_t \;\; (x \in \omega_i(t)) \,, \quad \mathbf{V}_t^{(N)}|_{\partial\omega_i(t)} = 0 \;.$$

Since

$$\int_{\omega_i(t)} \nabla \cdot \mathbf{u}_t^{(N)} \lambda_t \, dx = -\int_{\Sigma_i(t)} \mathbf{u}_t^{(N)} \cdot \mathbf{n} \, dS = 0 \;,$$

this problem is solvable, and in virtue of assumption 3)

$$\|\nabla \mathbf{V}_t^{(N)}\|_{L_2(\omega_i(t))} \le C_6 \, \|\mathbf{u}_t^{(N)}\|_{L_2(\omega_i(t))} \;.$$

This implies the estimates

$$\|\mathbf{U}_t^{(N)}\|_{L_2(\omega_i(t))} \le C_7 \|\mathbf{u}^{(N)}\|_{L_2(\omega_i(t))} \,, \qquad \|\nabla \mathbf{U}_t^{(N)}\|_{L_2(\omega_i(t))} \le C_8 \|\nabla \mathbf{u}^{(N)}\|_{L_2(\omega_i(t))} \,,$$

with C_7, C_8 independent of t.

After the substitution $\vec{\eta} = \mathbf{U}_t^{(N)}$ we can estimate separate terms in (2.14) as follows:

$$\int_{\Omega_N} \nabla \mathbf{u}^{(N)} : \nabla \mathbf{U}_t^{(N)} dx \geq \int_{\Omega_t} |\nabla \mathbf{u}^{(N)}|^2 dx - C_8 \int_{\omega_t} |\nabla \mathbf{u}^{(N)}|^2 dx ,$$

$$\left| \int_{\Omega_N} \sum_k (u_k^{(N)} + a_k^{(N)}) \mathbf{u}^{(N)} \cdot \frac{\partial \mathbf{U}_t^{(N)}}{\partial x_k} \, dx \right| = \left| \int_{\Omega_N} \sum_k (u_k^{(N)} + a_k^{(N)}) \mathbf{u}^{(N)} \cdot \frac{\partial (\mathbf{U}_t^{(N)} - \mathbf{u}^{(N)})}{\partial x_k} \, dx \right|$$
$$\leq C_9 \|\nabla \mathbf{u}^{(N)}\|^2_{L_2(\omega_t(t))} + C_{10} \|\nabla \mathbf{u}^{(N)}\|^3_{L_2(\omega_t(t))} ,$$

$$\left| \int_{\Omega_N} \sum_k u_k^{(N)} \mathbf{a} \cdot \frac{\partial \mathbf{U}_t^{(N)}}{\partial x_k} \, dx \right| \leq C_{11} \epsilon \int_{\Omega_N} \frac{|\mathbf{u}^{(N)}|}{\operatorname{dist}(x, \Gamma)} |\nabla \mathbf{U}_t^{(N)}| \, dx \tag{2.17}$$
$$\leq C_{12} \epsilon \int_{\Omega_{t+1}} |\nabla \mathbf{u}^{(N)}|^2 dx .$$

Here we have used the Hardy inequality. Finally,

$$|\mathcal{L}(\mathbf{U}_t^{(N)})| \leq C_{13} \|\nabla \mathbf{u}^{(N)}\|_{L_2(\Omega_{t+1})} \left(\|\mathbf{f}\|_{\mathcal{H}^*(\Omega_{t+1})} + \|\nabla \mathbf{a}\|_{L_2(\Omega_{t+1})} + \|\mathbf{a}\|^2_{L_4(\Omega_{t+1})} \right) .$$

With these estimates established, it is not hard of deduce from (1.13) the inequality

$$\|\nabla \mathbf{u}^{(N)}\|^2_{L_2(\Omega_t)} \leq C_{14} \|\nabla \mathbf{u}^{(N)}\|^2_{L_2(\omega_t)} + C_{15} \|\nabla \mathbf{u}^{(N)}\|^3_{L_2(\omega_t)} + C_{16} q(t) + C_{17}(t+1) . \tag{2.18}$$

Here $\omega_t = \omega_{1t} \cup \omega_{2t}$, and the constants C_{16} and C_{17} are proportional to $\|\!|\mathbf{f}|\!\|^2$ and to $|Q|^2(1+Q^2)$, respectively.

Let us introduce the function

$$y(t) = \int_{\Omega_N} \psi(t,x) \, |\nabla \mathbf{u}^{(N)}|^2 \, dx ,$$

where $\psi(t,x) = 1$ for $x \in \Omega_t$, $\psi(t,x) = 0$ for $x \in \Omega \backslash \Omega_{t+1}$, and $\psi(t,x) = t + 1 - y_1^{(i)}$ for $x \in \omega_i(t)$. Since $\frac{dy}{dt} = \int_{\omega_t} |\nabla \mathbf{u}^{(N)}|^2 \, dx$, (2.18) implies

$$y(t) \leq (C_{14}+1) y'(t) + C_{15} y'^{\frac{3}{2}}(t) + \frac{1}{2} \varphi(t) , \tag{2.19}$$

with $\varphi(t) = 2C_{16} q(t) + 2C_{17}(t+b)$, $b \geq 1$.

Now, we are going to apply Lemma 2.1. Inequality (2.10) follows from (2.15) if we take $2C_{16} \geq C_4 \|\!|\mathbf{f}|\!\|^2$, $2C_{17} \geq C_5 Q^2 (1+Q^2)$, and (2.10) reduces to

$$C_{16} q(t) + C_{17}(t+b) \geq (C_{14}+1) \left(2C_{16} q'(t) + 2C_{17} \right) + C_{15} (2C_{16} q'(t) + 2C_{17})^{\frac{3}{2}} .$$

This inequality will be satisfied in virtue of (2.7) if we require that

$$\mu_1 = 2(C_{14}+1) , \quad \mu_2 \geq 2^{\frac{3}{2}} C_{15} C_{16}^{\frac{1}{2}} , \quad \beta = C_{17}/C_{16} \tag{2.20}$$

and take b large enough.

By Lemma 2.1,

$$\|\nabla \mathbf{u}^{(N)}\|_{L_2(\Omega_t)} \leq y(t) \leq \varphi(t) ,$$

which proves estimate (2.8) for $\mathbf{u}^{(N)}$. Hence, there exists a subsequences $\mathbf{u}^{(N_i)}$ which is convergent weakly in $W_2^1(\Omega_t)$ and strongly in $L_4(\Omega_t)$ for arbitrary t. Clearly, a limiting vector field $\mathbf{u}$ satisfies inequality (2.8) and an integral identity (2.13). The theorem is proved.

Proof of Theorem 2.2: If $\mathbf{w}$ and $\mathbf{v} = \mathbf{w} + \mathbf{u}$ are two weak solutions of problem (1.4), (1.13) in Ω, then $\mathbf{u}$ satisfies the conditions $\int_\Sigma \mathbf{u}\cdot\mathbf{n}\, dS = 0$, $\mathbf{u}|_S = 0$ and the integral identity

$$\int_\Omega \left(\nu\nabla\mathbf{u} : \nabla\vec{\eta} - \sum_k (w_k\mathbf{u} + u_k\mathbf{w}) \cdot \frac{\partial\vec{\eta}}{\partial x_k} - \sum_k u_k\mathbf{u}\cdot\frac{\partial\vec{\eta}}{\partial x_k}\right) dx = 0\,, \tag{2.21}$$

with arbitrary $\vec{\eta} \in \mathcal{H}(\Omega)$ with a compact support. It is not hard to prove that $y(t) = \int_\Omega \psi(t,x)|\nabla\mathbf{u}|^2 dx$ satisfies inequality (2.19) without the term $\frac{1}{2}\varphi(t)$ in the right-hand side. This can be done by the same arguments as in the preceding theorem. But instead of (2.17) we have

$$\left|\int_\Omega \sum_k u_k\mathbf{w}\cdot\frac{\partial\mathbf{U}_t}{\partial x_k}\, dx\right| \le \left|\int_{\Omega_0} \sum_k u_k\mathbf{w}\cdot\frac{\partial\mathbf{U}_t}{\partial x_k}\, dx\right| + \sum_{m=0}^{\infty}\left|\int_{\omega_m} u_k\mathbf{w}\cdot\frac{\partial\mathbf{U}_t}{\partial x_k}\, dx\right|$$
$$\le C_{18}\delta\|\nabla\mathbf{u}\|^2_{L_2(\Omega_{t+1})}\,,$$

where $\mathbf{U}_t$ is determined by $\mathbf{u}$ as explained in Theorem 2.1. Since δ is small, we can easily arrive at (2.19) with $\varphi = 0$.

According to Lemma 2.2, $y(t) \ge C_{19}t^3$, but this contradicts the condition $t^{-3}y(t) \to 0$ as $t \to \infty$. Hence $y(t) = 0$ which proves the theorem.

Proof of Theorem 2.3: Suppose that the exit G_1 is cylindrical and let $\mathbf{w}$ be a Poiseuille flow in this cylinder with the same small flux Q. The difference $\mathbf{u} = \mathbf{v} - \mathbf{w}$ will satisfy (2.21) with $\vec{\eta} \in \mathcal{H}(\Omega)$ whose support is contained in G_1. If we take $\vec{\eta} = \mathbf{U}_t - \mathbf{U}_1$ in (2.21), we arrive at the inequality

$$z(t) \le (C_{14}+1)z'(t) + C_{15}z'^{\frac{3}{2}}(t) + C_{19}$$

for $z(t) = \int_{G_{1t+1}\backslash G_{11}} \psi(t,x)\,|\nabla\mathbf{u}|^2\, dx$. It is seen from the proof of lemma 2.2 that $z(t) - C_{19}$ can not be positive, hence, $\int_{G_1} |\nabla\mathbf{u}|^2\, dx < \infty$.

It follows that (2.21) is satisfied for all $\vec{\eta} \in \mathcal{H}(\Omega)$, and we can substitute there the test function $\vec{\eta} = \mathbf{u} - \mathbf{U}_t$ extended by zero into $\Omega\backslash G_1$. We can repeat once more all the arguments of Theorem 2.1. Since

$$\int_{G_1} \sum_k (u_k + w_k)\mathbf{u}\cdot\frac{\partial(\mathbf{u}-\mathbf{U}_t)}{\partial x_k}\, dx = \int_{\omega_{1t}} \sum_k (u_k + w_k)\mathbf{U}_t\cdot\frac{\partial(\mathbf{u}-\mathbf{U}_t)}{\partial x_k}\, dx \le$$
$$\le C_{20}\|\mathbf{u}+\mathbf{w}\|_{W_2^1(\omega_{1t})}\|\mathbf{U}_t\|_{W_2^1(\omega_t)}\|\nabla(\mathbf{u}-\mathbf{U}_t)\|_{L_2(\omega_t)} \le C_{21}\|\nabla\mathbf{u}\|^2_{L_2(\omega_t)}\,,$$

we arrive at the inequality

$$\xi(t) \le -\mu\xi'(t)\,, \qquad t \ge 2\,,$$

for the function

$$\xi(t) = \int_{G_1\backslash G_{1t}} \min(y_1^{(1)} - t, 1)\,|\nabla\mathbf{u}|^2\, dx\,.$$

Integration of this inequality gives $\xi(t) \le \xi(2)e^{-\frac{1}{\mu}(t-2)}$ which proves the theorem.

We observe at the conclusion that extensions and generalizations of the above results may be found in [36, 41, 42].

3 – Investigation of evolution problems in classes of vector fields with infinite energy integrals

In this section we consider problem (1.1)–(1.3), (1.15) giving up the condition that the "exits" G_j contain infinite cones, and also the condition of the boundedness of energy. We suppose that the domain Ω possesses the following properties:

1) $\Omega = \Omega_0 \cup G_1 \cup ... \cup G_m$, Ω_0 is bounded, G_j are unbounded "exits to infinity",

2) Every G_j can be represented as a union of bounded domains ω_{jk}: $G_j = \bigcup_{k=1}^{\infty} \omega_{jk}$. The domains $G_{jN} = \bigcup_{k=1}^{N} \omega_{jk}$. and $G_j \backslash G_{jN}$ are connected; a common part Σ_{jk} of the boundaries of ω_{jk} and ω_{jk+1} is a cross-section of G_j.

3) For every divergence free $\mathbf{u}(x)$, $x \in G_j$, vanishing on $\partial\Omega \cap \partial G_j$ and having a zero flux $\int_{\Sigma_j} \mathbf{u} \cdot \mathbf{n}\, dx = 0$, there exists a vector field $\mathbf{U}_k(x)$, $x \in \omega_{jk}$ satisfying the relations

$$\operatorname{div} \mathbf{U}_k = 0 (x \in \omega_{jk}), \quad \mathbf{U}_k|_{\Sigma_{jk}} = \mathbf{u}_k|_{\Sigma_{jk}}, \quad \mathbf{U}_k|_{\partial\omega_{jk} \backslash \Sigma_{jk}} = 0$$

and inequalities

$$\|\nabla \mathbf{U}_k\|_{L_2(\omega_{jk})} \le C_1 \|\nabla \mathbf{u}\|_{L_2(\omega_{jk})}, \quad \|\mathbf{U}_k\|_{L_2(\omega_{jk})} \le C_1 \|\mathbf{u}\|_{L_2(\omega_{jk})}. \tag{3.1}$$

The correspondence between $\mathbf{u}$ and $\mathbf{U}_k$ is linear.

4) Any vector field $\mathbf{u} \in W_2^1(\omega_{jk})$ vanishing on $\partial\omega_{jk} \cap S$ satisfies the inequalities

$$\begin{aligned} &\|\mathbf{u}\|_{L_6(\omega_{jk})} \le b_1 \|\nabla \mathbf{u}\|_{L_2(\omega_{jk})} \quad \text{for } n = 3, \\ &\|\mathbf{u}\|_{L_4(\omega_{jk})} \le b_2 \|\nabla \mathbf{u}\|_{L_2(\omega_{jk})}^{1-\frac{r}{4}} \|\mathbf{u}\|_{L_r(\omega_{jk})}^{\frac{r}{4}} \quad \text{for } n = 2, \end{aligned} \tag{3.2}$$

where $r \in [2, 4)$ and b_1, b_2 are independent of k. Analogous inequalities hold in the domain Ω_0.

5) In Ω there exist divergence free vector fields $\mathbf{A}_j$ satisfying the conditions

$$\mathbf{A}_j|_S = 0, \quad \int_{\Sigma_j} \mathbf{A}_j \cdot \mathbf{n}\, dS = 1, \quad j = 1, ..., m-1,$$

and having uniformly bounded norms $\|\mathbf{A}_j\|_{W_2^1(\Omega_0)}$, $\|\mathbf{A}_j\|_{W_2^1(\omega_{jk})}$ (vector fields of this type were constructed in Example 2).

Let us introduce the following notations

$$\Omega_k = \Omega_0 \cup G_{1k} \cup ... \cup G_{mk}, \quad Q_{kT} = \Omega_k \times (0, T);$$

$\phi(t)$: infinitely differentiable monotone function equal to 1 for $t \le \frac{1}{2}$ and to 0 for $t \ge 1$. For functions of the variable $t \in I_T \equiv (0,T)$ define a weighted space $L_2(I_T;\gamma)$, $\gamma \ge 0$, with the norm

$$\|u\|^2_{L_2(I_T;\gamma)} = \int_0^T e^{-2\gamma t}|u(t)|^2\,dt$$

and the space $D_0^{\frac{1}{2}}(I_T;\gamma)$ where the norm is defined by

$$\|u\|^2_{D_0^{\frac{1}{2}}(I_T;\gamma)} = \int_0^T e^{-2\gamma t}\left(\gamma|u|^2 + \frac{|u|^2}{t}\right)dt + \int_0^T e^{-2\gamma t}dt \int_0^t |u(t)-u(t-h)|^2\,\frac{dh}{h^2}\,.$$

For $T<\infty$, all the spaces $D_0^{\frac{1}{2}}(I_T;\gamma)$ coincide and the norms $\|u\|_{D_0^{\frac{1}{2}}(I_T;\gamma)}$ are equivalent. It is easy to verify that

$$\|u\|^2_{D_0^{\frac{1}{2}}(I_T;\gamma)} = \gamma\int_0^T e^{-2\gamma t}|u|^2\,dt + \int_{-\infty}^T e^{-2\gamma t}\,dt\int_0^\infty |u_0(t)-u_0(t-h)|^2\,\frac{dh}{h^2}\,,$$

where $u_0(t) = u(t)$ for $t>0$, $u_0(t)=0$ for $t<0$.

Let us define also some spaces of functions depending on $(x,t)\in Q_T$. We set $W_2^{1,0}(Q_T) = L_2(0,T;W_2^1(\Omega))$, $H^*(Q_T) = L_2(0,T;H^*(\Omega))$, $W_2^{1,1}(Q_T) = W_2^1(Q_T)$. Norms in these spaces are defined in a standard way, for instance,

$$\|u\|^2_{W_2^{1,0}(Q_T)} = \int_0^T \|u\|^2_{W_2^1(\Omega)}\,dt\,,$$

$$\|u\|^2_{W_2^{1,1}(Q_T)} = \|u\|^2_{W_2^{1,0}(Q_T)} + \|u_t\|^2_{L_2(Q_T)}\,.$$

We introduce also the spaces $L_2(Q_T;\gamma)$, $D_0^{0,\frac{1}{2}}(Q_T;\gamma) = L_2(\Omega;D_0^{\frac{1}{2}}(I_T;\gamma))$ and $D_0^{1,\frac{1}{2}}(Q_T;\gamma)$ with the norms

$$\|u\|^2_{L_2(Q_T;\gamma)} = \int_0^T e^{-2\gamma t}|u|^2\,dt\,,$$

$$\begin{aligned}\|u\|^2_{D_0^{0,\frac{1}{2}}(Q_T;\gamma)} &= \int_0^T e^{-2\gamma t}(\gamma\|u\|^2 + t^{-1}\|u\|^2)\,dt + \int_0^T e^{-2\gamma t}dt\int_0^t \|u(\cdot,t)-u(\cdot,t-h)\|^2\,\frac{dh}{h^2}\\ &= \gamma\int_0^T e^{-2\gamma t}\|u\|^2dt + \int_{-\infty}^T e^{-2\gamma t}dt\int_0^\infty \|u_0(\cdot,t)-u_0(\cdot,t-h)\|^2\frac{dh}{h^2}\,,\end{aligned} \tag{3.3}$$

$$\|u\|^2_{D_0^{1,\frac{1}{2}}(Q_T;\gamma)} = \|u\|^2_{D_0^{0,\frac{1}{2}}(Q_T;\gamma)} + \|\nabla u\|^2_{L_2(Q_T;\gamma)}\,, \tag{3.4}$$

where $\|u\| = \|u\|_{L_2(\Omega)}$. Norms corresponding to different $\gamma\ge 0$, are also equivalent in the case $T<\infty$. For $T=\infty$ they may be written in terms of the Laplace transform $\tilde{u}(x,t) = \int_0^\infty e^{-st}\,u(x,t)\,dt$ with $s=\gamma+i\xi_0$ which can also be considered as the Fourier transform of function $e^{-\gamma t}u(x,t)$ extended by zero into the half-space $t<0$. It can be shown (see [41]) that the norms (3.3), (3.4) are equivalent to $\int_{-\infty}^\infty |s|\,\|\tilde{u}(\cdot,s)\|^2\,d\xi_0$ and $\int_{-\infty}^\infty(|s|\|\tilde{u}\|^2 + \|\nabla\tilde{u}(\cdot,s)\|^2)\,d\xi_0$ respectively. This statement is also true for functions which need that vanish for $t<0$; we denote corresponding spaces by $D^{1,\frac{1}{2}}(\mathcal{G}_\infty,\gamma)$ and

$D^{0,\frac{1}{2}}(\mathcal{G}_\infty,\gamma)$, $\mathcal{G}_\infty = \Omega \times (-\infty,\infty)$. Norms in these spaces are also defined by formulas (3.3), (3.4) with the integration extended to all real $t \in (-\infty,\infty)$.

In the case $\gamma = 0$, we omit index γ in the definition of these spaces and norms.

Let $\wp = (1,2,...)$ and æ=(1,1,...). By $L_p(\Omega;\wp)$ and $L_p(\Omega;æ)$ we shall mean the spaces of function defined in Ω with finite norms

$$\|u\|^p_{L_p(\Omega;\wp)} = \sup_{k>0} k^{-1}\int_{\Omega_k} |u(x)|^p dx ,$$

$$\|u\|^p_{L_p(\Omega;æ)} = \max\left\{\int_{\Omega_0} |u(x)|^p dx,\ \sup_{k>0}\int_{\omega_{jk}} |u(x)|^p dx\right\} .$$

In a similar way are defined spaces $D_0^{1,\frac{1}{2}}(Q_T;\gamma;\wp)$, $D_0^{1,\frac{1}{2}}(Q_T;\gamma;æ)$, $H^*(Q_T;\gamma;\wp)$ etc.. For instance,

$$\|u\|^2_{D_0^{1,\frac{1}{2}}(Q_T;\gamma;\wp)} = \sup_{k>0} k^{-1}\|u(x)\|^2_{D_0^{1,\frac{1}{2}}(Q_{kT};\gamma)} ,$$

$$\|u\|^2_{D_0^{1,\frac{1}{2}}(Q_T;\gamma;æ)} = \max_{k>0}\left\{\|u\|^2_{D_0^{1,\frac{1}{2}}(Q_{0T};\gamma)}, \sup_{k>0}\|u\|^2_{D_0^{1,\frac{1}{2}}(\omega_{ik}\times(0,T);\gamma)}\right\} ,$$

$$\|u\|^2_{H^*(Q_T;\gamma;\wp)} = \sup_{k>0} k^{-1}\|u(x)\|^2_{H^*(Q_{kT};\gamma)} \equiv \sup_{k>0} k^{-1}\int_0^T e^{-2\gamma t}\|u(x)\|^2_{H^*(\Omega_k)} dt .$$

We shall study the problem

$$\begin{aligned}
&\mathbf{v}_t + (\mathbf{v}\cdot\nabla)\mathbf{v} - \nu\Delta\mathbf{v} + \nabla p = \mathbf{f}, \qquad \operatorname{div}\mathbf{v} = 0 ,\\
&\mathbf{v}|_{t=0} = \vec{\varphi}(x), \qquad \mathbf{v}|_{x\in S} = 0 ,\\
&\int_{\Sigma_j} \mathbf{v}\cdot\mathbf{n}\, dS = Q_j(t), \qquad j = 1,2,...,m-1 .
\end{aligned} \tag{3.5}$$

By a weak solution of this problem we mean a vector field $\mathbf{v}$ belonging to $L_2(0,T;W_2^1(\Omega'))$ in every bounded subdomain $\Omega' \subset \Omega$, satisfying the conditions

$$\mathbf{v}|_{x\in S} = 0, \qquad \int_{\Sigma_j} \mathbf{v}\cdot\mathbf{n}\, dS = Q_j(t),\ \ j = 1,2,...,m-1 , \tag{3.6}$$

and the integral identity

$$\begin{aligned}
\int_0^T dt \int_\Omega \left(-\mathbf{v}\cdot\vec{\eta}_t + \nu\nabla\mathbf{v} : \nabla\vec{\eta} - \sum_{k=1}^n v_k\mathbf{v}\cdot\frac{\partial\vec{\eta}}{\partial x_k}\right) dx =\\
= \int_0^T\int_\Omega \mathbf{f}\cdot\vec{\eta}\, dxdt + \int_\Omega \vec{\varphi}(x)\cdot\vec{\eta}(x,0)\, dx ,
\end{aligned} \tag{3.7}$$

for arbitrary solenoidal $\vec{\eta} \in W_2^{1,1}(Q_T)$ vanishing for $x \in S$ and $t = T$ and having a compact support for any fixed $t \in [0,T)$.

Theorem 3.1. *Suppose that $\Omega \in \mathbb{R}^n$, $n = 2,3$ satisfies the above conditions 1)–5), that $\mathbf{f} \in \mathcal{H}^*(Q_T;æ) \cap L_2(Q_T;æ)$, $\vec{\varphi} \in W_2^2(\Omega;æ) \cap \mathcal{H}^*(\Omega;æ)$, $Q_j \in W_2^1(0,T)$ and that the compatibility conditions hold*

$$\operatorname{div}\vec{\varphi} = 0, \qquad \vec{\varphi}|_{x\in S} = 0, \qquad \int_{\Sigma_j}\vec{\varphi}\cdot\mathbf{n}\, dS = Q_j(0),\ \ j = 1,2,...,m-1 .$$

If

$$\min(T^{\frac{1}{2}}, 1)M^2 \le \delta ,$$

where $\delta > 0$ *is a small constant and*

$$M^2 = \|\mathbf{f}\|^2_{\mathcal{H}^*(Q_T;æ)} + \|\mathbf{f}\|^2_{L_2(Q_T;æ)} + \sup_{x\in\Omega}|\varphi|^2 + \|\vec{\varphi}\|_{W_2^2(\Omega;æ)} + \|\vec{\varphi}\|^2_{\mathcal{H}^*(\Omega;æ)} + \sum_{j=1}^{m-1}\|Q_j\|^2_{W_2^1(0,T)} ,$$

then problem (3.5) has unique solution with the following properties (the case $T = \infty$ *is not excluded):* $\mathbf{v} - \vec{\varphi}\phi \in D_0^{1,\frac{1}{2}}(Q_T;æ)$, $\mathbf{v}_t \in L_2(Q_T;æ)$, $\nabla\mathbf{v} - \phi\nabla\vec{\varphi} \in D_0^{0,\frac{1}{2}}(Q_T;æ)$.

In the two-dimensional case it is possible to prove the solvability of problem (3.5) in a larger functional class and with less restrictions on the data.

Theorem 3.2. *Suppose that* $\Omega \subset \mathbb{R}^2$ *satisfies 1)–5),* $\mathbf{f} \in \mathcal{H}^*(\Omega;æ)$, $\vec{\varphi} \in W_2^1(\Omega;æ) \cap \mathcal{H}^*(\Omega;æ)$, $Q_j - \phi\int_{\Sigma_j}\vec{\varphi}\cdot\mathbf{n}\,dS \in D_0^{\frac{1}{2}}(0,T)$ *and the conditions* $\operatorname{div}\vec{\varphi} = 0$, $\vec{\varphi}|_{x\in S} = 0$ *hold.*
If

$$\|\mathbf{f}\|^2_{\mathcal{H}^*(Q_T;æ)} + \sum_{j=1}^{m}\Big\|Q_j - \phi\int_{\Sigma_j}\vec{\varphi}\cdot\mathbf{n}\;dS\Big\|^2_{D_0^{\frac{1}{2}}(0,T)} + \\ + \min(1,T)\big(\|\vec{\varphi}\|^2_{W_2^1(\Omega;æ)} + \|\vec{\varphi}\|^2_{\mathcal{H}^*(\Omega;æ)}\big) \le \delta , \qquad (3.8)$$

with a small $\delta > 0$, *then problem (3.5) has a weak solution such that* $\mathbf{v} - \phi\vec{\varphi} \in D_0^{1,\frac{1}{2}}(Q_T;æ)$. *The solution is unique in the class of vector fields with the following properties:* $\mathbf{v} - \phi\vec{\varphi} \in D_0^{1,\frac{1}{2}}(Q_T;æ)$ *and there exists such* $\rho > 0$ *that*

$$\max\Big(\int_\tau^{\tau+\rho}\|\mathbf{v}\|^4_{L_4(\Omega_0)}dt,\ \sup_k\int_\tau^{\tau+\rho}\|\mathbf{v}\|^4_{L_4(\omega_{jk})}dt\Big) \le \lambda, \quad \forall\tau \in (0, T-\rho) , \qquad (3.9)$$

with a certain small $\lambda > 0$.

Consider the following particular cases:

1) *Flow in a cylinder.*

The domain $\Omega = \mathbb{R}\times\Sigma$ has two "exits" to infinity, and additional conditions reduce to $\int_\Sigma v_1\,dS = Q_1(t)$. Conditions 1)–5) are absolutely evident. Since $\mathcal{H}^*(\Omega;æ) \subset L_2(\Omega;æ)$, the condition $\vec{\varphi} \in \mathcal{H}^*(\Omega;æ)$ is superfluous.

Problem (3.2) in a cylindrical tube was considered in the paper [29] by a slightly different techniques and with less restrictive hypothesis $\vec{\varphi} \in W_2^1(\Omega;æ)$, (the condition $\vec{\varphi} \in W_2^2(\Omega;æ)$ is too strong and may be relaxed.)

2) *Flow in* $\Omega_H \subset \mathbb{R}^2$ (see §1).

Additional condition has a form $\int_\Sigma v_1\,dS = Q_1(t)$ where $\Sigma = \{x_1 = 0,\ |x_2| < 1\}$. For $Q_1 \not\equiv 0$, the consideration of this problem in the class of vector fields with a finite energy is impossible. Let $\Omega_0 = \{x \in \Omega : |x| < 2\}$ $\omega_{1k} = \{2^k < |x| < 2^{k+1},\ x_1 > 0\}$, $\omega_{2k} = \{2^k < |x| < 2^{k+1},\ x_1 < 0\}$. All the hypotheses 3)–5) are easily verified. The construction of $\mathbf{U}_k$ reduces to the problem (1.8). The fact that the constant C_1 in

(1.11) may be taken common for all ω_{ik} is easily proved by a homothetic coordinate transformation. The same is true for estimates (3.1). Vector field $\mathbf{A}$ constructed in Example 1 decays at infinity as $|x|^{-1}$, and belongs to $W_2^1(\Omega_H;æ)$.

A detailed proof of Theorem 3.1 may be found in [41], therefore we concentrate our attention on Theorem 3.2. It is also formulated in [41] as Theorem 4.7.

The proof of Theorem 3.2 is based on the investigation of a linear problem which is carried out independently of the dimension of Ω.

Let $\Omega \subset \mathbb{R}^n$, $n=2,3$. Consider the problem of finding a vector field $\mathbf{v} \in L_2(0,T;W_2^1(\Omega'))$ (Ω' is an arbitrary bounded subdomain of Ω) satisfying the conditions

$$\operatorname{div}\mathbf{v}=0\,,\qquad \mathbf{v}|_{x\in S}=0\,,\qquad \int_{\Sigma_j}\mathbf{v}\cdot\mathbf{n}\,dS=0,\quad (j=1,2,...,m-1) \tag{3.10}$$

and the integral identity

$$\int_0^T dt\int_\Omega(-\mathbf{v}\cdot\vec{\eta}_t+\nu\nabla\mathbf{v}:\nabla\vec{\eta})\,dx=\int_0^T dt\int_\Omega\Big(\mathbf{f}\cdot\vec{\eta}+\mathbf{g}\cdot\vec{\eta}+\sum_{j=1}^n\mathbf{f}_j\cdot\vec{\eta}_{x_j}-\mathbf{f}_0\cdot\vec{\eta}_t\Big)\,dx\,, \tag{3.11}$$

with given $\mathbf{f}$, $\mathbf{f}_j$, $\mathbf{g}$, $\mathbf{f}_0$, and with arbitrary divergence free $\vec{\eta}\in W_2^{1,1}(Q_T)$ with a compact support vanishing for $x\in S$ and for $t=T$.

In addition, consider an auxiliary problem of determination of a vector field $\mathbf{v}^{(N)}\in D_0^{1,\frac12}(Q_{NT})$ ($Q_{NT}=\Omega_N\times(0,T)$) satisfying the condition

$$\operatorname{div}\mathbf{v}^{(N)}=0\,,\qquad \mathbf{v}^{(N)}|_{x\in\partial\Omega_N}=0$$

and the integral identity

$$\int_0^T dt\int_{\Omega_N}\Big(-\mathbf{v}^{(N)}\cdot\vec{\eta}_t+\nu\nabla\mathbf{v}^{(N)}:\nabla\vec{\eta}\Big)\,dx=$$

$$=\int_0^T dt\int_{\Omega_N}\Big(\mathbf{f}\cdot\vec{\eta}+\mathbf{g}\cdot\vec{\eta}+\sum_{j=1}^n\mathbf{f}_j\cdot\frac{\partial\vec{\eta}}{\partial x_j}-\mathbf{f}_0\cdot\vec{\eta}_t\Big)\,dx\,, \tag{3.12}$$

with arbitrary solenoidal $\vec{\eta}\in W_2^{1,1}(Q_{NT})$ vanishing when $x\in\partial\Omega_N$ and also when $t=T$.

We shall prove the following theorem.

Theorem 3.3. *Suppose that hypotheses 1)-3) are fulfilled and that* $\mathbf{f}\in\mathcal{H}^*(Q_T;\gamma;\wp)$, $\mathbf{g}$, $\mathbf{f}_j\in L_2(Q_T;\gamma;\wp)$, $\mathbf{f}_0\in D_0^{0,\frac12}(Q_T;\gamma;\wp)$ *with* $\gamma>0$, $T\le\infty$, $\mathbf{g}(x,t)=0$ *when* $t\le t_1$ *and* $t\ge t_2$ $(0<t_1<t_2\le T)$. *Then there exists a unique* $\mathbf{v}\in D_0^{1,\frac12}(Q_T;\gamma;\wp)$ *satisfying conditions (3.10) and integral identity (3.11). If* $\mathbf{f}\in\mathcal{H}^*(Q_T;\gamma;æ)$ $\mathbf{g}$, $\mathbf{f}_j\in L_2(Q_T;\gamma;\wp)$, $\mathbf{f}_0\in D_0^{0,\frac12}(Q_T;\gamma;\wp)$, *then* $\mathbf{v}\in D_0^{1,\frac12}(Q_T;\gamma;æ)$ *and* $\mathbf{v}(x,t)$ *satisfies the inequalities*

$$\|\mathbf{v}\|^2_{D_0^{1,\frac12}(Q_T;\gamma;\wp)}\le C_2\Big(\|\mathbf{f}\|^2_{\mathcal{H}^*(Q_T;\gamma;\wp)}+\|\mathbf{g}\|^2_{L_2(Q_T;\gamma;\wp)}+\sum_{j=1}^n\|\mathbf{f}_j\|^2_{L_2(Q_T;\gamma;\wp)}+\|\mathbf{f}_0\|^2_{D_0^{0,\frac12}(Q_T;\gamma;\wp)}\Big)\,, \tag{3.13}$$

$$\|\mathbf{v}\|^2_{D_0^{1,\frac12}(Q_T;\gamma;æ)}\le C_3\Big(\|\mathbf{f}\|^2_{\mathcal{H}^*(Q_T;\gamma;æ)}+\|\mathbf{g}\|^2_{L_2(Q_T;\gamma;æ)}+\sum_{j=1}^n\|\mathbf{f}_j\|^2_{L_2(Q_T;\gamma;æ)}+\|\mathbf{f}_0\|^2_{D_0^{0,\frac12}(Q_T;\gamma;æ)}\Big)\,, \tag{3.14}$$

with C_2, C_3 independent of T. Analogous assertions are true for an auxiliary problem.

In equalities (3.13), (3.14) are consequences of integral estimates "of Saint-Venant's type" obtained with the aid of the following lemma.

Lemma 3.1. *Let positive numbers y_k, $k = 1, ..., M \leq \infty$ satisfy the inequalities $y_k \leq y_{k+1}$ and*

$$y_k \leq b(y_{k+1} - y_k) + Bk\,, \quad b, B > 0\,, \tag{3.15}$$

If $y_M \leq B_1 M$ or $(\frac{b}{b+1})^N y_N \underset{N\to\infty}{\longrightarrow} 0$ (in the case $M = \infty$), then

$$y_k \leq C_4(B + B_1)k\,. \tag{3.16}$$

Proof: We rewrite (3.15) in the form $y_k \leq \mu y_{k+1} + B_2 k$ with $\mu = \frac{b}{b+1} \in (0,1)$ and $B_2 = \frac{B}{b+1}$. After iteration we obtain

$$y_k \leq B_2(k + \mu(k+1) + ... + (M-1)\mu^{M-1-k}) + \mu^{M-k} y_M$$

(in the case $M = \infty$ the last term vanishes).

Let $\theta \in (1, \frac{1}{\mu})$. Clearly, there exists a constant $C_5 \geq 1$ independent of k and $s \geq 1$ such that $k + s \leq C_5 \theta^s k$. Hence,

$$y_k \leq C_5(B_2 + B_1)k[1 + \mu\theta + ... + (\mu\theta)^{M-k}] \leq \frac{B_1 + B_2}{1 - \mu\theta} C_5 k\,.$$

Now, we prove *a priori* estimates for auxiliary problem.

Lemma 3.2. *Suppose that $\mathbf{v}^{(N)} \in D_0^{1,\frac{1}{2}}(Q_{N\infty};\gamma)$ satisfies the identity (3.12) with $T = \infty$. Then*

$$\|\mathbf{v}^{(N)}\|^2_{D_0^{1,\frac{1}{2}}(Q_{N\infty};\gamma;\wp)} \leq C_6\Big(\|\mathbf{f}\|^2_{\mathcal{H}^\bullet(Q_\infty;\gamma;\wp)} + \|\mathbf{g}\|^2_{L_2(Q_\infty;\gamma;\wp)} + \sum_{j=1}^{2} \|\mathbf{f}_j\|^2_{L_2(Q_\infty;\gamma;\wp)} + \|\mathbf{f}_0\|^2_{D_0^{0,\frac{1}{2}}(Q_\infty;\gamma;\wp)}\Big) \equiv C_6 \mathcal{F}_\wp\,, \tag{3.17}$$

$$\|\mathbf{v}^{(N)}\|^2_{D_0^{1,\frac{1}{2}}(Q_{N\infty};\gamma;æ)} \leq C_7\Big(\|\mathbf{f}\|^2_{\mathcal{H}^\bullet(Q_\infty;\gamma;æ)} + \|\mathbf{g}\|^2_{L_2(Q_\infty;\gamma;æ)} + \sum_{j=1}^{2} \|\mathbf{f}_j\|^2_{L_2(Q_\infty;\gamma;æ)} + \|\mathbf{f}_0\|^2_{D_0^{0,\frac{1}{2}}(Q_\infty;\gamma;æ)}\Big) \equiv C_7 \mathcal{F}_æ\,, \tag{3.18}$$

where $Q_{N\infty} = \Omega_N \times [0, \infty)$.

Proof: We replace $\vec{\eta}$ by $e^{-2\gamma t}\vec{\eta}$ in (3.12) and make the Laplace transform $\tilde{\mathbf{v}}(x,s) = \int_0^\infty \mathbf{v}(x,t)e^{-st}\,dt$ which can be considered as the Fourier transform of the function $\mathbf{v}(x,s)e^{-\gamma t}$ $\gamma = \text{Re}\ s$, extended by zero into the half-space $t < 0$. We extend also $\vec{\eta}(x,t)$ into this half-space in such a way that the Fourier transform of $\vec{\eta}(x,t)e^{-\gamma t}$ has a sense.

In virtue of the Parseval's formula we can rewrite (3.12) in the form

$$\int_{-\infty}^{\infty} d\xi_0 \int_{\Omega_N} \left(s\tilde{\mathbf{v}}^{(N)} \cdot \vec{\bar{\eta}} + \nu \nabla \tilde{\mathbf{v}}^{(N)} : \nabla \vec{\bar{\eta}} \right) dx =$$

$$= \int_{-\infty}^{\infty} d\xi_0 \int_{\Omega_N} \left(\tilde{\mathbf{f}} \cdot \vec{\bar{\eta}} + \tilde{\mathbf{g}} \cdot \vec{\bar{\eta}} + \sum_{j=1}^{2} \tilde{\mathbf{f}}_j \cdot \vec{\bar{\eta}}_{x_j} + s\tilde{\mathbf{f}}_0 \cdot \vec{\bar{\eta}} \right) dx \; ; \qquad (3.19)$$

here

$$s = \gamma + i\xi_0, \quad \tilde{\mathbf{f}} \cdot \vec{\bar{\eta}} = \tilde{f}_1 \cdot \bar{\eta}_1 + \tilde{f}_2 \cdot \bar{\eta}_2, \quad \nabla \tilde{\mathbf{v}}^{(N)} : \nabla \vec{\bar{\eta}} = \sum_{i,k=1}^{2} \frac{\partial \tilde{v}_i^{(N)}}{\partial x_k} \frac{\partial \bar{\eta}_i^{(N)}}{\partial x_k} \, .$$

Because of the equivalence of the norms $\|\vec{\eta}\|_{D^{1,\frac{1}{2}}(\mathcal{G}_{N,\infty},\gamma)}$ and $(\int_{-\infty}^{\infty}[|s|\|\vec{\tilde{\eta}}\|^2+\|\nabla\vec{\tilde{\eta}}\|^2]d\xi_0)^{\frac{1}{2}}$ ($\mathcal{G}_{N\infty} = \Omega_N \times \mathbb{R}$), (3.19) is meaningful for all $\vec{\eta} \in D^{1,\frac{1}{2}}(\mathcal{G}_{N,\infty},\gamma)$.

We set $\vec{\tilde{\eta}} = \tilde{\mathbf{v}}$, then $\vec{\tilde{\eta}} = i \text{ sgn } \xi_0 \tilde{\mathbf{v}}$ and compute real parts of both integrals in (3.19). The right-hand side may be easily estimated by the Cauchy inequality. In particular,

$$\left| \int_{-\infty}^{\infty} d\xi_0 \int_{\Omega_N} \tilde{\mathbf{g}} \cdot \vec{\bar{\eta}} \, dx \right| \le \left(\int_{-\infty}^{\infty} |s| \|\vec{\tilde{\eta}}\|^2 d\xi_0 \right)^{\frac{1}{2}} \left(\int_{-\infty}^{\infty} \|\tilde{\mathbf{g}}\|^2 \frac{d\xi_0}{|s|} \right)^{\frac{1}{2}} \qquad (3.20)$$

$$\le \frac{\sqrt{2\pi}}{\sqrt{\gamma}} \left(\int_{-\infty}^{\infty} |s| \|\vec{\tilde{\eta}}\|^2 d\xi_0 \right)^{\frac{1}{2}} \left(\int_{t_1}^{t_2} e^{-2\gamma t} \|\tilde{\mathbf{g}}\|^2 dt \right)^{\frac{1}{2}} ,$$

where $\| \cdot \| = \| \cdot \|_{L_2(\Omega_N)}$. After easy calculations we arrive at the inequality

$$\int_{-\infty}^{\infty} (|s| \|\tilde{\mathbf{v}}^{(N)}\|^2 + \nu \|\nabla \tilde{\mathbf{v}}^{(N)}\|^2) d\xi_0 \le C_8 \Big(\|\mathbf{f}\|^2_{\mathcal{H}^*(Q_{N\infty};\gamma)} + \|\mathbf{g}\|^2_{L_2(Q_{N\infty};\gamma)} \qquad (3.21)$$

$$+ \sum_{j=1}^{2} \|\tilde{\mathbf{f}}_j\|^2_{L_2(Q_{N\infty};\gamma)} + \|\mathbf{f}_0\|^2_{L_2(Q_{N\infty};\gamma)} \Big) \le C_8 N \mathcal{F}_{\wp} \, .$$

Next, we estimate $\int_{-\infty}^{\infty} d\xi_0 \int_{\Omega_k} (|s| \|\tilde{\mathbf{v}}^{(N)}\|^2 + \nu \|\nabla \tilde{\mathbf{v}}^{(N)}\|^2) \, dx$ with arbitrary k between 1 and $N-1$.

We substitute into (3.19) the test function $\vec{\tilde{\eta}} = \widetilde{\mathbf{U}}_k$ and $\vec{\tilde{\eta}} = i \text{ sgn } \xi_0 \widetilde{\mathbf{U}}_k$, where $\mathbf{U}_k$ is defined by $\mathbf{v}^{(N)}$ as explained in the property 3) of Ω. Since $\mathbf{U}_k$ depends on $\mathbf{v}^{(N)}$ as a linear function, estimates (3.1) hold for the transformed $\mathbf{v}^{(N)}$ and $\mathbf{U}_k$ as well, so we have

$$\int_{-\infty}^{\infty} d\xi_0 \int_{\omega_{ik}} (|s| \|\widetilde{\mathbf{U}}_k\|^2 + \nu \|\nabla \widetilde{\mathbf{U}}_k\|^2) \, dx \le C_1^2 \int_{-\infty}^{\infty} d\xi_0 \int_{\omega_{ik}} (|s| \|\tilde{\mathbf{v}}_k^{(N)}\|^2 + \nu \|\nabla \tilde{\mathbf{v}}_k^{(N)}\|^2) \, dx \, .$$

Repeating the arguments of Theorem 2.1, we arrive at the inequality (3.15) for $y_k = \int_{-\infty}^{\infty} d\xi_0 \int_{\Omega_k} (|s| |\tilde{\mathbf{v}}_k^{(N)}|^2 + \nu |\nabla \tilde{\mathbf{v}}_k^{(N)}|^2) \, dx$ with the constant B proportional to $\mathcal{F}_{\wp}$. Inequality (3.16) is equivalent to (3.13).

To prove (3.14), we set $\vec{\tilde{\eta}}(x,s) = \widetilde{\mathbf{U}}_{i,k+j} - \widetilde{\mathbf{U}}_{i,k-j}$, where $k \ge 0$ is a fixed integer, $j = 1, \dots, k-1$, $\widetilde{\mathbf{U}}_{i,k}(x,s) = \widetilde{\mathbf{U}}_k(x,s)$ for $x \in G_i$; $\widetilde{\mathbf{U}}_{ik} = \tilde{\mathbf{v}}^{(N)}(x,s)$ for $x \in \Omega \backslash G_i$. Let $\mathcal{G}_j = (G_{i,k+j+1} \backslash G_{i,k-j}) \times \mathbb{R}$. Clearly

$$\|\mathbf{g}\|^2_{L_2(\mathcal{G}_j;\gamma)} \le 2(j+1) \|\mathbf{g}\|^2_{L_2(Q_\infty;\gamma;æ)} \, ,$$

$$\|\mathbf{f}_0\|^2_{D_0^{0,\frac{1}{2}}(\mathcal{G}_j;\gamma)} \le 2(j+1) \|\mathbf{f}_0\|^2_{D_0^{0,\frac{1}{2}}(\mathcal{G}_j;\gamma;æ)} \, ,$$

It is also not hard to show (see [41], Lemma 3.1) that

$$\|\mathbf{f}\|^2_{\mathcal{H}^*(\mathcal{G}_j;\gamma)} \le C_9(j+1)\|\mathbf{f}\|^2_{\mathcal{H}^*(\mathcal{G}_j;\gamma;æ)} \ .$$

Repeating again arguments of Theorem 2.1, we show that

$$z_j = \int_{-\infty}^{\infty} \int_{G_{i,k+j+1}\setminus G_{i,k-j}} (|s|\|\tilde{\mathbf{v}}_k^{(N)}\|^2 + \|\nabla\tilde{\mathbf{v}}_k^{(N)}\|^2) \ dx d\xi_0$$

satisfy inequalities similar to (3.15), namely,

$$z_j \le b(z_{j+1} - z_j) + C_{10}\mathcal{F}_æ j \ ,$$

where $\mathcal{F}_æ$ is the sum of norms in the right-hand side of (3.14). In addition, (3.13) implies $z_k \le C_{11}\mathcal{F}_æ k$, hence, by Lemma 3.1, $z_j \le C_{12}\mathcal{F}_æ j$, $j = 1, ..., k-1$. Taking $j = 1$ we arrive at (3.14), and complete the proof of the lemma.

Remark. The condition $\gamma > 0$ was used only in (3.20), so in the case $\mathbf{g} = 0$ Lemma 3.2 and Theorem 3.3 hold for $\gamma \ge 0$.

Proof of Theorem 3.3: We establish first of all the solvability of auxiliary problem assuming that $T = \infty$ (if $T < \infty$, we may extend $\mathbf{f}$, $\mathbf{g}$, $\mathbf{f}_j$, $\mathbf{f}_0$, by zero for $t < 0$ and then as even functions of $t - T$ for $t > T$ which will preserve their differentiability properties). We may also assume that $\frac{\partial \mathbf{f}_0}{\partial t} \in L_2(Q_{N\infty})$. Let $\{\mathbf{a}_k(x)\}_{k=1,2,...}$ be a basis in $\mathcal{H}(\Omega_N)$ which is orthonormal in $L_2(\Omega_N)$ and let $\mathbf{v}_m^{(N)}(x,t) = \sum_{k=1}^m c_{km}(t)\mathbf{a}_k(x)$ be Galerkin approximate solutions of auxiliary problem whose coefficients $c_{km}(t)$ are found from the system of differential equations

$$\frac{dc_{km}(t)}{dt} = \int_{\Omega_N} \frac{\partial \mathbf{v}^{(N)}}{\partial t} \cdot \mathbf{a}_k(x) \ dx \tag{3.22}$$

$$= -\nu \int_{\Omega_N} \nabla \mathbf{v}^{(N)} : \nabla \mathbf{a}_k \ dx + \int_{\Omega_N} \left[(\mathbf{f} + \mathbf{g}) \cdot \mathbf{a}_k + \frac{\partial \mathbf{f}_0}{\partial t} \cdot \mathbf{a}_k + \sum_{j=1}^2 \mathbf{f}_j \cdot \frac{\partial \mathbf{a}_k}{\partial x_j} \right] dx$$

and initial condition $c_{km}(0) = 0$. After multiplication of (3.22) by smooth functions $d_{km}(t)$ and integration with respect to t we easily show that $\mathbf{v}^{(N)}$ satisfies the identities (3.12) and (3.19) with arbitrary $\vec{\eta}$ of the form $\vec{\eta} = \sum_{k=1}^m d_{km}(t)\mathbf{a}_k(x)$. In (3.19) we may set $\tilde{\vec{\eta}} = \tilde{\mathbf{v}}^{(N)}$ and $\tilde{\vec{\eta}} = i \ \mathrm{sgn} \ \xi_0 \tilde{\mathbf{v}}^{(N)}$ and prove estimate (3.21) for $\tilde{\mathbf{v}}^{(N)}$ which is uniform in m. A standard passage to the limit proves the solvability of auxiliary problem.

Estimates (3.17), and (3.18) for $\mathbf{v}^{(N)}$ which are uniform with respect to N permit us to pass to a weak limit as $N \to \infty$ in the identity (3.12) with a fixed $\vec{\eta}$ having a compact support. The limiting vector field will satisfy the identity (3.7) and the estimates (3.17), (3.18).

It remains to prove the uniqueness of the solution $\mathbf{v} \in D_0^{1,\frac{1}{2}}(Q_T, \wp)$. Let $T < \infty$ and let $\mathbf{v} \in D_0^{1,\frac{1}{2}}(Q_T, \wp)$ satisfy the identity (3.7) with zero right-hand side. If we replace $\vec{\eta}$ by $h(t)\vec{\eta}$ where $h \in C^\infty(\mathbb{R})$, $h(t) = 1$ for $t \le T - \delta$ and $h(t) = 0$ for $t \ge T - \frac{\delta}{2}$, we arrive at

$$\int_0^\infty dt \int_\Omega (-\mathbf{w} \cdot \vec{\eta}_t + \nu \nabla \mathbf{w} : \nabla \vec{\eta}) \ dx = \int_0^\infty dt \int_\Omega \mathbf{g}(x,t) \cdot \vec{\eta} \ dx \ ,$$

where $\mathbf{w} = \mathbf{v}h$ and $\mathbf{g} = \mathbf{v}h'$. As we have already seen it follows that

$$\int_{-\infty}^{\infty} d\xi_0 \int_{\Omega_k} (|s||\widetilde{\mathbf{w}}|^2 + \nu|\nabla\widetilde{\mathbf{w}}|^2)\, dx \le \frac{C_{11}}{\gamma} k \sup_{m\ge 0}\Big(\int_{T-\delta}^{T-\frac{\delta}{2}} e^{-2\gamma t}\|\mathbf{v}\|^2_{L_2(\Omega_m)}\, dt\; m^{-1}\Big),$$

hence,

$$\int_0^{T-\delta} e^{-2\gamma t}\|\mathbf{v}\|^2_{L_2(\Omega_k)}\frac{dt}{t} \le \frac{C_{11}T}{\gamma} e^{-2\gamma(T-\delta)} \sup_{m\ge 0}\Big(\int_0^T \|\mathbf{v}\|^2_{L_2(\Omega_m)}\frac{dt}{t}\; m^{-1}\Big)\,.$$

If we multiply this by $e^{2\gamma(T-\delta)}$ and let $\gamma \to \infty$ we see immediately that $\mathbf{v}(x,t) = 0$ for $t \in (T-\delta)$ with arbitrary $\delta > 0$. The proof is completed.

Let us turn to nonlinear problem. We preface the proof of Theorem 3.2 with an auxiliary proposition.

Lemma 3.3. *Any $v \in W_2^{\frac{1}{2}}(\mathbb{R}_+) \cap L_r(\mathbb{R}_+)$, $\mathbb{R}_+ = \{t > 0\}$, satisfies the inequality*

$$\|v\|_{L_q(\mathbb{R}_+)} \le C_{12}\Big(\int_0^\infty dt \int_0^\infty |v(t+h)-v(t)|^2\frac{dh}{h^2}\Big)^{\frac{1}{2}(1-\frac{r}{q})} \|v\|^{\frac{r}{q}}_{L_r(\mathbb{R}_+)}\,, \tag{3.23}$$

where $q \ge \max(2,r)$, $r \ge 1$.

Proof: We may suppose that v is continuously differentiable and make use of the identity proved by V. P. Il'in in [15], namely,

$$v(t) = \frac{1}{h}\int_0^h v(t+\xi)d\xi - \int_0^h \frac{d\xi}{\xi^2}\int_0^\xi [v(t+\xi)-v(t+s)]ds\,, \qquad \forall h > 0\,.$$

From this identity we easily obtain

$$|v(t)| \le \frac{1}{h}\int_0^h |v(t+\xi)|d\xi + \int_0^h \frac{d\xi}{\xi^{\frac{1}{2}}}\Big(\int_0^\infty |v(t+\xi)-v(t+\xi-s)|^2\frac{ds}{s^2}\Big)^{\frac{1}{2}}\,.$$

The terms in the right-hand side are convolutions of $|v(t)|$ and $(\int_0^\infty |v(t)-v(t-s)|^2\frac{ds}{s^2})^{\frac{1}{2}}$ with kernels $\frac{1}{h}\chi_h(t)$ and $\frac{1}{\sqrt{t}}\chi_h(t)$, respectively ($\chi_h(t)$ is a characteristic function of the interval $(0,h)$). Application of the Young's inequality leads to

$$\|v\|_{L_q(\mathbb{R}_+)} \le C_{13}h^{-\frac{1}{r}+\frac{1}{q}}\|v\|_{L_r(\mathbb{R}_+)} + C_{14}h^{\frac{1}{q}}\left(\int_0^\infty dt\int_0^\infty |v(t+s)-v(t)|^2\frac{ds}{s^2}\right)^{\frac{1}{2}}\,,$$

$\forall h > 0$. Minimizing the right-hand side with respect to h, we arrive at (3.23), and the lemma is proved.

Corollary. *Let $\Omega' = \Omega_0$ or $\Omega' = \omega_{ik}$, $\Omega \subset \mathbb{R}^2$. For arbitrary function $u \in D_0^{1,\frac{1}{2}}(Q'_T)$, $Q'_T = \Omega' \times (0,T)$, vanishing for $x \in S$, there holds the inequality*

$$\int_0^T \|u\|^4_{L_4(\Omega')}dt \le C_{15}\|\nabla u\|^2_{L_2(Q'_T)}\|u\|^2_{D_0^{1,\frac{1}{2}}(Q'_T)}\,. \tag{3.24}$$

Proof: In virtue of (3.2) and of the Minkowskii inequality,

$$\int_0^T \|u\|^4_{L_4(\Omega')} dt \le b_2^4 \int_0^T \|\nabla u\|^{4-r}_{L_2(\Omega')} \|u\|^r_{L_r(\Omega')} dt$$

$$\le b_2^4 \left(\int_0^T \|\nabla u\|^2 dt \right)^{\frac{4-r}{2}} \int_{\Omega'} dx \left(\int_0^T |u|^{\frac{2r}{r-2}} dt \right)^{\frac{r-2}{2}} .$$

Now we apply Lemma 3.3 to the function $u_0(x,t)$ which is defined for $t \in (-\infty, T)$ as zero extension of $u(x,t)$ into the domain $t < 0$. This leads to

$$\int_{\Omega'} dx \left(\int_0^T |u|^{\frac{2r}{r-2}} dt \right)^{\frac{r-2}{2}} \le C_{12}^2 \int_{\Omega'} \|u_0\|^{4-r}_{D_0^{\frac{1}{2}}(\mathbb{R}_T)} \|u_0\|^{2(r-2)}_{L_4(\mathbb{R}_T)} dx$$

$$\le C_{16} \|u\|^{4-r}_{D_0^{0,\frac{1}{2}}(Q'_T)} \|u\|^{2(r-2)}_{L_4(Q'_T)} , \qquad \mathbb{R}_T = (-\infty, T) .$$

From this and the preceding inequality the estimate (3.23) follows. The lemma is proved.

Let us give another variant of the estimate of the left-hand side in (3.24). Inequality (3.23) with $r = 2$ yields

$$\int_0^T \|u\|^4_{L_4(\Omega')} dt = \int_{\Omega'} dx \int_{-\infty}^T \|u_0\|^4_{L_4(\mathbb{R}_T)} dt \tag{3.25}$$

$$\le C_{12}^4 \int_{\Omega'} dx \int_{-\infty}^T dt \int_0^\infty |u_0(x,t) - u_0(x,t-h)|^2 \frac{dh}{h^2} \sup_{x \in \Omega'} \|u(x,\cdot)\|^2_{L_2(0,T)}$$

$$\le C_{12}^4 \int_0^T \|u(\cdot,t)\|^2_{L_\infty(\Omega')} dt \; \|u\|^2_{D_0^{0,\frac{1}{2}}(Q'_T)} .$$

Proof of Theorem 3.2: Suppose that $T = \infty$. Let

$$\mathbf{a}(x,t) = \sum_{j=1}^m (Q_j(t) - \phi(t) Q_{j0}) \mathbf{A}_j(x) , \qquad Q_{j0} = \int_\Sigma \vec{\varphi} \cdot \mathbf{n} \, dS ,$$

$$\mathbf{b}(x,t) = \vec{\varphi}(x)\phi(t) + \mathbf{a}(x,t) , \qquad \mathbf{u}(x,t) = \mathbf{v}(x,t) - \mathbf{b}(x,t) .$$

For a new unknown vector field $\mathbf{u}$, (3.6) and (3.7) imply

$$\operatorname{div} \mathbf{u} = 0 , \qquad \mathbf{u}|_{x \in S} = 0 , \qquad \int_{\Sigma_j} \mathbf{u} \cdot \mathbf{n} \, dS = 0 \tag{3.26}$$

and

$$\int_0^T dt \int_\Omega \left[-\mathbf{u} \cdot \vec{\eta}_t + \nu \nabla \mathbf{u} : \nabla \vec{\eta} - \sum_{k=1}^2 (b_k \mathbf{u} + u_k \mathbf{b} + u_k \mathbf{u}) \cdot \frac{\partial \vec{\eta}}{\partial x_k} \right] dx =$$

$$= \int_0^T dt \int_\Omega \left[(\mathbf{f} - \phi'_t(t) \vec{\varphi}) \cdot \vec{\eta} - \mathbf{a} \cdot \vec{\eta}_t - \nu \nabla \mathbf{b} : \nabla \vec{\eta} + \sum_{k=1}^2 b_k \mathbf{b} \cdot \frac{\partial \vec{\eta}}{\partial x_k} \right] dx \equiv \mathcal{L}[\vec{\eta}] .$$

Clearly, $\mathbf{a} \in D_0^{1,\frac{1}{2}}(Q_T, æ)$, $\vec{\varphi}\phi'_t \in L_2(Q_T, æ)$ (this vector field vanishes for $t \le \frac{1}{2}$ and $t \ge 1$), $\nabla \mathbf{b} \in L_2(Q_T; æ)$ and $b_k \mathbf{b} \in L_2(Q_T; æ)$, since in virtue of (3.24) and (3.2)

$$\|b_k \mathbf{b}\|^2_{L_2(Q_T;æ)} \le 4(\| |\mathbf{a}|^2 \|^2_{L_2(Q_T;æ)} + \| |\vec{\varphi}|^2 \phi^2 \|^2_{L_2(Q_T;æ)})$$

$$\le C_{17}(\|\mathbf{a}\|^4_{D_0^{1,\frac{1}{2}}(Q_T,æ)} + \|\vec{\varphi}\|^2_{W_2^1(\Omega,æ)} \min(1,T)) .$$

We construct $\mathbf{u}$ by successive approximations. We set $\mathbf{u}^{(0)} = 0$ and find $\mathbf{u}^{(m+1)}$ $(m \geq 0)$ successively as vector fields satisfying conditions (3.26) and integral identities

$$\int_0^T dt \int_\Omega \left[-\mathbf{u}^{(m+1)} \cdot \vec{\eta}_t + \nu \nabla \mathbf{u}^{(m+1)} : \nabla \vec{\eta}\right] dx =$$

$$= \mathcal{L}[\vec{\eta}] + \sum_{k=1}^{2} \int_0^T dt \int_\Omega (b_k \mathbf{u}^{(m)} + u_k^{(m)} \mathbf{b} + u_k^{(m)} \mathbf{u}^{(m)}) \cdot \frac{\partial \vec{\eta}}{\partial x_k} \, dx \,, \tag{3.27}$$

with arbitrary solenoidal $\vec{\eta} \in W_2^{1,1}(Q_T)$ vanishing for $x \in S$ and also for $t = T$ and having a compact support.

We have

$$\|u_k^{(m)} \mathbf{u}^{(m)}\|_{L_2(Q_T;æ)} \leq C_{17} \|\mathbf{u}^{(m)}\|^2_{D_0^{1,\frac{1}{2}}(Q_T;æ)}$$

and

$$\begin{aligned} \|b_k \mathbf{u}^{(m)}\|_{L_2(Q_T;æ)} &\leq \|a_k \mathbf{u}^{(m)}\|_{L_2(Q_T;æ)} + \|\varphi_k \phi \mathbf{u}^{(m)}\|_{L_2(Q_T;æ)} \\ &\leq C_{17}(\|\mathbf{a}\|_{D_0^{1,\frac{1}{2}}(Q_T;æ)} + \min(1, T^{\frac{1}{2}}) \|\vec{\varphi}\|_{W_2^{\frac{1}{2}}(Q_T;æ)}) \|\mathbf{u}^{(m)}\|_{D_0^{1,\frac{1}{2}}(Q_T;æ)} \,. \end{aligned}$$

The same estimate holds for $\|u_k^{(m)} \mathbf{b}\|_{L_2(Q_T;æ)}$, hence, in virtue of Theorem 3.3 (with $\gamma = 0$, see the Remark above), all the approximations are well defined and they satisfy the inequality

$$\|\mathbf{u}^{(m+1)}\|_{D_0^{1,\frac{1}{2}}(Q_T;æ)} \leq C_{18}(\|\mathbf{u}^{(m)}\|^2_{D_0^{1,\frac{1}{2}}(Q_T;æ)} + H \|\mathbf{u}^{(m)}\|_{D_0^{1,\frac{1}{2}}(Q_T;æ)}) + C_{19} |\mathcal{L}| \,, \tag{3.28}$$

with

$$H = \|\mathbf{a}\|_{D_0^{1,\frac{1}{2}}(Q_T;æ)} + \min(1, T^{\frac{1}{2}}) \|\vec{\varphi}\|_{W_2^{\frac{1}{2}}(\Omega;æ)} \,,$$

$$|\mathcal{L}| = \|\mathbf{f}\|_{\mathcal{H}^*(Q_T;æ)} + \left[\|\mathbf{a}\|_{D_0^{1,\frac{1}{2}}(Q_T;æ)} + (\|\vec{\varphi}\|_{W_2^{\frac{1}{2}}(\Omega;æ)} + \|\vec{\varphi}\|_{\mathcal{H}^*(Q_T;æ)}) \min(1, T^{\frac{1}{2}})\right] [1 + H] \,.$$

Let ξ be a minimal root of the quadratic equation $C_{18}\xi^2 + (C_{18}H - 1)\xi + C_{19}|\mathcal{L}| = 0$. If

$$C_{18} H < 1 \,, \quad 4 C_{18} C_{19} |\mathcal{L}| < (1 - C_{18} H)^2 \,, \tag{3.29}$$

then this root is positive and it is given by the formula

$$\xi = \frac{2 C_{19} |\mathcal{L}|}{1 - C_{18} H + \sqrt{(1 - C_{18}H)^2 - 4 C_{18} C_{19} |\mathcal{L}|}} \,.$$

Clearly, $\|\mathbf{u}^{(0)}\|_{D_0^{1,\frac{1}{2}}(Q_T;æ)} \leq \xi$, and from (3.28) we can obtain successively for all $\mathbf{u}^{(m)}$ a uniform estimate

$$\|\mathbf{u}^{(m+1)}\|_{D_0^{1,\frac{1}{2}}(Q_T;æ)} \leq \xi \,.$$

Conditions (3.29) hold if the constant δ in (3.8) is small.

To prove the convergence of sequence $\{\mathbf{u}^{(m)}\}$ we observe that the difference $\mathbf{w}^{(m+1)} = \mathbf{u}^{(m+1)} - \mathbf{u}^{(m)} = \mathbf{v}^{(m+1)} - \mathbf{v}^{(m)}$ ($\mathbf{v}^{(m)} = \mathbf{u}^{(m)} + \mathbf{b}$) satisfies the integral identity

$$\int_0^T dt \int_\Omega \left[-\mathbf{w}^{(m+1)} \cdot \vec{\eta}_t + \nu \nabla \mathbf{w}^{(m+1)} : \nabla \vec{\eta}\right] dx =$$

$$= \sum_{k=1}^{2} \int_0^T dt \int_\Omega \left(v_k^{(m-1)} \mathbf{w}^{(m)} + w_k^{(m)} \mathbf{v}^{(m)}\right) \cdot \frac{\partial \vec{\eta}}{\partial x_k} \, dx \, .$$

Therefore

$$\|\mathbf{w}^{(m+1)}\|_{D_0^{1,\frac{1}{2}}(Q_T;æ)} \leq C_4 \sum_{k=1}^{2} \left(\|v_k^{(m-1)} \mathbf{w}^{(m)}\|_{L_2(Q_T;æ)} + \|w_k^{(m)} \mathbf{v}^{(m)}\|_{L_2(Q_T;æ)}\right)$$
$$\leq 2C_4 C_{17} \|\mathbf{w}^{(m)}\|_{D_0^{1,\frac{1}{2}}(Q_T;æ)} (\xi + H)$$

and the sequence $\{\mathbf{u}^{(m)}\}$ is convergent in $D_0^{1,\frac{1}{2}}(Q_T; æ)$ if

$$2C_4 C_{17}(\xi + H) < 1 \, .$$

This condition is also satisfied if δ is small, and we can easily prove the existence of a weak solution by a standard passage to a limit in (3.17). The uniqueness is established by the same kind of arguments. The difference $\mathbf{w} = \mathbf{v} - \mathbf{v}'$ of two solutions satisfy the identity

$$\int_0^{T'} dt \int_\Omega (-\mathbf{w} \cdot \vec{\eta}_t + \nu \nabla \mathbf{w} : \nabla \vec{\eta}) \, dx = \sum_{k=1}^{2} \int_0^{T'} dt \int_\Omega (v_k \mathbf{w} + w_k \mathbf{v}') \cdot \frac{\partial \vec{\eta}}{\partial x_k} \, dx \, ,$$

with any $T' \leq T$, hence,

$$\|\mathbf{w}\|_{D_0^{1,\frac{1}{2}}(Q_T;æ)} \leq C_{21} \|\mathbf{w}\|_{D_0^{1,\frac{1}{2}}(Q_T;æ)} \cdot \tag{3.30}$$
$$\cdot \max\left\{\int_0^{T'} (\|\mathbf{v}\|_{L_4(\Omega_0)}^4 + \|\mathbf{v}'\|_{L_4(\Omega_0)}^4) dt, \sup_k \int_0^{T'} (\|\mathbf{v}\|_{L_4(\omega_{jk})}^4 + \|\mathbf{v}'\|_{L_4(\omega_{jk})}^4) dt\right\} .$$

Suppose that condition (3.9) holds. Then we can take T' so small that the integrals in the right-hand side are less than $2\lambda < \frac{1}{C_{21}}$, and (3.30) proves that $\mathbf{w} = 0$ for $t \leq T'$. Repeating this argument if necessary we show that $\mathbf{w} = 0$ for $t \in [0, T]$. The theorem is proved.

Remark. In the above proof all the constants were independent of T, so we have proved global solvability of problem (3.5) with small data. Condition (3.8) can be also satisfied at the expense of the magnitude of T, if we suppose that the data are slightly more regular with respect to t. For instance, since

$$\|\mathbf{f}\|_{\mathcal{H}^*(Q_T;æ)}^2 \leq T^{\frac{1}{q'}} \max\left\{\left(\int_0^T (\|\mathbf{f}\|_{\mathcal{H}^*(\Omega_0)}^{2q} dt\right)^{\frac{1}{q}}, \sup_k \left(\int_0^T (\|\mathbf{f}\|_{\mathcal{H}^*(\Omega_k)}^{2q} dt\right)^{\frac{1}{q}}\right\},$$

$$\left\|Q_j - \phi \int_{\Sigma_j} \vec{\varphi}(\cdot, 0) \cdot \mathbf{n} \, dS\right\|_{D_0^{\frac{1}{2}}(0,T)}^2 \leq C_{22} T^{2(r-\frac{1}{2})} \left\|Q_j - \phi \int_{\Sigma_j} \vec{\varphi} \cdot \mathbf{n} \, dS\right\|_{W_2^r(0,T)}^2 , \tag{3.31}$$

with arbitrary $q \in (1,\infty)$, $r \in (\frac{1}{2},1)$, the left-hand side in (3.8) is small for small T, if the norms in the right-hand side of (3.31) and $\|\vec{\varphi}\|^2_{W_2^1(\Omega;\text{æ})}$ are simply bounded.

We notice also that in the case of a finite T we need not require that $\vec{\varphi} \in \mathcal{H}^*(\Omega;\text{æ})$, since we may apply Theorem 3.3 with $\mathbf{g} = \vec{\varphi}\phi_t'$, $\gamma > 0$.

4 – Boundary and initial-boundary value problems for the Navier-Stokes equations with discontinuous boundary data

It is well known (see for instance [33]) that integral estimates of Saint-Venant's type can be applied to the investigation of solutions of different boundary and initial-boundary problems in the neighborhood of singular points of the boundary. In this sections we consider singular points generated by discontinuities of functions prescribing boundary conditions in the case of two spacial variables.

As a first example, we consider a stationary problem in a bounded domain $\Omega \subset \mathbb{R}^2$

$$-\nu\Delta\mathbf{v} + (\mathbf{v}\cdot\nabla)\mathbf{v} + \nabla p = \mathbf{f}\,, \qquad \operatorname{div}\mathbf{v} = 0 \;\; (x \in \Omega)\,, \qquad \mathbf{v}|_S = \mathbf{a}\,, \tag{4.1}$$

with a vector field $\mathbf{a}$ that can have jump discontinuities in some points $\xi_1, ..., \xi_m$, and these points can be also angular points of the contour S.

When Ω is a rectangle, and $\mathbf{a}$ assumes different constant values on its sides, this is a well known test problem in numerical simulations of the Navier-Stokes equations.

The solution of problem (4.1) cannot have a bounded Dirichlet integral, since discontinuous $\mathbf{a}$ has no extensions inside Ω with this property.

By a weak solution of problem (4.1) we mean a vector field $\mathbf{v}$ belonging to $W_2^1(\Omega')$ (Ω' is any subdomain of Ω bounded away from all the points $\xi_1, ..., \xi_m$) satisfying conditions $\operatorname{div}\mathbf{v} = 0$, $\mathbf{v}|_S = \mathbf{a}$ and the integral identity

$$\int_\Omega \left(\nu\nabla\mathbf{v} : \nabla\vec{\eta} - \sum_{k=1}^{2} v_k\mathbf{v}\cdot\frac{\partial\vec{\eta}}{\partial x_k}\right) dx = \int_\Omega \mathbf{f}\cdot\vec{\eta}\; dx\,,$$

with arbitrary $\vec{\eta} \in \mathcal{H}(\Omega)$ vanishing near all ξ_j, $j = 1, ..., m$.

The following theorem is proved in [44, 45].

Theorem 4.1. *Let $S = \bigcup_{i=1}^m S_i$ where S_j are C^2-smooth curves with endpoints ξ_i and ξ_{i+1} ($\xi_{m+1} \equiv \xi_1$) and let the angle between two such curves with the vertex ξ_i be equal to $\theta_i \in (0,2\pi]$. For arbitrary $\mathbf{f} \in \mathcal{H}^*(\Omega)$ and $\mathbf{a} \in L_\infty(S_i) \cap W_2^{\frac{1}{2}}(S_i)$, $i = 1, ..., m$ satisfying the necessary condition $\int_S \mathbf{a}\cdot\mathbf{n}\; dS = 0$ problem (4.1) has at least one weak solution such that*

$$\int_{\Omega_r} |\nabla\mathbf{v}|^2 dx \le C_1\left(\ln\frac{1}{r} + 1\right), \tag{4.2}$$

where $\Omega_r = \{x \in \Omega\colon |x - \xi_i| > r, i = 1, ..., m\}$ and $r \in (0,d)$, $d < 1$.

If

$$\sum_{i=1}^{n} \|\mathbf{a}\|^2_{W_2^{\frac{1}{2}}(S_i)} + \|\mathbf{a}\|^2_{L_\infty(S)} + \|\mathbf{f}\|^2_{\mathcal{H}^*(\Omega)} \le \beta^2$$

and β is small, then any weak solution satisfying (4.2) possesses the property

$$\int_{\omega_r} |\nabla \mathbf{v}|^2 dx \le C_2\beta^2 \tag{4.3}$$

($\omega_r = \Omega_{\frac{d}{2}} \backslash \Omega_r$) and in this class it is unique.

As a second example, we consider an evolution problem with a moving boundary that was formulated to the author by Professor R. Rannacher. Let $\Omega(t) \subset \mathbb{R}^2$ be a rectangle $0 < x_1 < l_1(t)$, $0 < x_2 < l_2 =$const where $l_1(t)$ is a given function with values $l_1(t) \in [L_1, L_2]$, $L_1 > 0$. Consider the initial-boundary value problem

$$\begin{aligned} &\mathbf{v}_t + (\mathbf{v}\cdot\nabla)\mathbf{v} - \nu\nabla^2\mathbf{v} + \nabla p = 0, \qquad \operatorname{div}\mathbf{v} = 0\,, \\ &\mathbf{v}|_{t=0} = \vec{\varphi}(x)\,, \qquad \mathbf{v}|_{x\in S(t)} = \mathbf{a}(x,t)\,, \end{aligned} \tag{4.4}$$

where $S(t) = \partial\Omega(t) = S_1 \cup S_2 \cup S_3 \cup S_4$, $\mathbf{a}(x,t) = 0$ for $x \in S_1 \cup S_3 = \{x_2 = 0,\ 0 < x < l_1(t)\} \cup \{x_2 = l_2,\ 0 < x_1 < l_1(t)\}$, $\mathbf{a}(x,t) = (l_1'(t), 0)$ for $x \in S_2 = \{x_1 = l_1(t),\ 0 < x_2 < l_2\}$, (this is the adherence boundary condition), and $\mathbf{a}(x,t)|_{x\in S_4}$ is a vector field with a support contained in the interval $(0, l_2)$ such that

$$\int_{S_1} a_1(x,t)dx_2 = \int_{S_1} l_1'(t)dx_2 = l_2 l_1'(t)\,.$$

This problem may be written as initial-boundary value problem in $\Omega(0) \equiv \Omega$ if we introduce new variables $y_1 = x_1\frac{l_1(0)}{l_1(t)}$, $y_2 = x_2$ and

$$w_1(y,t) = \frac{l_1(0)}{l_1(t)} v_1\left(y_1\frac{l_1(t)}{l_1(0)}, y_2, t\right), \qquad w_2(y,t) = v_2\left(y_1\frac{l_1(t)}{l_1(0)}, y_2, t\right).$$

In these variables (4.3) takes the form

$$\begin{aligned} &\Lambda^2(t)\left(\mathbf{w}_t + (\mathbf{w}\cdot\nabla)\mathbf{w} - \nu\hat{\nabla}^2\mathbf{w} - y_1\frac{\lambda'(t)}{\lambda(t)}\frac{\partial\mathbf{w}}{\partial y_1} + N\mathbf{w}\right) + \nabla p = \Lambda(t)\mathbf{f}(\lambda(t)y_1, y_2, t)\,, \\ &\operatorname{div}\mathbf{w} = 0\,, \qquad (y \in \Omega,\ t \in (0,T))\,, \\ &\mathbf{w}|_{t=0} = \vec{\varphi}(y)\,, \qquad \mathbf{v}|_{y\in S(0)} = \vec{\alpha}(y,t)\,. \end{aligned} \tag{4.5}$$

Here $\lambda(t) = \frac{l_1(t)}{l_1(0)}$, $\vec{\alpha}(y,t) = 0$ for $y \in S_1(0) \cup S_3(0)$; $\vec{\alpha}(y,t) = (l_1(0)\frac{\lambda'(t)}{\lambda(t)}, 0)$ for $y \in S_2(0)$; $\vec{\alpha}(y,t) = \frac{\mathbf{a}(y,t)}{\lambda(t)}$ for $y \in S_4(0)$. The condition $\int_{S(0)} \vec{\alpha}\cdot\mathbf{n}dS = 0$ is again satisfied. The matrices Λ and N are defined by

$$N = \begin{pmatrix} \frac{\lambda'(t)}{\lambda(t)} & 0 \\ 0 & 0 \end{pmatrix}, \quad \Lambda(t) = \begin{pmatrix} \lambda(t) & 0 \\ 0 & 1 \end{pmatrix}, \quad \Lambda^2(t) = \begin{pmatrix} \lambda^2(t) & 0 \\ 0 & 1 \end{pmatrix},$$

finally, $\hat{\nabla} = \left(\frac{1}{\lambda(t)}\frac{\partial}{\partial y_1}, \frac{\partial}{\partial y_2}\right) = \Lambda^{-1}\nabla$, hence $\hat{\nabla}^2\mathbf{w} = \frac{1}{\lambda^2}\frac{\partial^2\mathbf{w}}{\partial y_1^2} + \frac{\partial^2\mathbf{w}}{\partial y_2^2}$.

Since supp $\vec{\alpha}|_{x_1=0} \subset S_4(0)$, vector field $\vec{\alpha}$ has jump discontinuities only at the points $\xi_1 = (l_1(0), 0)$ and $\xi_2 = (l_1(0), l_2)$. Let $\Omega_r = \{x \in \Omega(0) \colon |x - \xi_1| > r, |x - \xi_2| > r\}$, $r \le d =$

$\min\left(1, \frac{l_1(0)}{3}, \frac{l_2}{3}\right)$, $\omega_r = \Omega_{\frac{r}{2}} \backslash \Omega_r$, $\Omega_k = \Omega_{2^{-k}d}$, $\omega_{ik} = \{x \in \Omega :\ 2^{-(k+1)}d < |x - \xi_i| \leq 2^{-k}d\}$, $i = 1, 2$, $Q_T = \Omega(0) \times (0, T)$, and let $L_2(Q_T; \wp)$, $D_0^{1,\frac{1}{2}}(Q_T; æ)$ etc. be spaces introduced above in §3.

Define a weak solution of problem (4.5) as a vector field $\mathbf{w} \in L_2(0, T; W_2^1(\Omega_r))$, $\forall r \in (0, d)$, satisfying the conditions

$$\operatorname{div} \mathbf{w} = 0\,, \qquad \mathbf{w}|_{y \in S(0)} = \vec{\alpha}(y, t)$$

and the integral identity

$$\int_0^T dt \int_\Omega \Big[-\mathbf{w} \cdot (\Lambda^2(t)\vec{\eta})_t + \nu \hat{\nabla}\mathbf{w} : \hat{\nabla}\Lambda^2\vec{\eta} - \sum_{k=1}^{2} w_k \mathbf{w} \cdot \frac{\partial}{\partial y_k} \Lambda^2 \vec{\eta} + N\mathbf{w} \cdot \Lambda^2 \vec{\eta} -$$
$$- y_1 \frac{\lambda'(t)}{\lambda(t)} \frac{\partial \mathbf{w}}{\partial y_1} \cdot \Lambda^2(t)\vec{\eta} \Big] dy = \int_0^T dt \int_\Omega \Lambda \mathbf{f} \cdot \vec{\eta}\ dy + \int_\Omega \vec{a} \cdot \vec{\eta}(x, 0)\ dy\ ,$$

for arbitrary divergence free $\vec{\eta} \in W_2^{1,1}(Q_T)$ vanishing for $y \in S(0)$, $t = T$, and in neighborhoods of ξ_1, ξ_2.

Let $\Gamma_{iT} = S_i(0) \times (0, T)$. Define the space $D_0^{\frac{1}{2},\frac{1}{4}}(\Gamma_{iT})$ as the set of functions $a(y, t)$ ($(y, t) \in \Gamma_{iT}$) with a finite norm

$$\|a\|^2_{D_0^{\frac{1}{2},\frac{1}{4}}(\Gamma_{iT})} = \int_0^T \|a\|^2 \frac{dt}{t} + \int_0^T dt \int_0^t \|a(\cdot, t) - a(\cdot, t - h)\|^2 \frac{dh}{h} + \int_0^T \|a\|^2_{W_2^{\frac{1}{2}}(S_i(0))} dt\ ,$$

where $\|\cdot\|$ is the norm in $L_2(S_1(0))$.

Theorem 4.2. *Let $T < \infty$, $\mathbf{f} \in L_2(Q_T; æ)$, $\vec{\varphi} \in W_2^1(\Omega; æ) \cap L_4(\Omega; æ)$, $\vec{\varphi}|_{S_i(0)} \in L_\infty(S_i(0))$, $\vec{\alpha} - \phi\vec{\varphi}|_{S_i(0)} \in D_0^{\frac{1}{2},\frac{1}{4}}(\Gamma_{iT})$, $i = 1, 2, 3, 4$. Suppose that $l_1(t)$ is a positive continuously differentiable function in $(0, T)$ and $\vec{\alpha} \in L_2(0, T; L_\infty(S(0)))$. If*

$$\|\mathbf{f}\|^2_{L_2(Q_T; æ)} + \|\vec{\varphi}\|^2_{W_2^1(\Omega; æ)} + \|\vec{\varphi}\|^2_{L_4(\Omega; æ)} + \|\vec{\varphi}\|^2_{L_\infty(S(0))} +$$
$$+ \int_0^T \|\vec{\alpha}\|^2_{L_\infty(S(0))} dt + \sum_{i=1}^{4} \|\vec{\alpha} - \phi\vec{\varphi}\|^2_{D_0^{\frac{1}{2},\frac{1}{4}}(\Gamma_{iT})} \leq \delta$$

with a small $\delta > 0$, then problem (4.5) has a weak solution $\mathbf{w}$ such that $\mathbf{w} - \phi\vec{\varphi} \in D_0^{1,\frac{1}{2}}(Q_T; æ)$. The solution is unique in the class of vector fields satisfying conditions $\mathbf{w} - \phi\vec{\varphi} \in D_0^{1,\frac{1}{2}}(Q_T; æ)$ and (3.9).

In the case $\vec{\varphi} = 0$ this theorem was proved in [46].

Before we proceed to prove Theorems 4.1 and 4.2, we should consider auxiliary problem of extension of discontinuous vector fields from S into Ω.

Lemma 4.1. *Suppose that on the interval $I_d = (0, d)$ functions φ_0 and φ_1 are prescribed such that $\varphi_0 \in W_2^{\frac{3}{2}}(I_d)$, $\varphi_0', \varphi_1 \in W_2^{\frac{1}{2}}(I_d) \cap L_\infty(I_d)$. In a square $K_d = \{|x_1| <$*

$d, |x_2| < d\}$ there exists a function $\phi \in W_2^2(K_d)$, with $\nabla\phi \in W_2^1(I_d) \cap L_\infty(K_d)$ satisfying the boundary conditions

$$\phi(x_1, 0) = \varphi_0(x_1)\,, \qquad \left.\frac{\partial \phi}{\partial x_2}\right|_{x_2=0} = \varphi_1(x_1) \qquad (0 < x_1 < d)$$

and the inequalities

$$\|\nabla\phi\|_{L_2(K_d)} + \left(\int_{K_d} \frac{|\phi(x) - \phi(0)|^2}{|x|^2} dx\right)^{\frac{1}{2}} \le C_3(\|\varphi_0'\|_{L_2(I_d)} + \|\varphi_1\|_{L_2(I_d)})\,, \tag{4.7}$$

$$\|\phi\|_{L_\infty(K_d)} + \|\nabla\phi\|_{L_\infty(K_d)} \le C_4(\|\varphi_0\|_{L_\infty(I_d)} + \|\varphi_0'\|_{L_\infty(I_d)}) + \|\varphi_1\|_{L_\infty(I_d)}\,, \tag{4.8}$$

$$\|\phi\|_{W_2^2(K_d)} \le C_5(\|\varphi_0\|_{W_2^{\frac{3}{2}}(I_d)} + \|\varphi_1\|_{W_2^{\frac{1}{2}}(I_d)})\,. \tag{4.9}$$

The correspondence between (φ_0, φ_1) and ϕ is linear.

Proof: First of all, we extend φ_0 and φ_1 from I_d to $I_{2d} = (0, 2d)$ with the preservation of class, i.e. in such a way that

$$\|\varphi_0\|_{W_2^{\frac{3}{2}}(I_{2d})} \le C_6\|\varphi_0\|_{W_2^{\frac{3}{2}}(I_d)}\,, \qquad \|\varphi_1\|_{W_2^{\frac{1}{2}}(I_{2d})} \le C_7\|\varphi_1\|_{W_2^{\frac{1}{2}}(I_d)}\,,$$

$$\|\varphi_0'\|_{L_\infty(I_{2d})} \le C_8\|\varphi_0'\|_{L_\infty(I_d)}\,, \qquad \|\varphi_1\|_{L_\infty(I_{2d})} \le C_9\|\varphi_1\|_{L_\infty(I_d)}\,.$$

Such an extension can be defined, for instance, by the well known Hestenes-Whitney formula. Now, we set:

$$\phi(x_1, x_2) = \int_{\frac{1}{2}}^1 K(t)\varphi_0(x_1 + tx_2)dt + x_2\int_{\frac{1}{2}}^1 K(t)\varphi_1(x_1 + tx_2)dt\,, \tag{4.10}$$

where $K \in C_0^\infty(\frac{1}{2}, 1)$ possesses the properties

$$\int_{\frac{1}{2}}^1 K(t)dt = 1\,, \qquad \int_{\frac{1}{2}}^1 tK(t)dt = 0\,. \tag{4.11}$$

This formula defines $\phi(x_1, x_2)$ in a smaller square $K_d' = \{0 < x_1 < d, 0 < x_2 < d\}$. Conditions (4.6) are consequences of (4.11).Differentiation of (4.10) gives

$$\frac{\partial \phi}{\partial x_i} = \int_{\frac{1}{2}}^1 K_i(t)\varphi_0'(x_1 + tx_2)dt + \delta_{2i}\int_{\frac{1}{2}}^1 K(t)\varphi_1(x_1 + tx_2)dt - \int_{\frac{1}{2}}^1 \frac{dK_i(t)}{dt}\varphi_1(x_1 + tx_2)dt\,, \tag{4.12}$$

where $K_1 = K(t)$, $K_2 = tK(t)$. Estimates (4.7) and (4.8) are evident consequences of (4.10) and (4.12) and of the formula

$$\phi(x_1, x_2) - \phi(0) = \int_{\frac{1}{2}}^1 K(t)[\varphi_0(x_1 + tx_2) - \varphi_0(0)]dt + x_2\int_{\frac{1}{2}}^1 K(t)\varphi_1(x_1 + tx_2)dt\,.$$

Finally,

$$\frac{\partial^2\phi}{\partial x_i\partial x_j} = -\int_{\frac{1}{2}}^{1}\frac{\partial K_{ij}(t)}{\partial t}[\varphi_0'(x_1+tx_2)-\varphi_0'(x_1)]\frac{dt}{x_2}$$
$$-\delta_{2i}\int_{\frac{1}{2}}^{1}\frac{\partial K_j(t)}{\partial t}[\varphi_1(x_1+tx_2)-\varphi_1(x_1)]\frac{dt}{x_2}$$
$$+\int_{\frac{1}{2}}^{1}\frac{\partial K_{ij}^{(1)}(t)}{\partial t}[\varphi_1(x_1+tx_2)-\varphi_1(x_1)]\frac{dt}{x_2}\,,$$

with $K_{i1}=K_i$, $K_{i2}=tK_i$, $K_{i1}^{(1)}=\frac{dK_i}{dt}$, $K_{i2}^{(1)}=t\frac{dK_i}{dt}$. From this representation formula it follows that

$$\left\|\frac{\partial^2\phi}{\partial x_i\partial x_j}\right\|_{L_2(K_d')} \le \int_{\frac{1}{2}}^{1}\left|\frac{\partial K_{ij}}{\partial t}\right| dt\left(\int_{K_d'}|\varphi_0'(x_1+tx_2)-\varphi_0'(x_1)|^2\frac{dx}{x_2^2}\right)^{\frac{1}{2}}$$
$$+\int_{\frac{1}{2}}^{1}\left(\delta_{2i}\left|\frac{\partial K_j}{\partial t}\right|+\left|\frac{\partial K_{ij}^{(1)}}{\partial t}\right|\right)dt\left(\int_{K_d'}|\varphi_1(x_1+tx_2)-\varphi_1(x_1)|^2\frac{dx}{x_2^2}\right)^{\frac{1}{2}}$$
$$\le C_{10}(\|\varphi_0'\|_{W_2^{\frac{1}{2}}(I_{2d})}+\|\varphi_1\|_{W_2^{\frac{1}{2}}(I_{2d})})\,,$$

hence,

$$\|\phi\|_{W_2^2(K_d')}\le C_{11}(\|\varphi_0\|_{W_2^{\frac{3}{2}}(I_d)}+\|\varphi_1\|_{W_2^{\frac{1}{2}}(I_d)})\,.$$

It remains to extend ϕ from K_d' to K_d, also with the preservation of class W_2^2. The lemma is proved.

Lemma 4.2. *Let $I^{(0)}$, $I^{(1)}$ be two segments on $\mathbb{R}^2$ with $|I^{(0)}|=|I^{(1)}|=d$ and with a common endpoint placed at the origin, and let D_θ be a circular sector bounded by $I^{(1)}$, $I^{(2)}$ and by a part of a circle $|x|=d$ ($\theta\in(0,2\pi]$ is a magnitude of the angle between $I^{(0)}$ and $I^{(1)}$).*

Suppose that on $I^{(i)}$ there are prescribed functions $\varphi_0^{(i)}\in W_2^{\frac{3}{2}}(I^{(i)})$ and $\varphi_1^{(i)}\in W_2^{\frac{1}{2}}(I^{(i)})\cap L_\infty(I^{(i)})$. Let $\varphi_0^{(i)\prime}\in L_\infty(I^{(i)})$ where $\varphi_0^{(i)\prime}$ is the derivative of $\varphi_0^{(i)}$ with respect to the arc length on $I^{(i)}$. Suppose also that the condition $\varphi_0^{(1)}(0)=\varphi_0^{(0)}(0)$ holds. Then there exists a function $\phi(x)$, $x\in D_\theta$ satisfying the boundary conditions

$$\phi|_{I^{(i)}}=\varphi_0^{(i)}\,,\qquad \left.\frac{\partial\phi}{\partial n_i}\right|_{I^{(i)}}=\varphi_1^{(i)}\,,\qquad i=0,1 \tag{4.13}$$

(n_i is a normal to $I^{(i)}$) and the estimates

$$\|\nabla\phi\|_{L_2(D_\theta)}+\left(\int_{D_\theta}|\phi(x)-\phi(0)|^2\frac{dx}{|x|^2}\right)^{\frac{1}{2}}\le C_{12}\sum_{i=0}^{1}(\|\varphi_0^{(i)\prime}\|_{W_2^{\frac{1}{2}}(I^{(i)})}+\|\varphi_1^{(i)}\|_{W_2^{\frac{1}{2}}(I^{(i)})})\,, \tag{4.14}$$

$$\|\phi\|_{L_\infty(D_\theta)}+\|\nabla\phi\|_{L_\infty(D_\theta)}\le C_{12}'N_\infty\,, \tag{4.15}$$

$$\left|\frac{\partial^2\phi(x)}{\partial x_i\partial x_j}\right|\le\frac{C_{13}N_\infty}{|x|}+\phi_{ij}\,, \tag{4.16}$$

where $N_\infty = \sum_{i=0}^{1}(\|\varphi_0^{(i)}\|_{L_\infty(I^{(i)})} + \|\varphi_0^{(i)\prime}\|_{L_\infty(I^{(i)})} + \|\varphi_1^{(i)}\|_{L_\infty(I^{(i)})})$, $\phi_{ij} \in L_2(D_\theta)$ and

$$\|\phi_{ij}\|_{L_2(D_\theta)} \le C_{14}\sum_{i=0}^{1}\left(\|\varphi_0^{(i)}\|_{W_2^{\frac{3}{2}}(I^{(i)})} + \|\varphi_1^{(i)}\|_{W_2^{\frac{1}{2}}(I^{(i)})}\right) .$$

Proof: In virtue of the preceding lemma, in D_θ there exist functions $\phi_i \in W_2^2(D_\theta)$ such that each ϕ_i satisfies the boundary conditions (4.13) on $I^{(i)}$, and can be estimated by inequalities (4.6)–(4.8) in terms of $\varphi_0^{(i)}$ and $\varphi_1^{(i)}$. Let $\phi = \lambda\left(\frac{x}{|x|}\right)\phi_0 + \left(1 - \lambda\left(\frac{x}{|x|}\right)\right)\phi_1$ where $\lambda(\omega)$ is a smooth function on a unit circle $|\omega| = 1$ such that $\lambda\left(\frac{x}{|x|}\right) = 1$ in the neighborhood of I_0 and $\lambda\left(\frac{x}{|x|}\right) = 0$ in the neighborhood of I_1. Then ϕ satisfies (4.13). Estimates for L_2- and L_∞-norms of ϕ are evident. Moreover, from the formula

$$\nabla\phi = \lambda\nabla\phi_0 + (1-\lambda)\nabla\phi_1 + \nabla\lambda\left(\frac{x}{|x|}\right)(\phi_0 - \phi_1)$$

and from the inequality

$$\begin{aligned}|\phi_0(x) - \phi_1(x)| &\le |\phi_0(x) - \phi_0(0)| + |\phi_1(0) - \phi_1(x)| \\ &\le |x|(\|\nabla\phi_0\|_{L_\infty(D_\theta)} + \|\nabla\phi_1\|_{L_\infty(D_\theta)}) ,\end{aligned}$$

it follows that

$$\begin{aligned}&\|\nabla\phi\|_{L_\infty(D_\theta)} \le C_{15}(\|\nabla\phi_0\|_{L_\infty(D_\theta)} + \|\nabla\phi_1\|_{L_\infty(D_\theta)}) \le C_{16}N_\infty , \\ &\|\nabla\phi\|_{L_2(D_\theta)} \le \|\nabla\phi_0\|_{L_2(D_\theta)} + \|\nabla\phi_1\|_{L_2(D_\theta)} \\ &\qquad + \max_\omega|\lambda'(\omega)|\left(\int_{D_\theta}\frac{|\phi_0(x) - \phi_0(0)|^2}{|x|^2}\,dx + \int_{D_\theta}\frac{|\phi_1(x) - \phi_1(0)|^2}{|x|^2}\,dx\right)^{\frac{1}{2}} \\ &\le C_{12}\sum_{i=0}^{1}(\|\varphi_0^{(i)\prime}\|_{L_2(I_d)} + \|\varphi_1^{(i)}\|_{L_2(I_d)}) .\end{aligned}$$

Finally, (4.16) follows from the formula

$$\begin{aligned}\frac{\partial^2\phi}{\partial x_i\partial x_j} &= \lambda\frac{\partial^2\phi_0}{\partial x_i\partial x_j} + (1-\lambda)\frac{\partial^2\phi_1}{\partial x_i\partial x_j} + \frac{\partial\lambda(\frac{x}{|x|})}{\partial x_i}\left(\frac{\partial\phi_0}{\partial x_j} - \frac{\partial\phi_1}{\partial x_j}\right) \\ &+ \frac{\partial\lambda}{\partial x_j}\left(\frac{\partial\phi_0}{\partial x_i} - \frac{\partial\phi_1}{\partial x_i}\right) + \frac{\partial^2\lambda}{\partial x_i\partial x_j}(\phi_0 - \phi_1)\end{aligned}$$

and from the estimates (4.8), (4.9) for ϕ_0 and ϕ_1. The lemma is proved.

Lemma 4.3. *Any* $\mathbf{a} \in W_2^{\frac{1}{2}}(S_i) \cap L_\infty(S_i)$, $i = 1, ..., m$ *satisfying the condition* $\int_S \mathbf{a}\cdot\mathbf{n}\,dS \equiv \sum_{i=1}^{m}\int_{S_i}\mathbf{a}\cdot\mathbf{n}\,dS = 0$ *can be extended into* Ω *in such a way that* $\operatorname{div}\mathbf{a} = 0$, $\mathbf{a} \in L_\infty(\Omega)$ *and the following estimates hold:*

$$\|\mathbf{a}\|_{L_\infty(\Omega)} \le C_7\|\mathbf{a}\|_{L_\infty(S)} , \tag{4.17}$$

$$|\mathbf{a}_{x_i}(x)| \le \frac{C_{18}\|\mathbf{a}\|_{L_\infty(S)}}{\min_j|x - \xi_j|} + |A_i(x)| , \tag{4.18}$$

$$\|A_i(x)\|_{L_2(\Omega)} \le C_{19}\sum_{j=1}^{m}\|\mathbf{a}\|_{W_2^{\frac{1}{2}}(S_j)} \tag{4.19}$$

The operator of the extension is linear.

Proof: We seek a vector field $\mathbf{a}(x)$, $x \in \Omega$ in the form $\mathbf{a} = (\frac{\partial \phi}{\partial x_2}, -\frac{\partial \phi}{\partial x_2})$ where ϕ satisfies the boundary conditions

$$\frac{\partial \phi}{\partial \tau} = \mathbf{a} \cdot \mathbf{n}\,, \qquad \frac{\partial \phi}{\partial n} = \mathbf{a} \cdot \vec{\tau} \equiv a_1 \qquad (x \in S_i).$$

(We assume that $\mathbf{n}$ is exterior normal to Ω and $\vec{\tau} = (-n_2, n_1)$). Hence

$$\phi|_S = \phi(\xi_1) + \int_{\xi_1}^{x} (\mathbf{a} \cdot \mathbf{n})\, ds \equiv a_0\,,$$

where $\xi_1 \in S$ and the integration is carried out along S. Clearly, $a_0 \in W_2^{\frac{3}{2}}(S_i)$, $a_1 \in W_2^{\frac{1}{2}}(S_i)$, $a_0, \frac{da_0}{ds}, a_1 \in L_\infty(S_i)$.

In Ω function ϕ may be defined by the formula

$$\phi = \sum_{i=1}^{m} \varsigma_i \phi_i\,,$$

where $\{\varsigma_i\}$ is a smooth "partition of unity" on S and in a neighborhood of S, supp ς_i is contained in a circle $|x - x_i| \le \rho$, $x_i \in S$. On the curve $S \cap \operatorname{supp} \varsigma_i$, ϕ_i should satisfy the boundary conditions

$$\phi_i = \varphi_0\,, \qquad \frac{\partial \phi_i}{\partial n} = \varphi_1\,.$$

Since $S_j \in C^2$, in the neighborhood of any point ξ_i there exists an invertible transformation mapping this neighborhood onto a circular sector D_{θ_i}, and the function ϕ_i may be constructed as described in Lemma 4.2.

The neighborhood of any interior point of S_j may be transformed into a semi-disk where this function may be constructed by the procedure described in Lemma 4.1. Estimate for ϕ follows easily from estimates (4.8), (4.9) or (4.15), (4.16) for ϕ_i. The lemma is proved.

Lemma 4.4. *Let $\Gamma_T = S \times (0,T)$, $\Gamma_{iT} = S_i \times (0,T)$. Arbitrary $\mathbf{b} \in D_0^{\frac{1}{2},\frac{1}{2}}(\Gamma_i) \cap L_2(0,T;L_\infty(S))$, $j = 1,...,m$, satisfying the condition $\int_S \mathbf{b} \cdot \mathbf{n}\, dS = 0$ can be extended into Q_T in such a way that $\mathbf{b} \in D_0^{1,\frac{1}{2}}(Q_T;æ) \cap L_2(0,T;L_\infty(\Omega))$, It satisfies the estimates*

$$\int_0^T \|\mathbf{b}\|^2_{L_\infty(\Omega)} dt \le C_{20} \int_0^T \|\mathbf{b}\|^2_{L_\infty(S)} dt\,, \tag{4.20}$$

$$\|\mathbf{b}\|^2_{D_0^{1,\frac{1}{2}}(Q_T;æ)} \le C_{21} \left(\int_0^T \|\mathbf{b}\|^2_{L_\infty(S)} dt + \sum_{j=1}^{m} \|\mathbf{b}\|^2_{D_0^{\frac{1}{2},\frac{1}{2}}(\Gamma_{jT})} \right). \tag{4.21}$$

Proof: We construct the extension of $\mathbf{b}$ into Ω as in Lemma 4.3, t plays a role of a fixed parameter. Inequality (4.20) follows from (4.17), and (4.18), (4.19) imply

$$\|\mathbf{b}_{x_i}\|^2_{L_2(Q_T;æ)} \le C_{22} \left(\int_0^T \|\mathbf{b}\|^2_{L_\infty(S)} dt + \sum_{j=1}^{m} \|\mathbf{b}\|^2_{D_0^{\frac{1}{2},\frac{1}{2}}(\Gamma_{jT})} \right).$$

Finally, when we apply (4.7), (4.14) to the differences $\phi_i(x,t) - \phi_i(x,t-h)$ we easily show that

$$\|\mathbf{b}\|^2_{D_0^{0,\frac{1}{2}}(Q_T;æ)} \le C_{23}\|\mathbf{b}\|^2_{D_0^{0,\frac{1}{2}}(\Gamma_T,æ)}$$

and this completes the proof of (4.21) and of the lemma.

Let us turn to the proof of Theorems 4.1 and 4.2.

Proof of Theorem 4.1: Let $\mathbf{a} = (\frac{\partial \phi}{\partial x_2}, -\frac{\partial \phi}{\partial x_2})$ be an extension of $\mathbf{a}$ from S into Ω whose existence was established in Lemma 4.3. To prove the solvability of problem (4.1) without any smallness hypotheses, we introduce another extension, namely, $\mathbf{a}_\delta = (\frac{\partial \phi\omega_\delta}{\partial x_2}, -\frac{\partial \phi\omega_\delta}{\partial x_1})$ where ω_δ is the so called "Hopf cut-off function", i.e. a smooth function equal to 1 in a very thin neighborhood of S and to zero outside larger neighborhood and possessing the property

$$|\nabla\omega_\delta(x)| \le \frac{\epsilon}{\operatorname{dist}(x,\delta)} \tag{4.22}$$

for an arbitrary ϵ fixed beforehand (see the construction of functions of such type in [24, 26, 39, 41]). Vector field $\mathbf{u} = \mathbf{v} - \mathbf{a}_\delta$ should satisfy the conditions

$$\operatorname{div}\mathbf{u} = 0\,, \qquad \mathbf{u}|_S = 0$$

and the integral identity

$$\int_\Omega \left[\nu\nabla\mathbf{u} : \nabla\vec{\eta} - \sum_{k=1}^{2}(a_{k\delta}\mathbf{u} + u_k\mathbf{a}_\delta + u_k\mathbf{u})\cdot\frac{\partial\vec{\eta}}{\partial x_k}\right]\,dx =$$

$$= \int_\Omega \left(\mathbf{f}\cdot\vec{\eta} - \nu\nabla\mathbf{a} : \nabla\vec{\eta} + \sum_{k=1}^{2} a_{k\delta}\mathbf{a}_\delta\cdot\frac{\partial\vec{\eta}}{\partial x_k}\right)\,dx\,.$$

The construction and estimate of $\mathbf{u}$ is carried out in the same way as in Theorem 2.1. We consider auxiliary problem which consists in finding the vector field $\mathbf{u}^{(\epsilon)} \in \mathcal{H}(\Omega_\epsilon)$ in the domain $\Omega_\epsilon = \{x \in \Omega\colon\ |x - \xi_j| > \epsilon, j = 1, ..., m\}$ which satisfies the integral identity

$$\mathcal{M}_\epsilon[\mathbf{u}^{(\epsilon)},\vec{\eta}] \equiv \int_{\Omega_\epsilon}\left[\nu\nabla\mathbf{u}^{(\epsilon)} : \nabla\vec{\eta} - \sum_{k=1}^{2}(a_{k\delta}\mathbf{u}^{(\epsilon)} + u_k^{(\epsilon)}\mathbf{a}_\delta + u_k^{(\epsilon)}\mathbf{u}^{(\epsilon)})\cdot\frac{\partial\vec{\eta}}{\partial x_k}\right]dx$$

$$= \int_{\Omega_\epsilon}\left(\mathbf{f}\cdot\vec{\eta} - \nu\nabla\mathbf{a} : \nabla\vec{\eta} + \sum_{k=1}^{2} a_{k\delta}\mathbf{a}_\delta\cdot\frac{\partial\vec{\eta}}{\partial x_k}\right)dx\,, \qquad \forall\vec{\eta} \in \mathcal{H}(\Omega_\epsilon)\ . \tag{4.23}$$

In virtue of estimate (4.22), the form $\mathcal{M}_\epsilon[\mathbf{u}^{(\epsilon)},\mathbf{u}^{(\epsilon)}]$ is positive definite, and the existence of $\mathbf{u}^{(\epsilon)}$ follows from the well known results [10, 24]. $\mathbf{u}^{(\epsilon)}$ satisfies the inequality

$$\begin{aligned}\|\nabla\mathbf{u}^{(\epsilon)}\|^2_{L_2(\Omega_\epsilon)} &\le C_{24}(\|\mathbf{f}\|^2_{\mathcal{H}^*(\Omega)} + \|\nabla\mathbf{a}_\delta\|^2_{L_2(\Omega_\epsilon)} + \|\mathbf{a}_\delta\|^4_{L_4(\Omega)}) \\ &\le C_{25}\left(\log\frac{1}{\epsilon} + 1\right).\end{aligned} \tag{4.24}$$

To estimate $\|\nabla\mathbf{u}^{(\epsilon)}\|^2_{L_2(\Omega_r)}$, $r \in (2\epsilon, d)$, we substitute into (4.23) the test function $\vec{\eta} = \mathbf{U}_r$ where $\mathbf{U}_r(x) = \mathbf{u}^{(\epsilon)}(x)$ for $x \in \Omega_r$, $\mathbf{U}_r(x) = 0$ for $x \in \Omega_\epsilon\backslash\Omega_{\frac{r}{2}}$, and in every domain $\omega_{ir} = \{x \in \Omega\colon\ \frac{r}{2} < |x - \xi_i| < r\}$, $\mathbf{U}_r$ satisfies the relations

$$\operatorname{div}\mathbf{U}_r = 0\,, \qquad \mathbf{U}_r|_{\gamma_{ir}} = \mathbf{u}^{(\epsilon)}|_{\gamma_{ir}}\,, \qquad \mathbf{U}_r|_{\partial\omega_{ir}\backslash\gamma_{ir}} = 0\,,$$

where $\gamma_{ir} = \{x \in \Omega : |x - \xi_i| = r\}$. $\mathbf{U}_r(x)$ may be constructed and estimated in ω_{ir} exactly in the same way as in Theorem 2.1. Repeating the arguments of this theorem, we arrive at the inequality

$$y(r) \le C_{25}[r(-y'(r)) + r^{\frac{5}{2}}(-y')^{\frac{3}{2}}] + C_{26}\left(\ln\frac{1}{2} + b\right), \qquad b > 1,$$

for the function $y(r) = \int_\Omega h(x,r)|\nabla \mathbf{u}^{(\epsilon)}|^2 dx$ where $h(x,r) = 1$ for $x \in \Omega_r$, $h(x,r) = 0$, $h(x,r) = \frac{2|x-\xi_i|}{r} - 1$ for $x \in \omega_{ir}$. From this inequality and from (4.23) we conclude that $y(r) \le 2C_{26}\left(\ln\frac{1}{2} + b\right)$, which permits to prove the existence of $\mathbf{u}$ by the same standard procedure as in Theorem 2.1.

We restrict ourselves with the proof of the first part of Theorem 4.1 and refer to papers [44, 45] for the proof of the second part; it also follows closely the corresponding arguments of Theorem 2.1.

Proof of Theorem 4.2: Following general scheme of §3,we introduce a new unknown vector field $\mathbf{u} = \mathbf{w} - \mathbf{z}$ where $\mathbf{z} = \mathbf{b} + \phi(t)\vec{\varphi}(x)$, $\mathbf{b}$ being the extension of $\mathbf{a}(x,t) - \phi(t)\vec{\varphi}(x)|_S$ into Ω (see Lemma 4.4). Clearly, $\mathbf{u}$ should satisfy the conditions $\mathbf{u}|_{y \in S(0)} = 0$, $\operatorname{div}\mathbf{u} = 0$ and the integral identity

$$\int_0^T dt \int_\Omega \Big[-\mathbf{u}\cdot(\Lambda^2\vec{\eta})_t + \nu\hat{\nabla}\mathbf{u} : \nabla\hat{\Lambda}^2\vec{\eta} - \sum_{k=1}^{2}(z_k\mathbf{u} + u_k\mathbf{z} + u_k\mathbf{u})\cdot\frac{\partial\Lambda^2\vec{\eta}}{\partial y_k} + N\mathbf{u}\cdot\Lambda^2\vec{\eta} -$$
$$- y_1\frac{\lambda'(t)}{\lambda(t)}\frac{\partial\mathbf{u}}{\partial y_1}\cdot\Lambda^2\vec{\eta}\Big]dy = \int_0^T dt\int_\Omega(\mathbf{F}\cdot\vec{\eta} + \sum_{j=1}^{2}\mathbf{F}_j\cdot\vec{\eta}_{y_j} - \mathbf{F}_0\cdot\vec{\eta}_t)dy \equiv \mathcal{F}[\vec{\eta}],$$

where

$$\begin{aligned}
&\mathbf{F} = \Lambda\mathbf{f} + \frac{d\Lambda^2}{dt}\mathbf{b} - \Lambda^2\vec{\varphi}\phi_t - \Lambda^2 N\mathbf{z} + y_1\frac{\lambda'(t)}{\lambda(t)}\Lambda^2\frac{\partial\mathbf{z}}{\partial y_1},\\
&\mathbf{F}_j = (\mathbf{F}_{1j}, \mathbf{F}_{2j}), \qquad \mathbf{F}_0 = \Lambda^2\mathbf{b},\\
&F_{1j} = -\nu\frac{1}{\lambda^2}\frac{\partial}{\partial y_1}(\Lambda^2\mathbf{z})_j + z_1(\Lambda^2\mathbf{z})_j, \qquad F_{2j} = -\nu\frac{\partial}{\partial y_2}(\Lambda^2\mathbf{z})_j + z_2(\Lambda^2\mathbf{z})_j,
\end{aligned} \tag{4.25}$$

and $\vec{\eta}$ is an arbitrary vector field from $W_2^{1,1}(Q_T)$ satisfying the conditions $\operatorname{div}\vec{\eta} = 0$, $\vec{\eta}|_{y\in S(0)} = 0$, $\vec{\eta}(x,T) = 0$ and vanishing in neighborhoods of ξ_1 and ξ_2.

Consider also the following auxiliary problem: find $\mathbf{u}^{(\epsilon)} \in D_0^{1,\frac{1}{2}}(Q_{\epsilon\infty})$, ($Q_{\epsilon\infty} = \Omega_\epsilon \times (0,\infty)$) such that $\operatorname{div}\mathbf{u}^{(\epsilon)} = 0$, $\mathbf{u}^{(\epsilon)}|_{y\in\partial\Omega_\epsilon} = 0$, and

$$\int_0^\infty dt\int_{\Omega_\epsilon}(-\mathbf{u}^{(\epsilon)}\cdot(\Lambda^2\vec{\eta})_t + \nu\hat{\nabla}\mathbf{u}^{(\epsilon)} : \hat{\nabla}\vec{\eta})dx = \mathcal{F}[\vec{\eta}], \tag{4.26}$$

with arbitrary solenoidal $\vec{\eta} \in W_2^{1,1}(Q_{\epsilon\infty})$ vanishing for $x \in \partial\Omega_\epsilon$ and with given $\mathbf{F} \in L_2(Q_\infty; æ)$, $\mathbf{F}_j \in L_2(Q_\infty; æ)$, $\mathbf{F}_0 \in D_0^{0,\frac{1}{2}}(Q_\infty; æ)$ and with $\lambda(t)$ extended into the half-axis $t > T$ with the preservation of class and of the sign.

Both problems are investigated by the same arguments as in §3. Let us show that these arguments can be carried out in the present case, when the coefficients in the left-hand side of (4.26) depend on t. We restrict ourselves with the proof of an estimate of a type (3.21) for the solution of auxiliary problem. First of all, we change $\vec{\eta}$ for $e^{-2\gamma t}\vec{\eta}$ and rewrite identity (4.26) in the form

$$\int_0^\infty dt \int_{\Omega_\epsilon} (-\Lambda^2\mathbf{u}^{(\epsilon)}e^{-\gamma t}\cdot(e^{-\gamma t}\vec{\eta})_t + \gamma e^{-2\gamma t}\Lambda^2\mathbf{u}^{(\epsilon)}\cdot\vec{\eta} -$$

$$-\frac{d\Lambda^2}{dt}\mathbf{u}^{(\epsilon)}e^{-2\gamma t}\cdot\vec{\eta} + \nu e^{-2\gamma t}\hat{\nabla}\mathbf{u}^{(\epsilon)} : \hat{\nabla}\vec{\eta})dx = \mathcal{F}[\vec{\eta}e^{-2\gamma t}] \,. \qquad (4.27)$$

This identity holds for $\vec{\eta} \in D^{1,\frac{1}{2}}(\mathcal{G}_{\epsilon\infty};\gamma)$ ($\mathcal{G}_{\epsilon\infty} = \Omega_\epsilon \times (-\infty,\infty)$) possessing the derivative $\vec{\eta}_t \in L_2(\mathcal{G}_{\epsilon\infty};\gamma)$, but we can get rid of the last assumption if we write the first term $J[\mathbf{u}^{(\epsilon)},\vec{\eta}]$ of the left-hand side in the form

$$J[\mathbf{u}^{(\epsilon)},\vec{\eta}] = \frac{1}{2\pi}\int_{-\infty}^\infty d\xi_0 \int_{\Omega_\epsilon} s(\widetilde{\Lambda^2\mathbf{u}^{(\epsilon)}})\cdot\vec{\eta}(x,s)dx\,, \qquad s = \gamma + i\xi_0\,.$$

The term $\int_0^\infty dt \int_{\Omega_\epsilon} \mathbf{F}_0\cdot\vec{\eta}_t e^{-2\gamma t}dx$ in the right-hand side may be interpreted in the same way. Hence, we can take $\vec{\eta} = \mathbf{u}^{(\epsilon)}$ in (4.27). To compute $J[\mathbf{u}^{(\epsilon)},\mathbf{u}^{(\epsilon)}]$ we approximate $\mathbf{u}^{(\epsilon)}$ by smooth vector fields $\mathbf{u}_m^{(\epsilon)}$ in the norm $D_0^{1,\frac{1}{2}}(Q_{\epsilon\infty};\gamma)$. Clearly,

$$J[\mathbf{u}^{(\epsilon)},\mathbf{u}^{(\epsilon)}] = \lim_{m\to\infty} J[\mathbf{u}_m^{(\epsilon)},\mathbf{u}_m^{(\epsilon)}] = \lim \int_0^\infty dt \int_{\Omega_\epsilon} (-\Lambda^2\mathbf{u}_m^{(\epsilon)})e^{-\gamma t}\cdot(e^{-\gamma t}\mathbf{u}_m^{(\epsilon)})_t\, dx$$

$$= \frac{1}{2}\int_0^\infty e^{-2\gamma t}dt \int_{\Omega_\epsilon} \frac{d\Lambda^2}{dt}\mathbf{u}^{(\epsilon)}\cdot\mathbf{u}^{(\epsilon)}\, dx$$

and after easy estimates, we arrive at the inequality

$$\int_0^\infty dt \int_{\Omega_\epsilon} (\gamma|\mathbf{u}^{(\epsilon)}|^2 + \nu|\nabla\mathbf{u}^{(\epsilon)}|^2)e^{-2\gamma t}\, dx \le C_{26}B^{\frac{1}{2}}\|\mathbf{u}^{(\epsilon)}\|_{D_0^{1,\frac{1}{2}}(Q_{\epsilon\infty};\gamma)}\,, \qquad (4.28)$$

where

$$B = \|\mathbf{F}\|^2_{L_2(Q_{\epsilon\infty};\gamma)} + \sum_{j=1}^2 \|\mathbf{F}_j\|^2_{L_2(Q_{\epsilon\infty};\gamma)} + \|\mathbf{F}_0\|^2_{D_0^{0,\frac{1}{2}}(Q_{\epsilon\infty};\gamma)}$$

and γ is large enough: $\gamma \ge \gamma_0 > 0$. Now, the substitution $\vec{\eta} = i \text{ sgn } \xi_0\tilde{\mathbf{u}}^{(\epsilon)}$ that has been made many times in §3, is equivalent to

$$e^{-\gamma t}\vec{\eta} = H\mathbf{u}_0^{(\epsilon)}e^{-\gamma t} = \frac{1}{\pi}\text{ V.p. } \int_{-\infty}^\infty \frac{\mathbf{u}_0^{(\epsilon)}(x,s)e^{-\gamma s}ds}{t-s}\,,$$

where $Hf = \frac{1}{\pi}\text{ V.p. } \int_{-\infty}^\infty \frac{f(s)ds}{t-s}$ is the Hilbert transform of $f(t)$ and $\mathbf{u}_0^{(\epsilon)}$ is zero extension of $\mathbf{u}^{(\epsilon)}$ into the domain $t<0$. For smooth $w(t)$ we have

$$\frac{dHw}{dt} = -\frac{1}{\pi}\text{ V.p. } \int_{-\infty}^\infty \frac{w(s)-w(t)}{(t-s)^2}ds\,,$$

hence, the expression

$$\int_{-\infty}^\infty v(t)\frac{dHw}{dt} = -\frac{1}{\pi}\int_{-\infty}^\infty v(t)dt\text{ V.p. } \int_{-\infty}^\infty \frac{w(s)-w(t)}{(s-t)^2}ds$$

$$= \frac{1}{\pi}\int_{-\infty}^\infty v(s)ds\text{ V.p. } \int_{-\infty}^\infty \frac{w(s)-w(t)}{(s-t)^2}dt$$

$$= \frac{1}{2\pi}\int_{-\infty}^\infty\int_{-\infty}^\infty (v(s)-v(t))(w(s)-w(t))\frac{dsdt}{(s-t)^2}$$

is meaningful for all $v, w \in W_2^{\frac{1}{2}}(\mathbb{R})$. Setting $e^{-\gamma t}\vec{\eta} = He^{-\gamma t}\mathbf{u}^{(\epsilon)}$ in (4.27), we obtain

$$\frac{1}{2\pi}\int_{-\infty}^{\infty}\int_{\Omega_\epsilon}(\Lambda^2(s)\mathbf{u}_0^{(\epsilon)}(x,s)e^{-\gamma s} - \Lambda^2(t)\mathbf{u}_0^{(\epsilon)}(x,t)e^{-\gamma t})\cdot$$

$$\cdot(\mathbf{u}_0^{(\epsilon)}(x,s)e^{-\gamma s} - \mathbf{u}_0^{(\epsilon)}(x,t)e^{-\gamma t})\frac{dsdt}{(s-t)^2} =$$

$$= \int_0^\infty dt\int_{\Omega_\epsilon}\Big(\gamma e^{-\gamma t}\Lambda^2\mathbf{u}^{(\epsilon)}\cdot He^{-\gamma t}\mathbf{u}_0^{(\epsilon)} - \frac{d\Lambda^2}{dt}\mathbf{u}^{(\epsilon)}e^{-\gamma t}\cdot H(e^{-\gamma t}\mathbf{u}_0^{(\epsilon)})$$

$$+\nu\hat{\nabla}\mathbf{u}^{(\epsilon)}e^{-\gamma t} : \hat{\nabla}H(e^{-\gamma t}\mathbf{u}_0^{(\epsilon)})\Big)\,dx - \mathcal{F}[e^{-\gamma t}H(e^{-\gamma t}\mathbf{u}_0^{(\epsilon)})]\,. \tag{4.29}$$

As the operator H is bounded in $L_2(\mathbb{R})$, it is not hard to conclude from (4.29) that

$$\|\mathbf{u}^{(\epsilon)}\|^2_{D_0^{0,\frac{1}{2}}(Q_{\epsilon\infty};\gamma)} \le C_{27}(\gamma\|\mathbf{u}^{(\epsilon)}\|^2_{L_2(Q_{\epsilon\infty};\gamma)} + \|\nabla\mathbf{u}^{(\epsilon)}\|^2_{L_2(Q_{\epsilon\infty};\gamma)}) + C_{28}B^{\frac{1}{2}}\|\mathbf{u}^{(\epsilon)}\|^2_{D_0^{1,\frac{1}{2}}(Q_{\epsilon\infty};\gamma)}\,.$$

From this inequality and from (4.28) it follows that

$$\|\mathbf{u}^{(\epsilon)}\|^2_{D_0^{1,\frac{1}{2}}(Q_{\epsilon\infty};\gamma)} \le C_{29}B$$

$$\le C_{30}(\|\mathbf{F}\|^2_{L_2(Q_T;\text{æ})} + \sum_{j=1}^{2}\|\mathbf{F}_j\|^2_{L_2(Q_T;\text{æ})} + \|\mathbf{F}_0\|^2_{D_0^{0,\frac{1}{2}}(Q_T;\text{æ})})\left(\ln\frac{1}{\epsilon}+1\right)$$

$$\equiv C_{30}A\left(\ln\frac{1}{\epsilon}+1\right).$$

This is an analogue of (3.21). It is clear that subsequent arguments leading to inequalities

$$\|\mathbf{u}^{(\epsilon)}\|^2_{D_0^{1,\frac{1}{2}}(Q_{r\infty};\gamma)} \le C_{31}A\left(\ln\frac{1}{r}+1\right)$$

$$\|\mathbf{u}^{(\epsilon)}\|^2_{D_0^{1,\frac{1}{2}}(Q_{\frac{\epsilon}{2}\infty}\setminus Q_{r\infty};\gamma)} \le C_{31}A \tag{4.30}$$

(analogous to (3.17), (3.18) also can be carried out in this way).

Now it is trivial to prove the solvability of the following linear problem: find divergence free $\mathbf{u} \in D_0^{1,\frac{1}{2}}(Q_{r\infty};\text{æ})$ vanishing for $x \in S$ and satisfying an integral identity

$$\int_0^\infty dt\int_\Omega[-\mathbf{u}(\Lambda^2\vec{\eta})_t + \nu\hat{\nabla}\mathbf{u} : \hat{\nabla}\vec{\eta}]\,dx = \mathcal{F}[\vec{\eta}]\,, \tag{4.31}$$

with arbitrary divergence free $\vec{\eta} \in W_2^{1,1}(Q_\infty)$ vanishing for $x \in S$ and in the neighborhood of ξ_1, ξ_2. The proof reduces to the passage to a weak limit in (4.30) and (4.31) as $\epsilon \to 0$.

The solvability of a nonlinear problem is established by the method of successive approximations according to the following scheme: $\mathbf{u}^{(0)} = 0$, $\mathbf{u}^{(m+1)}$ are solutions of a linear problem analogous to (4.31), namely, these functions satisfy the integral identity

$$\int_0^T dt\int_\Omega[-\mathbf{u}^{(m+1)}\cdot(\Lambda^2\vec{\eta})_t + \nu\hat{\nabla}\mathbf{u}^{(m+1)} : \nabla\Lambda^2\vec{\eta}]\,dx =$$

$$= \mathcal{F}[\vec{\eta}] + \int_0^T dt\int_\Omega\Big[\sum_{k=1}^{2}(z_k\mathbf{u}^{(m)} + u_k^{(m)}\mathbf{z} + u_k^{(m)}\mathbf{u}^{(m)})\cdot\frac{\partial\Lambda^2\vec{\eta}}{\partial y_k}$$

$$- N\mathbf{u}^{(m)}\cdot\Lambda^2\vec{\eta} + y_1\frac{\lambda'(t)}{\lambda(t)}\frac{\partial\mathbf{u}^{(m)}}{\partial y_1}\cdot\Lambda^2\vec{\eta}\Big]\,dy\,, \tag{4.32}$$

with $\mathbf{F}$, $\mathbf{F}_j$, $\mathbf{F}_0$ given by (4.25). Because of our assumptions and of estimate (3.25) which is applicable to domains ω_{1r} and ω_{2r} considered here, we have:

$$\|\mathbf{z}\|^2_{L_2(Q_T;æ)} + \|\nabla\mathbf{z}\|^2_{L_2(Q_T;æ)} \leq$$
$$\leq 2[\|\mathbf{b}\|^2_{L_2(Q_T;æ)} + \|\nabla\mathbf{b}\|^2_{L_2(Q_T;æ)} + \min(1,T)(\|\vec{\varphi}\|^2_{L_2(\Omega;æ)} + \|\nabla\vec{\varphi}\|^2_{L_2(\Omega;æ)})] ,$$

$$\|\mathbf{z}\|^4_{L_4(Q_T;æ)} \leq 8[\|\mathbf{b}\|^4_{L_4(Q_T;æ)} + \|\vec{\varphi}\|^4_{L_4(\Omega;æ)} \min(1,T)]$$
$$\leq C_{32}\Big[\|\mathbf{b}\|^4_{D_0^{1,\frac{1}{2}}(Q_T;æ)} + \Big(\int_0^T \|\mathbf{b}\|^2_{L_\infty(\Omega)}dt\Big)^2\Big] + 8\min(1,T)\|\vec{\varphi}\|^4_{L_4(\Omega;æ)} \cdot$$

As a consequence, we see that $\mathbf{F}$, $\mathbf{F}_j \in L_2(Q_T;æ)$, $\mathbf{F}_0 \in D_0^{0,\frac{1}{2}}(Q_T;æ)$. All the estimates of $\mathbf{u}^{(m+1)}$, the passage to the limit in (4.32) and the uniqueness theorem for the solution of nonlinear problem are proved according to the scheme of Theorem 3.2.

REFERENCES

[1] *C.I. Amick – Steady solutions of the Navier-Stokes equations in unbounded channels and pipes*, Ann. Scuola Norm. Sup. Pisa, **4** (1977), 473–573.

[2] *C.I. Amick – Properties of steady Navier-Stokes solutions for certain unbounded channels and pipes*, Nonlinear Analysis, Theory, Methods and Applications, **2** (1978), 689–720.

[3] *C.I. Amick & L.E. Fraenkel – Steady solutions of the Navier-Stokes equations representing plane flow in channels of various type*, Acta Math., **144**, No. 1–2 (1980), 83–152.

[4] *J. Babushka & A.K. Aziz – Survey lectures on the mathematical foundations of the finite elements method. The mathematical formulations of the finite elements method with application to partial differential equations*, Acad. Press, N.Y., (1972), 5-359.

[5] *K.I. Babenko – On stationary solutions of the problem of flow past a body of a viscous incompressible fluid*, Math. USSR Sbornik, **20** (1973), 1–25.

[6] *T. Beale – Large-time regularity of viscous surface waves*, Arch. Rat. Mech. Anal, **84** (1984), 307–352.

[7] *M.E. Bogovskii – Solutions of some problems of vector analysis related to operators div and grad*, Proc. Semin. S.L. Sobolev, **1** (1980), 5–40.

[8] *W. Borchers & K. Pileckas – Existence, uniqueness and asymptotics of steady jets*, Univ. of Paderborn, Department of Math. Preprint, 1–62.

[9] *R. Finn – On the exterior stationary problem for the Navier-Stokes equations and associated perturbation problems*, Arch. Rat. Mech. Anal, **19** (1965), 363–406.

[10] *H. Fujita – On the existence and regularity of the steady-state solutions of the Navier-Stokes equations*, J. Fac. Sci. Univ. Tokyo, **9** (1961), 59–102.

[11] *J.G. Heywood – On uniqueness questions in the theory of viscous flow*, Acta Math., **136** (1976), 61–102.

[12] *Chang Huakang – Existence and uniqueness of solutions of the steady Navier-Stokes equations in unbounded domains*, Proc. Int. Conf. on the Navier-Stokes equations in Oberwofach (Aug. 1991), to appear.

[13] *C.O. Horgan & L.T. Wheeler – Spatial decay estimates for the Navier-Stokes equations with application to the problem of entry flow*, SIAM. Journ. Appl. Math., **35** (1978), 97–116.

[14] *C.O. Horgan – Plane entry flows and energy estimates for the Navier-Stokes equations*, Arch. Rat. Mech. Anal., **68** (1978), 359–381.

[15] *V.P. Il'in – On a theorem of Hardy and Littlewood*, Trudy. Math. Inst. Steklov, **53** (1959), 128–144.

[16] *G.A. Iosif'ian – Analogue of Saint-Venant's principle and the uniqueness of the solutions of the first boundary value problems for the Stokes system in domains with non-compact boundaries*, Doklady Acad Sci. USSR, **242** (1978), 36–39.

[17] *G.A. Iosif'ian – On the Saint-Venant's Principle for incompressible viscous flow*, Uspekhi Mat. Nauk., **34** (1979), 191–192.

[18] *L.V. Kapitanskii – Coincidence of the spaces $\overset{\circ}{J}{}_2^1\ (\Omega)$ and $\hat{J}_2^1(\Omega)$ for plane domains Ω having exits at infinity*, Zapiski Nauchn. Semin. LOMI, **110** (1981), 74–81.

[19] *L.V. Kapitanskii & K.I. Pileckas – On spaces of solenoidal vector fields and boundary value problems for the Navier-Stokes equations in domains with noncompact boundaries*, Trudy Math. Inst. Steklov, **159** (1983).

[20] *L.V. Kapitanskii & K.I. Pileckas – Certain problems of vector analysis*, Zapiski Nauchn. Semin. LOMI, **138** (1984), 65–85.

[21] *I.K. Knowles – A Saint-Venant's principle in the two-dimensional linear theory of elasticity*, Arch. Rat. Mech. Anal., **21** (1966), 1–22.

[22] *I.K. Knowles – A Saint-Venant's principle for a class of second order elliptic boundary value problems*, SAMP, **18** (1967), 473–490.

[23] *O.A. Ladyzhenskaya – Investigation of the Navier-Stokes equations in the case of steady-state flow of an incompressible fluid*, Uspekhi Mat. Nauk., **13** (1958), 219–220; **14** (1959), 75–97.

[24] *O.A. Ladyzhenskaya – Mathematical theory of a viscous incompressible flow*, Moscow, (1961), second edition, M (1970), 288p.

[25] *O.A. Ladyzhenskaya & V.A. Solonnikov – On some problems of vector analysis and generalized formulations of boundary value problems for the Navier-Stokes equations*, Zapiski Nauchn. Semin. LOMI, **59** (1976), 81–116.

[26] *O.A. Ladyzhenskaya & V.A. Solonnikov – On the solvability of boundary and initial-boundary value problems for the Navier-Stokes equations in domains with noncompact boundaries*, Vestink Leningr. Univ., Ser. Mat., Mech. Astr., **13**, No. 3 (1977), 39–48.

[27] *O.A. Ladyzhenskaya & V.A. Solonnikov – Determination of solutions of boundary value problem for the Stokes and Navier-Stokes equations with an infinite Dirichlet integral*, Zapiski Nauchn. Semin. LOMI, **96** (1980), 117-160.

[28] *O.A. Ladyzhenskaya & V.A. Solonnikov – On initial-boundary value problem for the linearized Navier-Stokes system in domains with noncompact boundaries*, Trudy Math. Inst. Steklov, **159** (1983), 37–40.

[29] *O.A. Ladyzhenskaya, V.A. Solonnikov & H. True – Résolution des équations de Stokes et Navier-Stokes dans les tuyaux infinis*; C.R. Acad. Sci. Paris, **292**, Ser I (1981), 251–254.

[30] *S. Nazarov & K. Pileckas – On noncompact free boundary problems for the plane stationary Navier-Stokes equations*, to appear.

[31] *O.A. Oleinik & G.I. Iosifian – Analogue of Saint-Venant's principle and the uniqueness of solutions of boundary value problems in unbounded domains for parabolic equations*, Uspekhi Mat. Nauk, **31** (1976), 142–166.

[32] *O.A. Oleinik & G.I. Iosifian – Boundary value problems for second order elliptic equations in unbounded domains and Saint-Venant's principle*, Ann. Scuola Norm. Sup. Pisa, **IV** (1977), 269–290.

[33] *O.A. Oleinik, G.A. Iosifian & I.N. Tavhelidze - On the asymptotics of solutions of the biharmonic equation in neighborhoods of irregular points of the boundary and at infinity,* Proc. Moscow Math. Soc., **142** (1981), 160–175.

[34] *K.I. Pileckas - On the unique solvability of boundary value problems for the Stokes system of equations in domains with noncompact boundaries,* Trudy Math. Inst. Steklov, **147** (1980), 115-123.

[35] *K.I. Pileckas - Solvability of a problem on a plane motion of a viscous incompressible liquid with a noncompact free boundary,* Diff. Equations and Their Application, Vilnius, **30** (1981), 57–96.

[36] *K.I. Pileckas - On the existence of solutions of the Navier-Stokes equations with an infinite dissipation of energy in a class of domains with noncompact boundaries,* Zapiski Nauchn. Semin. LOMI, **110** (1981), 180–202.

[37] *K.I. Pileckas - On spaces of solenoidal vectors,* Trudy Math. Inst. Steklov, **159** (1983).

[38] *K.I. Pileckas - On the problem of a flow of a heavy viscous incompressible liquid with a noncompact free boundary,* Liet. Mat. Rinkinys, **28** (1988), 315–333.

[39] *V.A. Solonnikov & K.I. Pileckas - On some spaces of divergence free vector fields and on the solvability of a boundary value problem for the system of Navier-Stokes equations in domains with noncompact boundaries,* Zapiski Nauchn. Semin. LOMI, **73** (1977), 136–157.

[40] *V.A. Solonnikov - On the solvability of boundary and initial-boundary value problems for the Navier-Stokes system in domains with noncompact boundaries,* Pacific J. of Math., **93** (1981), 443–458.

[41] *V.A. Solonnikov - Stokes and Navier-Stokes equations in domains with noncompact boundaries,* Nonlinear PDE and Their Applications, Collège de France Seminar, H. Brezis and J.L. Lions (editors), **IV** (1983), 240–349.

[42] *V.A. Solonnikov - On solution of stationary Navier-Stokes equation with an infinite Dirichlet integral,* Zapiski Nauchn. Semin. LOMI, **115** (1982), 257–263.

[43] *V.A. Solonnikov - Solvability of the problem of effluence of a viscous incompressible fluid into an infinite open basin,* Trudy. Math. Inst. Steklov, **179** (1989), 193–225.

[44] *V.A. Solonnikov - On a boundary value problem for the Navier-Stokes equations with discontinuous boundary data,* Rendiconti di Matematica, Ser VII, **10** (1990), 757–772.

[45] *V.A. Solonnikov - On boundary value problem for the Navier-Stokes equations with discontinuous boundary data,* Proc. of Conf. on PDE in Trento, (1990), to appear.

[46] *V.A. Solonnikov - On an initial-boundary value problem for the Navier-Stokes equations discontinuous boundary data in the case of two spacial variables,* Zapiski Nauchn. Semin. LOMI, to appear.

[47] *Y. Teramoto - The initial value problem for viscous incompressible flow down an inclined plane,* Hiroshima Math. J., **15** (1985), 619–643.

[48] *R.A. Toupin - Saint-Venant's principle,* Arch Rat. Mech. Anal., **18** (1965), 83–96.

V.A. Solonnikov,
LOMI Steklov Math. Institute, Fontanka 27,
191011 StPetersburg - RUSSIA

Stability Problems in Electrohydrodynamics, Ferrohydrodynamics and Thermoelectric Magnetohydrodynamics

BRIAN STRAUGHAN

1 – Introduction

In this series of lectures we present stability results for problems in fluid dynamics when interaction effects with electric, magnetic, and thermal fields are taken into account. These subjects have been broadly divided into three categories and we here examine basic problems in each of the three categories, namely, electrohydrodynamics (EHD), ferrohydrodynamics (FHD), and magnetohydrodynamics (MHD), although in the last case we consider only the relatively new and technologically important subject of thermoelectric magnetohydrodynamics (TEMHD).

The analysis we employ involves either the classical linear instability technique (Chandrasekhar [8]), or the more recent nonlinear energy stability method (Straughan [49]). We do not intend reviewing either of these ideas since they are covered in detail in [8] and [49]; the results described here are all new. However, to understand a certain peculiarity of the energy method for the nonlinear classes of problem to be studied here, we firstly describe some aspects of this technique in the context of penetrative convection in a micropolar fluid.

In the context of magnetizable fluids (FHD), Rosensweig [43], pp. 249–263, shows how a micropolar fluid fits within the framework of FHD. Hence the work now described is in keeping with the remainder of the article, and so we now include an exposition of a nonlinear stability analysis for convection in a micropolar fluid. After the section dealing with the micropolar fluid we study in section 3 the nonlinear stability of the thermal convection problem in the electrohydrodynamic setting. The penultimate section, 4, primarily presents a new nonlinear energy stability analysis for the thermal ferrohydrodynamic convection model of Blennerhassett *et al.* [4]. Finally in section 5, we present findings on the thermal convection problem in thermoelectric magnetohydrodynamics, and we review extensions of these ideas. Throughout standard indicial notation as employed in the continuum machanics literature is used. Thus $\partial/\partial t$ or subscript $,t$ denotes partial differentiation with respect to time, repeated indices imply summation over 1 to 3, etc.

2 – Convection in an isotropic micropolar fluid with a nonlinear buoyancy law

We study the theory of micropolar fluids developed by Eringen [11], a recent very readable and useful generalization is contained in Eringen [12]. In connection with this practical studies of micropolar fluids applied to lubrication problems are given by

Khonsari [17], Khonsari & Brewe [18], Prakash & Sinha [32] and Tipei [50].

If we assume the microinertia moment tensor has form

$$j_{ik} = j\delta_{ik}$$

for j a constant, then the equations of Eringen [11] for an incompressible, isotropic thermomicropolar fluid are:

$$v_{i,i} = 0, \tag{2.1}$$
$$\rho(v_{i,t} + v_j v_{i,j}) = \rho f_i + t_{ki,k}, \tag{2.2}$$
$$\rho j(\nu_{i,t} + v_j \nu_{i,j}) = \rho \ell_i + \epsilon_{ikh} t_{kh} + m_{ki,k}, \tag{2.3}$$
$$\rho(\epsilon_{,t} + v_i \epsilon_{,i}) = t_{kh} b_{kh} + m_{kh}\nu_{h,k} + q_{k,k} + \rho h, \tag{2.4}$$

where $b_{kh} = v_{h,k} - \epsilon_{khr}\nu_r$, v_i, ν_i, $\epsilon(T)$ are fluid velocity, particle spin vector and internal energy, T is temperature, $\rho, t_{ki}, f_i, \ell_i, m_{ki}, q_k$ and h are density (presumed constant except in the body force ρf_i term), stress tensor, body force, body couple, couple stress tensor, heat flux vector and heat supply, respectively, (2.1)-(2.4) representing, respectively, balance of mass, momentum, moment of momentum, and energy.

The constitutive equations of Eringen [11] are:

$$t_{kh} = -\pi\delta_{kh} + \mu(v_{k,h} + v_{h,k}) + \bar{\kappa}(v_{h,k} - \epsilon_{kha}\nu_a), \tag{2.5}$$
$$m_{kh} = \bar{\alpha}\nu_{r,r}\delta_{kh} + \bar{\beta}\nu_{k,h} + \gamma\nu_{h,k} + \alpha\epsilon_{khm}T_{,m}, \tag{2.6}$$
$$q_k = \kappa T_{,k} + \beta\epsilon_{khm}\nu_{h,m}, \tag{2.7}$$

where π is the pressure and for the general case of Eringen's [11] theory, $\mu, \bar{\kappa}, \bar{\alpha}, \bar{\beta}, \gamma, \alpha, \kappa, \beta$ are functions of T. We assume $\mu, \bar{\kappa}, \bar{\alpha}, \bar{\beta}, \gamma, \alpha, \kappa$ are constant and ρ $(= \rho_0)$ is also constant except in the body force term in (2.1) where we adopt the rule:

$$\rho = \rho_0[1 - A(T - T_0)^2]. \tag{2.8}$$

For simplicity we examine the case of zero body couple and heat supply, namely, $\ell_i \equiv h \equiv 0$.

Chapter 9 of Straughan [49] explains how the balance of energy equation reduces to

$$\rho_0 c\dot{T} = \kappa\Delta T + \hat{\beta}\epsilon_{khm}\nu_{h,m}T_{,k}, \tag{2.9}$$

for $\hat{\beta}$ *constant,* where $c = -\psi_{TT}$ is likewise constant.

Thermodynamics, Eringen [11], shows the coefficients in (2.5)–(2.7) must satisfy:

$$\begin{gathered} 2\mu + \bar{\kappa} \geq 0, \quad 2\mu + \kappa \geq 0, \quad \kappa \geq 0, \quad \bar{\kappa} \geq 0, \quad \mu \geq 0, \\ 3\bar{\alpha} + \bar{\beta} + \gamma \geq 0, \quad \bar{\beta} + \gamma \geq 0, \quad \frac{2\kappa}{T}(\gamma - \bar{\beta}) \geq \left(\alpha - \frac{\beta}{T}\right)^2. \end{gathered} \tag{2.10}$$

We assume $\hat{\beta} = \beta T^{-1} > 0$.

We suppose the fluid is contained in the layer $z \in (0, d)$ with gravity in the negative z-direction, and the planes $z = 0, d$ are kept at constant temperatures T_1, T_2. The steady solution whose stability we investigate is:

$$\bar{T} = -Bz + T_1, \quad \bar{v}_i = 0, \quad \bar{\nu}_i = 0, \tag{2.11}$$

where

$$B = \frac{T_1 - T_2}{d}$$

is the temperature gradient.

The stability investigation of (2.11) begins by defining the perturbations by:

$$T = \bar{T} + \theta, \qquad v_i = \bar{v}_i + u_i, \qquad \nu_i = \bar{\nu}_i + \nu_i.$$

The perturbations are now non-dimensionalized according to the scalings:

$$\begin{aligned}
&x_i = x_i^* d, \qquad u_i = u_i^* U, \qquad U = \frac{\mu}{d\rho_0}, \qquad Pr = \frac{\mu c}{\kappa},\\
&P = \frac{U\mu}{d}, \qquad t^* = \frac{t\mu}{\rho_0 d^2}, \qquad \nu_i = \nu_i^* N, \qquad N = \frac{U}{d},\\
&b = \frac{\hat{\beta}}{d^2 c\rho_0}, \qquad \theta = \theta^* T^\sharp, \qquad \varsigma = \frac{T_0 - T_1}{T_2 - T_1},\\
&Ra = R^2 = \frac{gAB^2 d^5 c\rho_0^2}{\kappa\mu}, \qquad T^\sharp = U\Big(\frac{Pr}{gAd}\Big)^{1/2},\\
&K = \frac{\bar{\kappa}}{\mu}, \qquad \Gamma = \frac{\gamma}{\mu d^2}, \qquad G = \frac{\bar{\alpha} + \bar{\beta}}{\mu d^2},
\end{aligned}$$

where Ra is the Rayleigh number and K, Γ, G are non-dimensional micropolar coefficients. The non-dimensionalized perturbation equations become (omitting all stars):

$$\begin{aligned}
u_{i,t} + u_k u_{i,k} &= -\pi_{,i} - 2R(\varsigma - z)\theta k_i + \Delta u_i + K(\Delta u_i - \epsilon_{kir}\nu_{r,k}) + Pr\theta^2 k_i,\\
u_{i,i} &= 0,\\
Pr(\theta_{,t} + u_i\theta_{,i}) &= -Rw + \Delta\theta - bR\xi + bPr\epsilon_{khm}\nu_{h,m}\theta_{,k},\\
j(\nu_{i,t} + u_a\nu_{i,a}) &= K(\epsilon_{ikh}u_{h,k} - 2\nu_i) + G\nu_{a,ai} + \Gamma\Delta\nu_i,
\end{aligned} \tag{2.12}$$

where $w = u_3$ and $\xi = (\mathrm{curl}\,\vec{\nu})_3$.

Equations (2.12) hold in the layer $z \in (0,1)$ and the boundary conditions on the perturbations are

$$u_i = \theta = \nu_i = 0 \qquad \text{on} \qquad z = 0, 1$$

which are in accordance with two fixed boundaries at the surfaces $z = 0, 1$. We assume the solution exhibits a periodic structure in the x, y-plane.

The linear instability problem for (2.12) is studied in detail in Lindsay & Straughan [22] where many numerical results are reported. This aspect is not discussed here.

Nonlinear energy stability when $b \neq 0$. When $b \neq 0$ we may develop a conditional nonlinear energy stability analysis. This may be considerably improved by employing a weighted energy, if $b = 0$. However, suppose now $b > 0$. We form three separate energy identities from (2.12) by consecutively multiplying $(2.12)_1$, $(2.12)_3$, $(2.12)_4$ by u_i, θ, ν_i and integrating over V. Upon using the boundary conditions we may obtain:

$$\begin{aligned}
\frac{d}{dt}\frac{1}{2}\|\mathbf{u}\|^2 = &-2R < f\theta w > - \|\nabla\mathbf{u}\|^2 - K\|\mathrm{curl}\,\mathbf{u}\|^2\\
&- K\epsilon_{kir} < \nu_{r,k}u_i > + Pr < \theta^2 w >,
\end{aligned} \tag{2.13}$$

$$\frac{d}{dt}\frac{1}{2}Pr\|\theta\|^2 = -R < \theta w > -bR < \xi\theta > -\|\nabla\theta\|^2, \tag{2.14}$$

$$\frac{d}{dt}\frac{1}{2}j\|\vec{\nu}\|^2 = K\epsilon_{ikh} < u_{h,k}\nu_i > -2K\|\vec{\nu}\|^2 - G\|\nu_{i,i}\|^2 - \Gamma\|\nabla\vec{\nu}\|^2. \tag{2.15}$$

The most general approach to nonlinear energy stability is now to form

$$(2.13) + \lambda_2(2.14) + \lambda_1(2.15)$$

for coupling parameters λ_1, λ_2 (> 0) to be selected. Hence, define

$$E(t) = \frac{1}{2}\|\mathbf{u}\|^2 + \frac{1}{2}j\lambda_1\|\vec{\nu}\|^2 + \frac{1}{2}\lambda_2 Pr\|\theta\|^2, \tag{2.16}$$

and we find

$$\begin{aligned}\frac{dE}{dt} =& - R[2 < f\theta w > +\lambda_2 < \theta w >] - bR\lambda_2 < \xi\theta > \\ &- \|\nabla\mathbf{u}\|^2 - \lambda_2\|\nabla\theta\|^2 \\ &- \lambda_1(K\|\vec{\nu}\|^2 + G\|\nu_{i,i}\|^2 + \Gamma\|\nabla\vec{\nu}\|^2) \\ &- K\|\vec{\omega}\|^2 + (1+\lambda_1)K < \vec{\nu}\cdot\vec{\omega} > -\lambda_1 K\|\vec{\nu}\|^2 \\ &+ Pr < \theta^2 w >,\end{aligned} \tag{2.17}$$

where we have set $\vec{\omega} = \operatorname{curl}\mathbf{u}$. Since there is symmetry in the micropolar and velocity terms in $(2.12)_1$ and $(2.12)_4$ it is tempting to choose $\lambda_1 = 1$ in (2.16). However, as we now show, parametric differentiation leads to the conclusion that the optimal value of λ_1 is not 1. This must be offset against the fact that the thermodynamic inequalities of Eringen [11], viz. (2.10), are derived under the premise of what amounts to $\lambda_1 = 1$. Thus, any analysis based on (2.10) with $\lambda_1 \neq 1$ must also ensure the dissipation on the right hand side of (2.10) may be employed as a stabilization ingredient. This is a subtle point, and any cavalier attempt to proceed with $\lambda_1 \neq 1$ may break down. Certainly the analysis for $\lambda_1 \neq 1$ is numerically more complex, requiring care in the multi-dimensional maximization/minimization routine

$$\max_{\lambda_1,\lambda_2}\min_{a^2} R_E^2(\lambda_1,\lambda_2;a^2)$$

which arises. The above numerical problem is solved by Broyden's method in Lindsay & Straughan [22] who show that the energy critical Rayleigh number is sensitive to λ_1 and so since we need to ensure the dissipation contributes to stabilization but are simultaneously restricted by (2.10), this is a non-trivial point which is further discussed in Lindsay & Straughan [22].

The essential content of the argument we have just discussed may be seen easily with the aid of an example when there are three fields present.

Consider u, v, w satisfying

$$\begin{aligned}\frac{\partial u}{\partial t} &= \Delta u + \alpha w + \beta v, \\ \frac{\partial v}{\partial t} &= \Delta v + \beta u + \gamma w, \\ \frac{\partial w}{\partial t} &= \Delta w + \alpha u,\end{aligned} \tag{2.18}$$

defined on the domain $\Omega \times (0,\infty)$, with Ω a bounded domain in $\mathbf{R}^3$. The coefficents α, β, γ we take as constants. On the boundary of Ω, Γ, we assume that

$$u = v = w = 0. \tag{2.19}$$

To investigate the stability of the zero solution to (2.18) by the energy method, we might begin with the "energy":

$$E(t) = \frac{1}{2}\|u\|^2 + \frac{1}{2}\lambda_1\|v\|^2 + \frac{1}{2}\lambda_2\|w\|^2, \tag{2.20}$$

where λ_1, λ_2 are positive coupling parameters to be selected.

It is instructive to rewrite (2.18) as

$$\frac{\partial}{\partial t}\begin{pmatrix} u \\ v \\ w \end{pmatrix} = \begin{pmatrix} \Delta & 0 & 0 \\ 0 & \Delta & 0 \\ 0 & 0 & \Delta \end{pmatrix}\begin{pmatrix} u \\ v \\ w \end{pmatrix} + \begin{pmatrix} 0 & \beta & \alpha \\ \beta & 0 & \gamma \\ \alpha & 0 & 0 \end{pmatrix}\begin{pmatrix} u \\ v \\ w \end{pmatrix},$$

from which we see that the operator on the right hand side is symmetric if $\gamma = 0$. Since the skew-symmetry only arises in the v-equation in (2.18) it is tempting to argue that it is sufficient to take $\lambda_2 = 1$ in (2.20). However, we now show by parametric differentiation, that both λ_1 and λ_2 are not equal to 1 at the "best" value for stability. What this shows is that for problems involving three or more fields, the energy stability analysis is likely to involve multi-dimensional optimization.

Next, derive energy identities from (2.18), (2.19), namely:

$$\begin{aligned} \frac{1}{2}\frac{d}{dt}\|u\|^2 &= -\|\nabla u\|^2 + \alpha < uw > + \beta < uv >, \\ \frac{1}{2}\frac{d}{dt}\|v\|^2 &= -\|\nabla v\|^2 + \beta < uv > + \gamma < vw >, \\ \frac{1}{2}\frac{d}{dt}\|w\|^2 &= -\|\nabla w\|^2 + \alpha < uw > . \end{aligned}$$

Thus, employing (2.20) we find

$$\begin{aligned} \frac{dE}{dt} = &- \left(\|\nabla u\|^2 + \lambda_1\|\nabla v\|^2 + \lambda_2\|\nabla w\|^2\right) \\ &+ \alpha(1+\lambda_2) < uw > + \beta(1+\lambda_1) < uv > + \gamma\lambda_1 < vw > . \end{aligned} \tag{2.21}$$

If we define

$$I = \alpha(1+\lambda_2) < uw > + \beta(1+\lambda_1) < uv > + \gamma\lambda_1 < vw >,$$

and

$$D = \|\nabla u\|^2 + \lambda_1\|\nabla v\|^2 + \lambda_2\|\nabla w\|^2,$$

and then define Λ by

$$\Lambda = \max \frac{I}{D}, \tag{2.22}$$

it follows from (2.21) that

$$\frac{dE}{dt} \le -D(1-\Lambda),$$

and hence using Poincaré's inequality, nonlinear stability follows provided $\Lambda < 1$. It is not our goal here to actually calculate Λ, merely to deduce some features associated with it. The optimal stability threshold will occur when $\Lambda = 1$. Hence, put

$$\phi = \sqrt{\lambda_1}v, \qquad \psi = \sqrt{\lambda_2}w,$$

in (2.22), and then the Euler-Lagrange equations for the maximizing solution become, upon setting $\Lambda = 1$:

$$\begin{aligned} &A\psi + B\phi + 2\Delta u = 0,\\ &Bu + C\psi + 2\Delta\phi = 0,\\ &Au + C\phi + 2\Delta\psi = 0, \end{aligned} \tag{2.23}$$

where we have put

$$A = \frac{\alpha(1+\lambda_2)}{\sqrt{\lambda_2}}, \qquad B = \frac{\beta(1+\lambda_1)}{\sqrt{\lambda_1}}, \qquad C = \gamma\sqrt{\frac{\lambda_1}{\lambda_2}}.$$

To employ parametric differentiation we let (u^1,ϕ^1,ψ^1), (u^2,ϕ^2,ψ^2), be solutions to (2.23) for $(\lambda_1^{(1)},\lambda_2^{(1)})$ and $(\lambda_1^{(2)},\lambda_2^{(2)})$, respectively. Then, multiply each of (2.23) with superscript 1, respectively by u^2,ϕ^2,ψ^2 and integrate over Ω. The process is repeated with 1 and 2 reversed. After dividing by either $\lambda_1^{(2)}-\lambda_1^{(1)}$ or $\lambda_2^{(2)}-\lambda_2^{(1)}$ and taking the limit $\lambda_i^{(2)} \to \lambda_i^{(1)}$ we may derive

$$\begin{aligned} &\frac{\partial A}{\partial\lambda_2} < u\psi > + \frac{\partial C}{\partial\lambda_2} < \phi\psi > = 0,\\ &\frac{\partial B}{\partial\lambda_1} < u\phi > + \frac{\partial C}{\partial\lambda_1} < \phi\psi > = 0. \end{aligned}$$

From these equations it follows that

$$\begin{aligned} &\alpha\Big(\frac{\lambda_2-1}{2\lambda_2^{3/2}}\Big) < u\psi > - \frac{\gamma\sqrt{\lambda_1}}{2\lambda_2^{3/2}} < \phi\psi > = 0,\\ &\beta\Big(\frac{\lambda_1-1}{2\lambda_1^{3/2}}\Big) < u\phi > + \frac{\gamma}{2\sqrt{\lambda_1\lambda_2}} < \phi\psi > = 0. \end{aligned} \tag{2.24}$$

Equations (2.24) hold on the nonlinear stability threshold $\Lambda = 1$. Since we do not expect the eigenfunctions u,ϕ,ψ to be orthogonal to each other, it is evident from (2.24) that $\lambda_1 = \lambda_2 = 1$ only when $\gamma = 0$, i.e. there is symmetry in (2.18). If $\gamma \neq 0$, then we must expect $\lambda_1,\lambda_2 \neq 1$ and the nonlinear stability problem, therefore, will involve an optimization procedure in λ_1,λ_2, to determine the sharpest result.

It is essentially the above idea which is present in the micropolar problem.

We return to the micropolar problem and proceed with a formal analysis of (2.17). Define I and D by

$$I = -(2 < f\theta w > + \lambda_2 < \theta w >) - b\lambda_2 < \xi\theta >, \tag{2.25}$$

$$
\begin{aligned}
\mathcal{D} =& \|\nabla \mathbf{u}\|^2 + \lambda_2 \|\nabla \theta\|^2 \\
&+ \lambda_1 (K\|\vec{\nu}\|^2 + G\|\nu_{i,i}\|^2 + \Gamma \|\nabla \vec{\nu}\|^2) \\
&+ K\|\vec{\omega}\|^2 - K(1+\lambda_1) < \vec{\nu}\cdot\vec{\omega} > + \lambda_1 K \|\vec{\nu}\|^2 .
\end{aligned}
\tag{2.26}
$$

Then (2.17) is

$$
\frac{dE}{dt} = RI - \mathcal{D} + Pr < \theta^2 w > . \tag{2.27}
$$

If we define R_E by

$$
\frac{1}{R_E} = \max_{\mathcal{H}} \frac{I}{\mathcal{D}}, \tag{2.28}
$$

with $\lambda_1, \lambda_2\,(>0)$ chosen so that $\mathcal{D}$ is positive-definite, then provided $R < R_E$ we may establish a conditional nonlinear stability result as follows. For $R < R_E$ we deduce from (2.27) that

$$
\frac{dE}{dt} \le -a\mathcal{D} + Pr < \theta^2 w >, \tag{2.29}
$$

where

$$
a \stackrel{\text{def}}{=} \frac{R_E - R}{R_E} \, (>0).
$$

Upon using the Cauchy-Schwarz inequality in (2.29) we see that

$$
\frac{dE}{dt} \le -a\mathcal{D} + Pr \|\theta^2\| \|w\|, \tag{2.30}
$$

and we then employ the Sobolev embedding inequality

$$
\|\theta^2\| \le c_1 \|\nabla \theta\|^2,
$$

to find

$$
\frac{dE}{dt} \le -\mathcal{D}(a - \sqrt{2} Pr\, c_1 E^{1/2}). \tag{2.31}
$$

Conditional nonlinear stability then follows from (2.31) provided

$$
(A) \quad R < R_E; \qquad\qquad (B) \quad E(0) < \frac{a^2}{2c_1^2 Pr^2} :
$$

details of the proof of conditional stability from an inequality like (2.31) may be found in Straughan [49], chapter 2.

Condition (B) is a restriction on the size of the initial data while (A) requires resolution of (2.28).

The Euler-Lagrange equations arising from (2.28) are:

$$
\begin{aligned}
& R_E F \theta k_i - 2\Delta u_i - 2K \epsilon_{kji} \omega_{k,j} + K \Lambda_1 \epsilon_{kji} \nu_{k,j} = \pi_{,i}, && (2.32)\\
& - R_E b \Lambda \epsilon_{3si} \theta_{,s} k_i + 4K\nu_i - 2\Gamma \Delta \nu_i - 2G \nu_{a,ai} - K\Lambda_1 \omega_i = 0, && (2.33)\\
& R_E F w + R_E b \Lambda \xi - 2\Delta\theta = 0, && (2.34)
\end{aligned}
$$

where F, Λ, Λ_1 are defined by

$$
F = \frac{2f + \lambda_2}{\sqrt{\lambda_2}}, \quad \Lambda = \sqrt{\frac{\lambda_2}{\lambda_1}}, \quad \Lambda_1 = \frac{1+\lambda_1}{\sqrt{\lambda_1}}. \tag{2.35}
$$

By using parametric differentiation on (2.32)–(2.34) we find, with a denoting λ_1 or λ_2 :

$$\Big\langle\Big(R_E\frac{\partial F}{\partial a}+F\frac{\partial R_E}{\partial a}\Big)\theta w\Big\rangle+b<\xi\theta>\Big(R_E\frac{\partial \Lambda}{\partial a}+\Lambda\frac{\partial R_E}{\partial a}\Big)$$
$$-K<\vec{\nu}\cdot\vec{\omega}>\frac{\partial \Lambda_1}{\partial a}=0. \tag{2.36}$$

Hence, using (2.35), (2.36) yields

$$\frac{\partial R_E}{\partial \lambda_1}<F\theta w>+b\Big(\frac{\partial R_E}{\partial \lambda_1}\Lambda-\frac{1}{2}R_E\frac{\sqrt{\lambda_2}}{\lambda_1^{3/2}}\Big)<\xi\theta>$$
$$-K<\vec{\nu}\cdot\vec{\omega}>\Big(\frac{1}{\sqrt{\lambda_1}}-\frac{1+\lambda_1}{2\lambda_1^{3/2}}\Big)=0, \tag{2.37}$$

$$\frac{\partial R_E}{\partial \lambda_2}<(2f+\lambda_2)\theta w>+R_E\Big\langle\Big(1-\frac{f}{\lambda_2}\Big)\theta w\Big\rangle$$
$$+\frac{b}{\sqrt{\lambda_1}}<\xi\theta>\Big(\frac{1}{2}R_E+\lambda_2\frac{\partial R_E}{\partial \lambda_2}\Big)=0. \tag{2.38}$$

Equation (2.37) is of much use since at the optimal value of λ_1, $\frac{\partial R_E}{\partial \lambda_1}=0$ and then (2.37) yields:

$$bR_E\sqrt{\lambda_2}<\xi\theta>+K<\vec{\nu}\cdot\vec{\omega}>(\lambda_1-1)=0. \tag{2.39}$$

This clearly demonstrates that when $b=0$, $\lambda_1=1$ is the best value to utilize. When, however, $b\neq 0$, (2.39) also shows λ_1 will be different from 1 at optimality.

A similar argument, evaluating (2.38) when $\frac{\partial R_E}{\partial \lambda_2}=0$ gives:

$$<(\lambda_2-2f)\theta w>+b<\xi\theta>\frac{\lambda_2}{\sqrt{\lambda_1}}=0.$$

If $b=0$ this suggests

$$\lambda_2=2f_{average}$$

will be a good value to start a numerical search.

The problem for general λ_1, λ_2 is resolved in Lindsay & Straughan [22]. We here choose $\lambda_1=1$ and set $\lambda_2=\lambda$, thus we have the assistance of Eringen's inequalities (2.10). The energy, $E(t)$, is now:

$$E(t)=\frac{1}{2}(\|\mathbf{u}\|^2+j\|\vec{\nu}\|^2)+\frac{1}{2}\lambda Pr\|\theta\|^2,$$

and this leads to the energy equation:

$$\frac{dE}{dt}=RI-\mathcal{D}+Pr<\theta^2 w>, \tag{2.40}$$

where the production term I and the dissipation $\mathcal{D}$ are in this case given by

$$I=-[2<f\theta w>+\lambda<\theta w>+b\lambda<\xi\theta>],$$

$$\mathcal{D} = \|\nabla \mathbf{u}\|^2 + K\|\vec{\nu}\|^2 + G\|\nu_{i,i}\|^2 + \Gamma\|\nabla\vec{\nu}\|^2 + \lambda\|\nabla\theta\|^2 + K\|\vec{\omega} - \vec{\nu}\|^2.$$

Nonlinear conditional stability follows from (2.40) in the same manner as the development from (2.27). The relevant Euler-Lagrange equations in the energy maximum problem are now:

$$\begin{aligned}
&(1+K)\Delta u_i + K\epsilon_{ijk}\nu_{k,j} - R_E F\phi k_i = \pi_{,i},\\
&\Delta\phi - R_E(Fw + \frac{1}{2}b\lambda^{1/2}\xi) = 0,\\
&\Gamma\Delta\nu_i + G\nu_{a,ai} + K(\epsilon_{ijk}u_{k,j} - 2\nu_i) + \frac{1}{2}\lambda^{1/2}b\epsilon_{3ji}R_E\phi_{,j} = 0,
\end{aligned} \tag{2.41}$$

where u_i is solenoidal, π is a Lagrange multiplier and where we have put

$$F = \frac{2f+\lambda}{2\sqrt{\lambda}}, \qquad \phi = \sqrt{\lambda}\theta.$$

Equations (2.41) are now reduced to the system:

$$\begin{aligned}
&\Delta\xi = \frac{K}{\Gamma}\Delta w + \frac{2K}{\Gamma}\xi - \frac{1}{2\Gamma}\lambda^{1/2}bR_E\Delta^*\phi,\\
&\Delta\phi = R_E(Fw + \frac{1}{2}b\lambda^{1/2}\xi),\\
&(1+K)\Delta^2 w - R_E F\Delta^*\phi + K\Big\{\frac{K}{\Gamma}\Delta w + \frac{2K}{\Gamma}\xi - \frac{1}{2\Gamma}\lambda^{1/2}bR_E\Delta^*\phi\Big\} = 0.
\end{aligned} \tag{2.42}$$

Equations (2.42) are in a form convenient for numerical solution, after using normal modes. The critical Rayleigh number is found from (2.42) with the optimization

$$Ra_E = \max_{\lambda}\min_{a^2} R_E^2(a^2;\lambda).$$

We do not discuss the stability problem when $b = 0$, but observe that when $b = 0$ the nonlinear stability results are considerably improved by Lindsay and Straughan [22] by employing a spatial weight. The stability then obtained is valid for *arbitrarily large initial perturbations.*

In addition to the above micropolar problem having a connection with FHD in its own right, it is useful in that the many coupling parameter method is also necessary in EHD, FHD, and TEMHD. Indeed, the multi-dimensional constrained optimization problem solved in [22] is the first case I know of where such a calculation is performed in nonlinear energy stability theory. However, it is important in that many problems in EHD, MHD, and TEMHD require exactly such an approach if accurate, quantitative nonlinear stability bounds are to be found.

3 – Thermal convection in a dielectric fluid

Throughout the next three sections we shall deal with an incompressible fluid. When we are interested in solving a convection problem involving a fluid when electromagnetic effects are present then we have to solve for the fields v_i, p, T, E_i and H_i, these being, respectively, velocity, pressure, temperature, electric field and magnetic field. Therefore, equations governing these fields have to be prescribed. The most general approach involves a thermodynamic prescription from the outset, cf. Müller [30], chapter 9. While this is the correct way to proceed, it leaves one with a very complicated system of equations to solve and little progress has been made in the field of thermal convection following this line of attack. Instead, people have preferred to write down the Navier-Stokes equations, a balance of energy equation, and Maxwell's equations, and incorporate the electromagnetic effects into the fluid behaviour through the body force term. An account of this is given in [49] and we do not dwell on it here.

We begin with an incompressible Newtonian fluid for which the momentum and continuity equations are:

$$\frac{\partial v_i}{\partial t} + v_j v_{i,j} = -\frac{1}{\rho} p_{,i} + \nu \Delta v_i + \alpha g T \delta_{i3} + f_i, \tag{3.1}$$

$$v_{i,i} = 0, \tag{3.2}$$

in which the $\alpha g T$ term is due to the buoyancy force while f_i denotes the externally supplied body force. The pressure here incorporates that part of the buoyancy force due to the reference temperature. The incorporation of electromagnetic effects is done through the f_i term.

An equation of energy balance is adopted which would have form

$$\rho\Big(\frac{\partial \epsilon}{\partial t} + v_i \epsilon_{,i}\Big) = -q_{i,i} + t_{ij} d_{ij} + \rho r, \tag{3.3}$$

where $\epsilon, q_i, t_{ij}, d_{ij}$ and r are the internal energy density, heat flux, stress tensor, symmetric part of the velocity gradient and the externally supplied heat source. A general procedure would now take account of the fact that quantities like the internal energy density and heat flux may depend directly on the electric and magnetic fields as well as temperature and temperature gradient. This aspect is discussed further in [49], chapter 12. We here follow *all* the papers I have seen on convection in dielectric fluids which simply write the balance of energy equation as the usual heat conduction equation in a moving body. This is clearly suspect and at least dimensional arguments should be given to remove the extra terms. However, the literature follows such an approach and often gives satisfactory agreement with experiment. Thus, we take as balance of energy the equation

$$\frac{\partial T}{\partial t} + v_i T_{,i} = \kappa \Delta T, \tag{3.4}$$

where we have also neglected heat supply.

To equations (3.1), (3.2) and (3.4) must be appended Maxwell's equations governing the behaviour of the electric and magnetic fields. These are:

$$\frac{\partial B_i}{\partial t} = -\epsilon_{ijk} E_{k,j}\,, \tag{3.5}$$

$$\epsilon_{ijk} H_{k,j} = j_i + \frac{\partial D_i}{\partial t}\,, \tag{3.6}$$

$$D_{i,i} = Q\,, \tag{3.7}$$

$$B_{i,i} = 0, \tag{3.8}$$

$$\frac{\partial Q}{\partial t} + j_{i,i} = 0, \tag{3.9}$$

where the terms D_i, B_i, j_i and Q are the electric displacement, magnetic induction, current and free charge, respectively. Equation (3.5) is Faraday's law, (3.6) is Ampère's law with the electric displacement correction of Maxwell, (3.7) and (3.8) are called the laws of Gauss, while (3.9) represents conservation of charge. In EHD, FHD or MHD it is usual to employ Maxwell's equations in some limiting form rather than their full form (3.5)-(3.9).

Convection in a dielectric fluid is discussed in chapter 11 of [49], and in particular the early contributions of Roberts [40] and Turnbull [51] are discussed at length. We cite several pertinent papers dealing with this topic: Castellanos *et al.* [5,6], Deo & Richardson [10], González & Castellanos [14], González *et al.* [15], Hoburg [16], Lacroix *et al.* [19], McCluskey & Atten [23], Martin & Richardson [24], Mohamed & El Shehawey [25], Rodriguz-Luis *et al.* [42], Rosensweig *et al.* [45], Sneyd [48] and Worraker & Richardson [53]. These articles cover various aspects of convection in dielectric fluids, by linear and weakly nonlinear stability methods. The paper by Deo & Richardson [10] uses the energy method but no temperature field is included, i.e. they consider only the isothermal situation. Therefore, we here begin a study of the nonlinear energy stability of convection in a dielectric fluid under non-isothermal conditions. I have not seen such an analysis before.

The problem considered now is one where an incompressible dielectric liquid is contained between two perfectly conducting rigid planar electrodes of infinite horizontal extent. The lower electrode at $z = 0$ is maintained at a fixed temperature Θ_0 which is greater than that of the collecting electrode at $z = d$ of constant temperature Θ_d. A homogeneous injection of unipolar charge of mobility K is present between the electrodes. The emitter at $z = 0$ is held at fixed potential ϕ_0 while the collector at $z = d$ is held at fixed potential ϕ_d. The precise situation is that considered by Rodriguez-Luis *et al.* [42] who developed a linearized instability analysis neglecting charge diffusion. Since we here develop a fully nonlinear energy analysis we include a charge diffusion term as did Deo & Richardson [10] in their investigation into the isothermal problem. As is pointed out in [10], the linear results without charge diffusion consistently exceed the experimental ones and so there is also a physical motivation for the inclusion of charge diffusion.

Rodriguez-Luis *et al.* [42] adopt a standard reduction of Maxwell's equations which neglects (3.6) and (3.8), (3.7) and (3.9) are retained while (3.5) is simplified to

$$\epsilon_{irs} E_{r,s} = 0$$

which allows the introduction of an electrical potential ϕ, via

$$E_i = -\phi_{,i}.$$

The electric field is introduced into the momentum equation (3.1) via the f_i term with a Coulombic force with representation

$$f_i \propto Q E_i.$$

The current j_i in (3.9) is here taken to have form

$$j_i = qv_i + KqE_i - D_c q_{,i}, \tag{3.10}$$

where, in line with [42], we have used q rather than Q to denote the charge, D_c is a (constant) charge diffusion coefficient and K is a mobility coefficient. In [42] K is a linear function of temperature but we here treat it as constant so as not to obscure the essentials of the nonlinear energy stability analysis. The inclusion of a temperature-dependent mobility K will not cause any conceptual difficulties, but will certainly complicate the mathematical analysis. The electric displacement D_i is related to the electric field E_i by the relation

$$D_i = \epsilon E_i, \tag{3.11}$$

where ϵ is the electrical permittivity. In [42] ϵ is a linear function of temperature but we again treat it as constant. Again, we do not expect any serious conceptual problems with this simplification, but the analysis is certainly clearer.

We now adopt the *non-dimensional* formulation of Rodriguez-Luis *et al.* [42] and then the governing equations for the dielectric convection problem are:

$$\frac{\partial v_i}{\partial t} + v_j \frac{\partial v_i}{\partial x_j} = -p_{,i} + \Delta v_i + \frac{Ra}{Pr}\Theta\delta_{i3} + \frac{CT^2}{M^2} qE_i, \tag{3.12}$$

$$v_{i,i} = 0, \tag{3.13}$$

$$Pr\Big(\frac{\partial \Theta}{\partial t} + v_j \frac{\partial \Theta}{\partial x_j}\Big) = \Delta\Theta, \tag{3.14}$$

$$\frac{\partial q}{\partial t} + v_j \frac{\partial q}{\partial x_j} + ReK(qE_i)_{,i} = S\Delta q, \tag{3.15}$$

$$E_{i,i} = \frac{C}{\epsilon} q, \tag{3.16}$$

$$E_i = -\phi_{,i}, \tag{3.17}$$

where we note that from (3.16), (3.17)

$$q = -\frac{\epsilon}{C}\Delta\phi. \tag{3.18}$$

Equations (3.12)-(3.17) are, respectively, the momentum equation, continuity equation, balance of energy equation, conservation of charge equation, and essentially the Maxwell equations (3.7) and (3.5). In these equations $v_i, p, \Theta, q, E_i, \phi$ denote velocity, pressure, temperature, charge, electric field, and electric potential, while the parameters Ra, Pr, C, T, M, Re, S are non-dimensional quantities being respectively, the Rayleigh number, Prandtl number, injection strength, applied voltage measure, mobility ratio, Reynolds number (based on ionic velocity), and charge diffusion coefficient.

Given the boundary conditions, equations (3.12)-(3.17) posess the steady state:

$$\bar{v}_i \equiv 0, \quad \bar{\Theta} = -z, \tag{3.19}$$

with $\bar{p}(z)$ determined from (3.12) and $\bar{\phi}(z), \bar{\mathbf{E}} = (0,0,\bar{E}_3(z)), \bar{q}(z)$ are found from (3.16), (3.17) and (3.15), the last which may be integrated to yield

$$\bar{\phi}'' + \frac{ReK}{2S}(\bar{\phi}')^2 = c_1 z + c_2, \tag{3.20}$$

where c_1, c_2 are constants determined from the boundary conditions. Thus, the basic state is a nonlinear function of z.

Before proceeding to the perturbation equations it is convenient to introduce a partial symmetrization by the transformation

$$\theta = \frac{R}{Pr}\Theta,$$

where $R = \sqrt{Ra}$. We now introduce perturbations $u_i, \pi, \theta, e_i, \phi, q$ by

$$v_i = \bar{v}_i + u_i, \qquad p = \bar{p} + \pi, \qquad \theta = \bar{\theta} + \theta,$$

$$E_i = \bar{E}_i + e_i, \qquad \phi = \bar{\phi} + \phi, \qquad q = \bar{q} + q.$$

The (nonlinear) perturbation equations then become:

$$\frac{\partial u_i}{\partial t} + u_j \frac{\partial u_i}{\partial x_j} = -\pi_{,i} + \Delta u_i + R\theta\delta_{i3} + \frac{CT^2}{M^2}[(\bar{q}+q)e_i + q\bar{E}_i], \tag{3.21}$$

$$u_{i,i} = 0, \tag{3.22}$$

$$Pr\Big(\frac{\partial \theta}{\partial t} + v_j \frac{\partial \theta}{\partial x_j}\Big) = Rw + \Delta\theta, \tag{3.23}$$

$$\frac{\partial q}{\partial t} + u_j(\bar{q}+q)_{,j} = S\Delta q - ReK[(\bar{q}+q)e_i + q\bar{E}_i]_{,i}, \tag{3.24}$$

$$e_{i,i} = \frac{C}{\epsilon}q, \quad e_i = -\phi_{,i}, \quad \Big(q = -\frac{\epsilon}{C}\Delta\phi\Big). \tag{3.25}$$

These equations hold on the three-dimensional spatial region $z \in (0,1)$ and the boundary conditions are that $u_i, \pi, \theta, q, \phi, e_i$ have an (x,y) planform which forms a repetitive plane tiling shape, typically hexagonal, and, on $z = 0,1$:

$$u_i = 0, \quad \phi = 0, \quad q = 0, \quad \theta = 0. \tag{3.26}$$

(I believe it should be possible to proceed with just these boundary conditions, but we later need an inequality which requires also $e_i = 0$ on $z = 0,1$. This is unfortunate, but is in keeping with the treatment of Deo & Richardson [10].)

Four separate integral (energy) identities are formed from (3.21)-(3.26). We multiply (3.21) by u_i, (3.23) by θ, and (3.24) by q, and then integrate over a period cell V, to obtain with the aid of (3.26) (no boundary conditions on $z = 0,1$ on e_i):

$$\begin{aligned}\frac{1}{2}\frac{d}{dt}\|\mathbf{u}\|^2 = &- \|\nabla\mathbf{u}\|^2 + R < \theta w > \\ &+ \frac{CT^2}{M^2}[< \bar{q}e_i u_i > + < \bar{E}_3 wq >] + \frac{CT^2}{M^2} < qe_i u_i >,\end{aligned} \tag{3.27}$$

$$\frac{1}{2}Pr\frac{d}{dt}\|\theta\|^2 = R < w\theta > -\|\nabla\theta\|^2, \tag{3.28}$$

$$\begin{aligned}\frac{1}{2}\frac{d}{dt}\|q\|^2 =& - S\|\nabla q\|^2 - \Big\langle\frac{d\bar{q}}{dz}wq\Big\rangle - ReK\Big\langle q^2\Big(\frac{C}{\epsilon}\bar{q} + \frac{1}{2}\frac{d\bar{E}_3}{dz}\Big)\Big\rangle \\ & - ReK\Big\langle\frac{d\bar{q}}{dz}e_3q\Big\rangle - \frac{1}{2}ReK\frac{C}{\epsilon} < q^3 >,\end{aligned} \tag{3.29}$$

where $\|\cdot\|$ and $<\cdot>$ denote the $L^2(V)$ norm and integration over V, respectively.

A rigorous nonlinear stability analysis is now possible using these three identities, as we show below. However, it may be of benefit for a sharper analysis to derive additionally an equation for $\mathbf{e}$, and to this end we multiply (3.24) by ϕ and integrate over V using (3.26) to find:

$$\begin{aligned}\frac{1}{2}\frac{d}{dt}\|\mathbf{e}\|^2 =& - \frac{CS}{\epsilon}\|q\|^2 - ReK < \bar{q}e^2 > - < \bar{q}e_iu_i > \\ & - ReK < \bar{E}_3qe_3 > - < qu_ie_i > -ReK < qe^2 >,\end{aligned} \tag{3.30}$$

where $e^2 = e_ie_i$.

The first approach. For positive coupling parameters λ_1, λ_2 we define $E(t)$ by

$$E(t) = \frac{1}{2}\|\mathbf{u}\|^2 + \frac{1}{2}\lambda_1Pr\|\theta\|^2 + \frac{1}{2}\lambda_2\|q\|^2. \tag{3.31}$$

Then from (3.27)-(3.29) we obtain an energy identity

$$\frac{dE}{dt} = I - D + N, \tag{3.32}$$

where I contains quadratic perturbation terms, D is the (positive) dissipation, and N is a nonlinear term cubic in the perturbations,

$$\begin{aligned}I =& R(1+\lambda_1) < w\theta > + \frac{CT^2}{M^2}[< \bar{q}e_iu_i > + < \bar{E}_3wq >] \\ & - \lambda_2\Big\langle\frac{d\bar{q}}{dz}\big(wq + ReKe_3q\big)\Big\rangle - \lambda_2ReK\Big\langle q^2\Big(\frac{C}{\epsilon}\bar{q} + \frac{1}{2}\frac{d\bar{E}_3}{dz}\Big\rangle,\end{aligned} \tag{3.33}$$

$$D = \|\nabla\mathbf{u}\|^2 + \lambda_1\|\nabla\theta\|^2 + \lambda_2S\|\nabla q\|^2, \tag{3.34}$$

$$N = \frac{CT^2}{M^2} < qe_iu_i > -\frac{1}{2}\lambda_2ReK\frac{C}{\epsilon} < q^3 > . \tag{3.35}$$

Define now

$$\frac{1}{\Lambda} = \max_{\mathcal{H}} \frac{I}{D}, \tag{3.36}$$

where $\mathcal{H}$ is the space of admissible solutions. The stability problem reduces to two calculations. The first consists in showing Λ^{-1} is meaningful and then determining the maximizing solution which leads to the nonlinear stability boundary relationship between R, C, T, M, Re, K and S. We do not calculate the Euler-Lagrange equations for (3.36) here as their solution involves an extensive numerical calculation involving optimization in λ_1, λ_2, with a $z-$dependent basic state. However, we do observe that the maximum in

(3.36) is meaningful. To do this is is sufficient to note that u_i, θ, q are bounded in (3.33) by the D terms in (3.34) using Poincaré's inequality. We must further show the e_i terms are likewise bounded. To do this we note that using (3.25), the x, y periodicity, and the fact that $\phi = 0$ on $z = 0, 1$,

$$< q\phi >= \frac{\epsilon}{C}\|\nabla\phi\|^2 = \frac{\epsilon}{C}\|\mathbf{e}\|^2,$$

and then with the aid of the Cauchy-Schwarz inequality:

$$\begin{aligned} \frac{\epsilon}{C}\|\mathbf{e}\|^2 &\le \|q\|\|\phi\|, \\ &\le \pi^{-1}\|q\|\|\nabla\phi\| = \pi^{-1}\|q\|\|\mathbf{e}\|, \end{aligned} \tag{3.37}$$

where Poincaré's inequality has been employed. From (3.37)

$$\|\mathbf{e}\| \le \frac{C}{\pi\epsilon}\|q\| \le \frac{C}{\pi^2\epsilon}\|\nabla q\|, \tag{3.38}$$

where Poincaré's inequality is again employed, noting $q = 0$ on $z = 0, 1$. Since $\|\mathbf{e}\|$ is bounded by $\|\nabla q\|$, I may be dominated by D and the maximum in (3.36) makes sense.

We leave the calculation of the nonlinear stability parameter boundary and return to the second part of the stability calculation. From (3.32) and (3.36) we know

$$\frac{dE}{dt} \le -D\Big(1 - \frac{1}{\Lambda}\Big) + N. \tag{3.39}$$

We require $\Lambda^{-1} < 1$ and an inequality of form

$$N \le cE^{\frac{1}{2}}D, \tag{3.40}$$

for a constant c, for then (3.39) leads to the inequality

$$\frac{dE}{dt} \le -aD + cE^{\frac{1}{2}}D,$$

where $a = 1 - \Lambda^{-1}\,(> 0)$. It is then easy to show with the aid of Poincaré's inequality that $E \to 0$ as $t \to \infty$, see e.g. Straughan [49], chapter 2, and so nonlinear stability follows.

We show how to establish (3.40). From (3.35) we use the Cauchy-Schwarz inequality to obtain

$$N \le \frac{1}{2}\lambda_2 ReK\frac{C}{\epsilon}\|q\|\|q^2\| + \frac{CT^2}{M^2}\|\mathbf{u}\|\|q\mathbf{e}\|. \tag{3.41}$$

Next, use the Sobolev inequality to deduce there is a constant c_1 such that

$$\|q^2\| \le c_1\|\nabla q\|^2. \tag{3.42}$$

After this, observe by further use of the Cauchy-Schwarz inequality:

$$\|q\mathbf{e}\| \le \|q^2\|^{\frac{1}{2}} < e^4 >^{1/4}. \tag{3.43}$$

We now assume $e_i = 0$ on $z = 0, 1$. (This is the only place such an assumption is made, and I do not see how to apply the appropriate Sobolev inequality without it.) Then we have the Sobolev inequality

$$< e^4 > \leq \|e\| < e_{i,j} e_{i,j} >^{3/2} .$$

Since $e_i = -\phi_{,i}$,

$$< e_{i,j} e_{i,j} > = < (e_{i,i})^2 > = \frac{C^2}{\epsilon^2} \|q\|^2,$$

and then

$$< e^4 > \leq \frac{C^{3/2}}{\epsilon^{3/2}} \|e\| \|q\|^3.$$

Hence,

$$\begin{aligned} \|q\mathbf{e}\| &\leq \frac{C^{3/8}}{\epsilon^{3/8}} \|q^2\|^{\frac{1}{2}} \|e\|^{1/4} \|q\|^{3/4}, \\ &\leq \frac{C^{5/8}}{\epsilon^{5/8} \pi^{1/4}} \|q^2\|^{\frac{1}{2}} \|q\|, \end{aligned}$$

using (3.38). Further use of inequality (3.42) then yields

$$\|q\mathbf{e}\| \leq \frac{C^{5/8} c_1^{\frac{1}{2}}}{\pi^{1/4} \epsilon^{5/8}} \|\nabla q\| \|q\|. \tag{3.44}$$

Hence, putting (3.42) and (3.44) in (3.41) we deduce $\exists$ positive constants k_1, k_2 such that

$$N \leq k_1 \|q\| \|\nabla q\|^2 + k_2 \|\mathbf{u}\| \|q\| \|\nabla q\|. \tag{3.45}$$

Recalling the form for E, (3.31), and using Poincaré's inequality in (3.45), (3.40) follows.

Thus, a rigorous nonlinear stability criterion is established.

The second approach. This approach basically proceeds as in the first one except identity (3.30) is also included. Hence, define now $E(t)$ by

$$E(t) = \frac{1}{2} \|\mathbf{u}\|^2 + \frac{1}{2} \lambda_1 Pr \|\theta\|^2 + \frac{1}{2} \lambda_2 \|q\|^2 + \frac{1}{2} \lambda_3 \|\mathbf{e}\|^2, \tag{3.46}$$

where $\lambda_1, \lambda_2, \lambda_3$ are coupling parameters at our disposal. The energy equation is in this case obtained from (3.27)-(3.30) as

$$\frac{dE}{dt} = I - D + N, \tag{3.47}$$

where I, D and N are now given by

$$\begin{aligned} I =& R(1 + \lambda_1) < w\theta > + \frac{CT^2}{M^2} [< \bar{q} e_i u_i > + < \bar{E}_3 w q >] \\ & - \lambda_2 \Big\langle \frac{d\bar{q}}{dz} \big(wq + ReK e_3 q \big) \Big\rangle - \lambda_2 ReK \Big\langle q^2 \Big(\frac{C}{\epsilon} \bar{q} + \frac{1}{2} \frac{d\bar{E}_3}{dz} \Big) \Big\rangle \\ & - \lambda_3 ReK < \bar{E}_3 q e_3 > - \lambda_3 < \bar{q} u_i e_i >, \end{aligned} \tag{3.48}$$

$$D = \|\nabla \mathbf{u}\|^2 + \lambda_1\|\nabla\theta\|^2 + \lambda_2 S\|\nabla q\|^2 + \lambda_3\frac{CS}{\epsilon}\|q\|^2\lambda_3 ReK < \bar{q}e^2 >, \tag{3.49}$$

$$N = \frac{CT^2}{M^2} < qe_iu_i > -\frac{1}{2}\lambda_2 ReK\frac{C}{\epsilon} < q^3 > -\lambda_3 < u_ie_iq > -\lambda_3 ReK < qe^2 > . \tag{3.50}$$

It should be noted that the last term in D is included because $\bar{q}$ is the charge in the steady state and so is physically a positive quantity. Deo & Richardson [10] adopt a procedure where they use the positivity of $\bar{q} + q$ to facilitate their stability analysis. Doubtless such an approach could be followed here also.

Again we now define

$$\frac{1}{\Lambda} = \max_{\mathcal{H}} \frac{I}{D}, \tag{3.51}$$

where $\mathcal{H}$ is the space of admissible solutions. The stability problem reduces to two calculations as before: that of finding the maximising solution and hence the nonlinear stability parameter boundary, and the controlling of the cubic nonlinearities. The only essential new nonlinearity in N in (3.50) is the $< qe^2 >$ term. There are various ways to proceed with this term. One such is to use the Cauchy-Schwarz inequality to obtain

$$< qe^2 > \leq \|q\|\|e^2\|,$$

and then use the inequality

$$\|e^2\| \leq \frac{C^{3/2}}{\epsilon^{3/2}}\|e\|\|q\|^3.$$

Thus

$$< qe^2 > \leq \|q\|^4\|e\|,$$

and remembering the form of D it is easy to show

$$N \leq cE^{\frac{1}{2}}D,$$

and so the stability inequality may be established. The maximum problem for I/D is not solved here, not because it presents any conceptual difficulties, simply that it is an involved numerical procedure which also involves optimization over the λ_i, and its inclusion would render this article too long.

Before closing this section we remark on not unrelated work of Morro [26], Morro *et al.* [27], and Morro & Parodi [28,29]. These writers develop a beautiful theory for studying the equilibrium solution to nonlinear problems for dielectric media by using a Lagrangian variational approach. Their work is very much motivated by genetic problems and they treat specific studies. They remark in [29] for example that, *High fields in dielectric media are responsible for electric nonlinear effects. Their determination is becoming more and more important in several contexts, such as electrochemistry, biophysics, optics and molecular electronics.* They also pay particular attention to the importance of nonlinear effects near a polyelectrolyte macroion in a water solution. There is a large variation between the electrical permittivity of water near a DNA macroion and that in the bulk of the fluid and the understanding of this is crucial to explaining the behaviour of the ionic charges surrounding the macroion. The analysis of [26]-[29] is entirely based on a static approach. A dynamic analogue of their work investigating the dynamic nonlinear stability of the equilibrium configurations would be very desirable.

4 – Thermal convection in a magnetic fluid

The topic of this section is in some sense at the opposite end of the physical spectrum to that of section 3. We consider a layer of magnetizable fluid contained betwen two planes $z = \pm\frac{1}{2}d$, with prescribed constant temperatures T_U, T_L at $z = \pm\frac{1}{2}d$. Intriguing convection patterns can be set up in this problem, including a fascinating labyrinthine one, see Rosensweig [43], p. 208, Rosensweig *et al.* [45]; the book by Rosensweig [43] is an extremely stimulating account of the beautiful subject of magnetic fluids. The thermal convection problem in ferrohydrodynamics is reviewed in Rosensweig [43], p. 228, and in Straughan [49], where various models are discussed in chapter 12, section 2. A nonlinear energy stability analysis for a further model with a force representation due to Korteweg & Helmholtz is presented in [49], section 12.3. Various other investigations of the above problem by linear and weakly nonlinear analyses may be found in Blennerhasset *et al.* [4], Curtis [9], Finlayson [13], Lalas & Carmi [20], Rosensweig [43,44], Rosensweig *et al.* [45], Shliomis [47], and for a similar problem in a porous medium by Vaidyanathan *et al.* [52]. Work on another class of ferrohydrodynamic problems of interest, particularly in the context of rotating neutron stars, was instigated by Muzikar & Pethick [31] and Roberts [41], and the thermal convection problem for a model based on their theory is investigated by Abdullah & Lindsay [1,2]. In this paper we restrict attention wholly to the recent model of Blennerhasset *et al.* [4] who use linear and weakly nonlinear methods. Our work is complimentary to theirs as it develops a rigorous fully nonlinear three-dimensional energy stability analysis. We employ the notation of [4].

The ferrohydrodynamic Bénard problem again uses (3.1) and (3.2), and Blennerhasset *et al.* [4] use (3.4), although other writers use a version of (3.3), see e.g. the account in [49], section 12.2. The electric field effects are in some sense neglected and Maxwell's equations (3.5)-(3.9) are employed in the limit

$$\epsilon_{ijk}H_{k,j} = 0, \qquad B_{i,i} = 0. \tag{4.1}$$

However, this is a special case and there will be instances when this is inadequate: a fuller account of this may be found in Rosensweig [43], p. 91.

The body force term f_i in (3.1) is given by Rosensweig [43] as

$$f_i = \mu_0 M H_{,i}\,, \tag{4.2}$$

where μ_0 is the permeability of free space, and where M is the magnitude of the magnetization vector $\mathbf{M}$, H being the magnitude of the magnetic field $\mathbf{H}$. Blennerhasset *et al.* [4] replace (4.2) by

$$f_i = \mu_0 M_j H_{i,j}. \tag{4.3}$$

The equivalence of (4.2), (4.3) is easily shown if $\mathbf{M}$ and $\mathbf{H}$ are collinear, as in Rosensweig [43], p. 110. For then, $\mathbf{M} = (M/H)\mathbf{H}$ and

$$M_j H_{i,j} = \frac{M}{H} H_j H_{i,j}\,.$$

By standard arguments of Tensor calculus,

$$\begin{aligned}
H_j H_{i,j} =& H_j H_{j,i} - \epsilon_{ijk} H_j (\epsilon_{krs} H_{s,r}),\\
=& H_j H_{j,i},\\
=& \frac{1}{2} H^2_{,i},\\
=& H H_{,i},
\end{aligned}$$

where in the first line $(4.1)_1$ has been employed, and so

$$M_j H_{i,j} = M H_{,i}.$$

Blennerhasset *et al.* [4] restrict attention to the situation of a strong applied field and adopt the relationship

$$M_i = [M_0 - K(T - T_0)] \frac{H_i}{H}, \tag{4.4}$$

where K is called the pyromagnetic coefficient.

The steady state of [4] is one which

$$\bar{T} = T_0 - \beta z d, \quad T_0 = \frac{1}{2}(T_L + T_U), \quad \beta = \frac{T_L - T_U}{d}, \tag{4.5}$$

although they allow β to be positive (heating below) or negative (heating above). The steady magnetization and magnetic field are given by

$$\bar{M}_i = (M_0 + K\beta z d)\delta_{i3}, \quad \bar{H}_i = (H_0 - K\beta z d)\delta_{i3}. \tag{4.6}$$

Their non-dimensionalization writes the perturbations to the temperature $\bar{T}$ and magnetic field $\bar{H}_i$, Θ, h_i, as

$$\begin{aligned}
T &= \bar{T} + d\beta\Theta,\\
H_i &= \bar{H}_i + K\beta d h_i = \bar{H}_i + K\beta d \phi_{,i}\,,
\end{aligned}$$

ϕ being the potential of the perturbation magnetic field. Since in [4] attention is restricted to the limit of strong applied magnetic field, they take

$$M_i = \bar{M}_i - d\beta K \Theta \delta_{i3}. \tag{4.7}$$

(The negative sign in (4.7) is positive in [4], but I think it should be as in (4.7).) In this case the perturbation quantity arising from the momentum equation (3.1) differs from that of the classical Bénard problem by the $\mathbf{M} \cdot \nabla \mathbf{H}$ term which is

$$\begin{aligned}
M_j H_{i,j} =& (\bar{M}_j - \delta_{j3} d\beta K\Theta)(\bar{H}_{i,j} + K\beta d \phi_{,ij})\\
=& \bar{M}_j \bar{H}_{i,j} - \bar{H}_{i,z} d\beta K \Theta + \bar{M}_j K\beta d \phi_{,ij} - (d\beta K)^2 \Theta \phi_{,iz}.
\end{aligned}$$

Only the last three terms figure in the perturbation equation (the first belonging to the steady state equation) and so the terms arising from the perturbation of $\mathbf{M} \cdot \nabla \mathbf{H}$ are, using (4.6)

$$\begin{aligned}
M_j H_{i,j\,(\text{perturbation})} =& (d\beta K)^2 \Theta \delta_{i3} + \bar{M} K\beta d \phi_{,iz} - (d\beta K)^2 \Theta \phi_{,iz},\\
=& (d\beta K)^2 \Theta \delta_{i3} - (d\beta K)^2 \Theta \phi_{,iz} - (K\beta d)^2 \delta_{i3} \phi_{,z} + K\beta d (\bar{M}\phi_{,z})_{,i}.
\end{aligned}$$

The last term is incorporated into the pressure and then using the non-dimensionalization of [4] which uses

$$Ra = \frac{g\alpha\beta d^4}{\kappa\nu}, \qquad N = \frac{\mu_0 K^2\beta^2 d^4}{\kappa\nu\rho_0},$$

where Ra is the Rayleigh number and N is a non-dimensional magnetic number, the perturbation momentum equation becomes:

$$Pr^{-1}\Big(\frac{\partial u_i}{\partial t} + u_j\frac{\partial u_i}{\partial x_j}\Big) = -p_{,i} + \Delta u_i + (Ra + N)\Theta\delta_{i3} - N\phi_{,z}\delta_{i3} - N\Theta\phi_{,zi}. \tag{4.8}$$

Due to incompressibility the velocity u_i is solenoidal. The temperature perturbation equation of [4] is

$$\frac{\partial\Theta}{\partial t} + u_i\frac{\partial\Theta}{\partial x_i} = w + \Delta\Theta, \tag{4.9}$$

w being u_3.

It is convenient for our purpose to introduce the transformation

$$\theta = R\Theta, \tag{4.10}$$

where $R = \sqrt{Ra}$, and then the perturbation equations of momentum, continuity, and balance of energy, are:

$$Pr^{-1}\Big(\frac{\partial u_i}{\partial t} + u_j\frac{\partial u_i}{\partial x_j}\Big) = -p_{,i} + \Delta u_i + (R + \frac{N}{R})\theta\delta_{i3} - N\phi_{,z}\delta_{i3} - \frac{N}{R}\theta\phi_{,zi}, \tag{4.11}$$

$$u_{i,i} = 0, \tag{4.12}$$

$$\frac{\partial\theta}{\partial t} + u_i\frac{\partial\theta}{\partial x_i} = \pm Rw + \Delta\theta, \tag{4.13}$$

where the positive sign is taken when heating from below and the negative sign when heating from above. (Actually, in the heated above case Ra is negative and we have to work with $\sqrt{|Ra|}$.)

There is one further perturbation equation which arises from $(4.1)_2$. A derivation of this employing a Boussinesq approximation may be found in [49], pp. 182, 183. In the notation of [4], recollecting the transformation (4.10), the relevant equation is

$$\Delta\phi = \frac{1}{R}\theta_{,z}. \tag{4.14}$$

The complete system of equations is (4.11)-(4.14) which hold on the three-dimensional spatial region $z \in (-\frac{1}{2}, \frac{1}{2})$. The boundary conditions we adopt are those of [4] which are

$$u_i = 0, \qquad \theta = 0, \qquad z = \pm\frac{1}{2}, \tag{4.15}$$

$$\phi_{,z} \pm a\phi = 0, \qquad z = \pm\frac{1}{2}, \tag{4.16}$$

and that u_i, θ, p and ϕ have a periodic x, y shape. The period cell of this shape is denoted by V. In the boundary condition (4.15) a is effectively a wave number and a derivation of such a condition may be found in Roberts [40], Straughan [49], p. 165.

The nonlinear analysis begins by multiplying (4.11) by u_i, (4.13) by θ, and integrating over V, to find with the use of the boundary conditions (4.15), and (4.12),

$$\begin{aligned}\frac{1}{2Pr}\frac{d}{dt}\|\mathbf{u}\|^2 = &- \|\nabla\mathbf{u}\|^2 + \Big(R + \frac{N}{R}\Big) < \theta w > \\ &- N < w\phi_{,z} > - \frac{N}{R} < \theta u_i \phi_{,iz} >,\end{aligned} \tag{4.17}$$

$$\frac{1}{2}\frac{d}{dt}\|\theta\|^2 = \pm R < w\theta > - \|\nabla\theta\|^2, \tag{4.18}$$

where $\|\cdot\|$ and $< \cdot >$ denote the $L^2(V)$ norm and integration over V, respectively.

To simplify things we shall henceforth replace (4.16) by $\phi = 0$ on $z = \pm\frac{1}{2}$, although we stress it is not necessary. In the more general case of (4.16) we have to work with ϕ defined not only in V, but also in $\hat{V}$, the hexagonal cylinder exterior to V, and employ continuity conditions on ϕ at $z = \pm\frac{1}{2}$. With the above assumption we multiply (4.14) by ϕ and integrate over V to find

$$0 = -\|\nabla\phi\|^2 - \frac{1}{R} < \phi\theta_{,z} > . \tag{4.19}$$

The idea is to form (4.17) + λ_1(4.18) + λ_2(4.19), for coupling parameters $\lambda_1, \lambda_2 (> 0)$ to be judiciously chosen. Hence, defining E, I, D and $\mathcal{N}$ by

$$E = \frac{1}{2Pr}\|\mathbf{u}\|^2 + \frac{\lambda_1}{2}\|\theta\|^2, \tag{4.20}$$

$$I = \Big(R + \frac{N}{R} \pm \lambda_1 R\Big) < \theta w > - N < w\phi_{,z} > - \frac{\lambda_2}{R} < \phi\theta_{,z} >, \tag{4.21}$$

$$D = \|\nabla\mathbf{u}\|^2 + \lambda_1\|\nabla\theta\|^2 + \lambda_2\|\nabla\phi\|^2, \tag{4.22}$$

$$\mathcal{N} = -\frac{N}{R} < \theta u_i \phi_{,iz} >, \tag{4.23}$$

we obtain

$$\frac{dE}{dt} = I - D + \mathcal{N}. \tag{4.24}$$

Define now Λ by

$$\frac{1}{\Lambda} = \max_{\mathcal{H}} \frac{I}{D}, \tag{4.25}$$

where $\mathcal{H}$ is the space of admissible solutions. Then we write

$$\frac{dE}{dt} \leq -D\Big(1 - \frac{1}{\Lambda}\Big) + \mathcal{N}. \tag{4.26}$$

The first requirement is that $\Lambda^{-1} < 1$, which we return to later. Then, put $a = 1 - \Lambda^{-1}$ ($>$ 0) and so

$$\frac{dE}{dt} \leq -aD + \mathcal{N}. \tag{4.27}$$

The second step is to control the cubic nonlinearity $\mathcal{N}$. To do this we use the Cauchy-Schwarz inequality twice, and then the Sobolev inequality to derive

$$\begin{aligned}
\mathcal{N} \leq & \frac{N}{R} < |\mathbf{u}|^2\theta^2 >^{\frac{1}{2}} \|\nabla\phi_{,z}\|, \\
\leq & \frac{N}{R} < |\mathbf{u}|^4 >^{1/4} < \theta^4 >^{1/4} \|\nabla\phi_{,z}\|, \\
\leq & \frac{Nc_1}{R} \|\nabla\mathbf{u}\| < \theta^4 >^{1/4} \|\nabla\phi_{,z}\|.
\end{aligned} \tag{4.28}$$

An estimate is now obtained for the term $\|\nabla\phi_{,z}\|$. Squaring (4.14) we show

$$\|\Delta\phi\|^2 = \frac{1}{R^2}\|\theta_{,z}\|^2. \tag{4.29}$$

By direct calculation

$$\|\nabla\phi_{,z}\|^2 = < \phi_{,xz}^2 > + < \phi_{,yz}^2 > + < \phi_{,zz}^2 > . \tag{4.30}$$

From a straightforward integration by parts

$$\int_V \phi_{,xx}\phi_{,zz}dV = -\int_V \phi_{,x}\phi_{,xzz}dV + \int_V \frac{\partial}{\partial x}(\phi_{,x}\phi_{,zz})dV.$$

In the case of a hexagonal cell the last term when evaluated on the boundary yields contributions which add up to zero (due to periodicity). Then, a further integration by parts shows

$$\int_V \phi_{,xx}\phi_{,zz}dV = -\int_V \phi_{,x}\phi_{,xzz}dV = \int_V \phi_{,xz}\phi_{,xz}dV - \int_V \frac{\partial}{\partial z}(\phi_{,x}\phi_{,xz})dV.$$

Since $\phi = 0$ on $z = \pm\frac{1}{2}$, the last term is zero and

$$< \phi_{,xz}^2 > = < \phi_{,xx}\phi_{,zz} > . \tag{4.31}$$

A similar calculation shows

$$< \phi_{,yz}^2 > = < \phi_{,yy}\phi_{,zz} > . \tag{4.32}$$

Using the arithmetic-geometric mean inequality we have

$$\begin{aligned}
\|\nabla\phi_{,z}\|^2 \leq & < \phi_{,zz}^2 > +2 < \phi_{,xz}^2 > +2 < \phi_{,yz}^2 > \\
& + 2 < \phi_{,xx}\phi_{,yy} > + < \phi_{,xx}^2 > + < \phi_{,yy}^2 >, \\
= & < \phi_{,xx}^2 > + < \phi_{,yy}^2 > + < \phi_{,zz}^2 > \\
& + 2 < \phi_{,xx}\phi_{,yy} > +2 < \phi_{,xx}\phi_{,zz} > +2 < \phi_{,zz}\phi_{,yy} >, \\
= & \|\Delta\phi\|^2,
\end{aligned} \tag{4.33}$$

where (4.31), (4.32) have also been employed. Hence, combining (4.33) and (4.29) we derive

$$\|\nabla\phi_{,z}\| \leq \frac{1}{R}\|\theta_{,z}\|. \tag{4.34}$$

If we now use (4.34) in (4.28), and the definition of gradient, we see that

$$\mathcal{N} \leq \frac{Nc_1}{R^2} < \theta^4 >^{1/4} \|\nabla \mathbf{u}\| \|\nabla \theta\|. \tag{4.35}$$

We want an inequality of form

$$\mathcal{N} \leq cE^{\alpha} D,$$

for some $\alpha \geq 0$. In view of the forms of E and D, (4.20) and (4.22), this would not appear possible *directly* from (4.35). Instead, we must consider a generalized approach and construct an energy identity for $\|\theta^2\|$. (Generalized energy arguments are described in much detail in [49].) To this end multiply (4.13) by θ^3 and integrate over V. The result is

$$\frac{1}{4}\frac{d}{dt}\|\gamma\|^2 = \pm R < w\theta^3 > -\frac{3}{4}\|\nabla\gamma\|^2, \tag{4.36}$$

where we have put $\gamma = \theta^2$. The first term on the right of (4.36) is manipulated as follows, using the Cauchy-Schwarz, Sobolev and Poincaré inequalities:

$$\begin{aligned} \pm R < w\theta^3 > &\leq R\|w\| < \theta^2\gamma^2 >^{\frac{1}{2}}, \\ &\leq R\|w\| < \theta^4 >^{1/4} < \gamma^4 >^{1/4}, \\ &\leq \frac{Rc_1}{\pi} < \theta^4 >^{1/4} \|\nabla \mathbf{u}\| \|\nabla\gamma\|. \end{aligned} \tag{4.37}$$

We now define a *generalized energy* $\mathcal{E}(t)$ by

$$\mathcal{E}(t) = E(t) + \frac{\xi}{4} < \theta^4 >, \tag{4.38}$$

for $\xi(> 0)$ another coupling parameter to be selected. Then combining (4.27), (4.35)-(4.37) we find

$$\begin{aligned} \frac{d}{dt}\mathcal{E} \leq &- aD - \frac{3}{4\xi}\|\nabla\gamma\|^2 + \frac{Nc_1}{R^2}\|\nabla \mathbf{u}\|\|\nabla\theta\| < \theta^4 >^{1/4} \\ &+ \frac{\xi Rc_1}{\pi}\|\nabla \mathbf{u}\|\|\nabla\gamma\| < \theta^4 >^{1/4}. \end{aligned} \tag{4.39}$$

Define $\mathcal{D}$ by

$$\mathcal{D} = aD + \frac{3\xi}{4}\|\nabla\gamma\|^2. \tag{4.40}$$

Then, using the definitions of $\mathcal{D}$ and $\mathcal{E}$ it is not difficult to obtain a constant c such that (4.39) leads to

$$\frac{d}{dt}\mathcal{E} \leq -\mathcal{D} + c\mathcal{D}\mathcal{E}^{1/4}. \tag{4.41}$$

If

$$\mathcal{E}^{1/4}(0) < \frac{1}{c}, \tag{4.42}$$

then it is easy to show $\mathcal{E}(t) \to 0$ as $t \to \infty$ and nonlinear stability is established, cf. [49], chapter 2.

The nonlinear stability criterion just derived requires not only the initial data restriction (4.42) but also the restriction on Λ defined in (4.25), namely $\Lambda^{-1} < 1$. The critical value of Λ is $\Lambda = 1$ and with this value the maximum problem in (4.25) leads to a system of equations for the critical stability parameter boundary, involving R, N, λ_1 and λ_2. The Euler-Lagrange equations for this critical case are:

$$\begin{aligned}
\left[\frac{R(1\pm\lambda_1)+(N/R)}{\sqrt{\lambda_1}}\right]\theta\delta_{i3} - \frac{N}{\sqrt{\lambda_2}}\phi_{,z}\delta_{i3} + 2\Delta u_i &= \pi_{,i},\\
\left[\frac{R(1\pm\lambda_1)+(N/R)}{\sqrt{\lambda_1}}\right]w + \frac{1}{R}\sqrt{\frac{\lambda_2}{\lambda_1}}\phi_{,z} + 2\Delta\theta &= 0,\\
\frac{N}{\sqrt{\lambda_2}}w_{,z} - \frac{1}{R}\sqrt{\frac{\lambda_2}{\lambda_1}}\theta_{,z} + 2\Delta\phi &= 0,
\end{aligned} \tag{4.43}$$

where π is a Lagrange multiplier which arises since u_i is solenoidal.

At the time of writing I have not been able to make any useful deductions from (4.43) using parametric differentiation. The numerical solution of (4.43) has yet to be performed but will involve solving an eighth order eigenvalue problem and then optimizing in λ_1 and λ_2.

5 – Thermoelectric magnetohydrodynamics

In this section we turn to a technologically important topic, but one which has evidently not seen much mathematical analysis.

Stability analyses in MHD without thermoelectric effects are many, see e.g. Roberts [39], Straughan [49], and the references therein. In particular, the original nonlinear energy stability studies of Rionero [33-38] and not unrelated calculations of Backus [3] and Chandrasekhar [7] are worth highlighting. Nevertheless, it has been fashionable in papers on nonlinear energy stability for the Bénard problem in MHD to prescribe the components of the magnetic field itself on the boundary of the fluid. The physically correct conditions are those of Chandrasekhar [8], p. 163, which involve continuity of magnetic field and zero normal component of current density when the medium adjoining the fluid is electrically non-conducting, or zero normal component of magnetic field and zero tangential current when the medium adjoining the fluid is a perfect conductor: nonlinear energy stability analyses which employ these boundary conditions are less common.

Rather than discuss energy stability work in MHD we look at problems in MHD with thermoelectric effects. The basic equations are given in Shercliff [46], see also Landau *et al.* [21], pp. 97-100. Shercliff [46] treats Hartmann flow and points out the relevance of TEMHD in liquid metal use, such as lithium, in nuclear reactors. I have not seen the Bénard problem discussed in the context of TEMHD, as is now done.

There are two fundamental differences with MHD. The first is the generalization of Ohm's law

$$\frac{\mathbf{j}}{\sigma} = \mathbf{E} + \mathbf{v}\times\mathbf{B} - \alpha\,\mathrm{grad}\,T, \tag{5.1}$$

where α is called the *absolute thermoelectric power* and in general is a function of temperature, T. The second is the relation between the amount of heat evolved per unit time and volume, Q, in the conductor, and the electromagnetic quantities, cf. Landau *et al.* [21]. p. 99,

$$
\begin{aligned}
Q =& - \operatorname{div} \mathbf{q} \\
=& \operatorname{div}(\kappa \operatorname{grad} T) + \frac{j^2}{\sigma} - T\mathrm{j} \cdot \operatorname{grad} \alpha,
\end{aligned}
\tag{5.2}
$$

where the first term is the usual Fourier heat conduction term, the second is the Joule heating, while the third is due to the thermoelectric effect. In fact, the last term gives rise to the *Thomson effect*

$$
-T\mathrm{j} \cdot \operatorname{grad} \alpha = -T\frac{d\alpha}{dT}\mathrm{j} \cdot \operatorname{grad} T, \tag{5.3}
$$

where $-Td\alpha/dT$ is called the *Thomson coefficient.*

To develop a theory for the Bénard problem in TEMHD we take the fluid to be an incompressible Newtonian one. Then the relevant equations may be deduced from Shercliff [46], as

$$
\frac{\partial v_i}{\partial t} + v_j v_{i,j} = -\frac{1}{\rho}p_{,i} + \frac{1}{\rho}(\mathrm{j} \times \mathbf{B})_i + \nu\Delta v_i + g\alpha_T T k_i, \tag{5.4}
$$

$$
v_{i,i} = 0, \tag{5.5}
$$

$$
-\frac{\partial B_i}{\partial t} = (\operatorname{curl} \mathbf{E})_i, \tag{5.6}
$$

$$
\mu \mathrm{j} = \operatorname{curl} \mathbf{B}, \tag{5.7}
$$

$$
\operatorname{div} \mathbf{B} = 0, \tag{5.8}
$$

$$
c\Big(\frac{\partial T}{\partial t} + v_i T_{,i}\Big) = \operatorname{div}(\kappa \operatorname{grad} T) + \frac{j^2}{\sigma} - T\frac{d\alpha}{dT}\mathrm{j} \cdot \operatorname{grad} T, \tag{5.9}
$$

where equation (5.9) is the balance of energy in which the viscous stress power contribution has been neglected since we are primarily interested in the thermoelectric effect.

The generalized Ohm's law (5.1) is used to remove $\mathbf{E}$ from Faraday's law (5.6) and then (5.7) is additionally employed to derive the evolutionary equation for the magnetic induction $\mathbf{B}$ as

$$
B_{i,t} = \frac{1}{\mu\sigma}\Delta B_i + (\operatorname{curl}(\mathbf{v} \times \mathbf{B}))_i, \tag{5.10}
$$

where we observe the thermoelectric term vanishes since

$$
\operatorname{curl}(\alpha(T)\nabla T) \equiv 0.
$$

Using (5.7) again we derive the governing equations for v_i, B_i, p, T as

$$
\frac{\partial v_i}{\partial t} + v_j v_{i,j} = -\frac{1}{\rho}p_{,i} + \frac{1}{\rho\mu}(\operatorname{curl} \mathbf{B} \times \mathbf{B})_i + \nu\Delta v_i + g\alpha_T T k_i, \tag{5.11}
$$

$$v_{i,i} = 0, \tag{5.12}$$

$$\frac{\partial B_i}{\partial t} + v_j B_{i,j} = v_{i,j} B_j + \frac{1}{\mu\sigma}\Delta B_i, \tag{5.13}$$

$$B_{i,i} = 0, \tag{5.14}$$

$$\frac{\partial T}{\partial t} + v_i T_{,i} = \kappa \Delta T + \frac{1}{\mu^2 \sigma}|\mathrm{curl}\,\mathbf{B}|^2 - \frac{T}{\mu}\frac{d\alpha}{dT}\mathrm{curl}\mathbf{B}\cdot \mathrm{grad}\, T, \tag{5.15}$$

where we have taken κ constant, α_T is the thermal expansion coefficient, and the specific heat c is absorbed in the other coefficients.

The spatial geometry in which (5.11)-(5.15) hold is the infinite three-dimensional strip $z \in (0, d)$. We suppose the steady state is one in which the magnetic field $\bar{\mathbf{H}}$ is constant and in the $z-$direction, and the temperatures are prescribed constants on $z = 0, d$, namely, $T = T_1, T_2$ on $z = 0, d$, respectively. The steady state whose stability we investigate is then,

$$\bar{v}_i \equiv 0, \qquad \bar{T} = -\beta z + T_1, \qquad \bar{H}_i = \delta_{i3}\bar{H} = \frac{\bar{B}}{\mu}, \tag{5.16}$$

$\bar{H}$ and $\bar{B}$ being constants, and the temperature gradient β is given by

$$\beta = \frac{T_1 - T_2}{d}\ (> 0),$$

with the steady pressure $\bar{p}(z)$ given by (5.11).

We seek a perturbation about this steady state of form

$$v_i = \bar{v}_i + u_i, \quad T = \bar{T} + \theta, \quad p = \bar{p} + \pi,$$

$$B_i = \bar{B}_i + b_i, \quad (\text{or } H_i = \bar{H}_i + h_i).$$

The perturbation equations arising from (5.11)-(5.15) are now linearized about the steady state (5.16). Employing the magnetic field h_i rather than the induction b_i (to be consistent with Chandrasekhar [8]), the linearized equations may be written as:

$$\begin{aligned} u_{i,t} &= -\frac{1}{\rho}\pi_{,i} + \frac{\mu}{\rho}\bar{H}(h_{i,z} - h_{3,i}) + \nu\Delta u_i + g\alpha_T \theta \delta_{i3}, \\ h_{i,t} &= \bar{H}u_{i,z} + \eta \Delta h_i, \\ \theta_{,t} &= \kappa\Delta\theta + \beta w + f(\bar{T})(h_{2,1} - h_{1,2}), \end{aligned} \tag{5.17}$$

where $\eta = 1/\sigma\mu$ and

$$f(\bar{T}) = \beta \bar{T}\frac{d\alpha}{dT}\Big|_{\bar{T}}. \tag{5.18}$$

To completely analyse this system requires the addition of equations (103), (104) of Chandrasekhar [8],

$$\begin{aligned} \frac{\partial \xi}{\partial t} &= \eta \Delta \xi + \bar{H}\frac{\partial \varsigma}{\partial z}, \\ \frac{\partial \varsigma}{\partial t} &= \nu \Delta \varsigma + \frac{\mu}{\rho}\bar{H}\frac{\partial \xi}{\partial z}, \end{aligned} \tag{5.19}$$

where $\xi = h_{2,1} - h_{1,2}$ and $\varsigma = u_{2,1} - u_{1,2}$ are the third components of the current density and vorticity, respectively.

Boundary conditions. The boundary conditions may be derived as in [8]. For case A of [8] where the medium adjoining the fluid is electrically non-conducting then

$j_3 = 0$ and $\mathbf{h}$ is continuous across $z = 0, d$,

which because of (5.7) is equivalent to

$$\xi = 0 \text{ and } \mathbf{h} \text{ continuous, on } z = 0, d. \tag{5.20}$$

When the medium adjoining the fluid is a perfect conductor, case B of [8], then

$$h_3 = 0, \quad E_1 = E_2 = 0 \qquad \text{on} \qquad z = 0, d.$$

From Ohm's law (5.1),

$$\mathbf{j} = \sigma(\mathbf{E} + \mathbf{v} \times \mathbf{B} - \alpha \nabla T).$$

Since $v_i = 0$ and T is constant *on the boundary*, this means $j_1 = j_2 = 0$ on $z = 0, d$, also. Then using (5.7) $j_{i,i} = 0$ and so $\partial j_3 / \partial z = 0$ on $z = 0, d$. Thus the same boundary conditions as Chandrasekhar [8], p. 163, apply.

A complete solution of (5.17) subject to the correct boundary conditions is not attempted here. We observe, however, that the thermoelectric effect manifests itself in $(5.17)_3$ via the $f(\bar{T})\xi$ term. Of course, a form for α must be known for practical calculations. Also, since $\bar{T}$ is a function of z $f(\bar{T})$ will likewise depend on z. Hence, system (5.17) would appear to require numerical solution. As in [8], the case of complex growth rate will doubtless be of importance and cannot be neglected.

We also observe that for other applications one should perhaps consider convection when a current flows in the simultaneous presence of an electric field, a magnetic field, and a temperature gradient. This will be an extremely rich field and may be referred to as involving *thermogalvanomagnetic phenomena*, cf. Landau *et al.* [21], p. 101. In that case one replaces (5.1) and (5.2) by

$$\mathbf{E} = \frac{\mathbf{j}}{\sigma} + \alpha \operatorname{grad} T + R\mathbf{H} \times \mathbf{j} + N\mathbf{H} \times \operatorname{grad} T, \tag{5.21}$$

$$\mathbf{q} = (\phi + \alpha T)\mathbf{j} - \kappa \operatorname{grad} T + NT\mathbf{H} \times \mathbf{j} + L\mathbf{H} \times \operatorname{grad} T. \tag{5.22}$$

In addition to the thermelectric effect there are present the *Hall effect* (the $R\mathbf{H} \times \mathbf{j}$ term), the *Nernst effect* (the $N\mathbf{H} \times \operatorname{grad} T$ term), and the *Leduc-Righi effect* (the $L\mathbf{H} \times \operatorname{grad} T$ term). Rionero [38] has studied the nonlinear energy stability of MHD incorporating the Hall effect, without the other thermomagnetoelectric effects. Clearly, this is a rich area mathematically and physically.

We finally observe that convection in electrolytes is yet another worthwhile area for study. Some models for electrolytes have already been discussed in section 3. Basically, one has then to add a concentration field into the plethora of effects already present. Some details may be found in Landau *et al.* [21], p. 102.

ACKNOWLEDGEMENT - This research was presented at the Summer School on Mathematical Topics in Fluid Dynamics, Lisbon, September 1991. I am indebted to Professor Jose Francisco Rodrigues for the opportunity to deliver lectures at that school.

REFERENCES

[1] Abdullah, A.A. & Lindsay, K.A. Bénard convection in a non-linear magnetic fluid. *Acta Mech.* **85** (1990), 27–42.

[2] Abdullah, A.A. & Lindsay, K.A. Bénard convection in a non-linear magnetic fluid under the influence of a non-vertical magnetic field. *Continuum Mech. Thermodyn.* **3** (1991), 13–25.

[3] Backus, G. The axisymmetric self-excited fluid dynamo. *Astrophys. J.* **125** (1957), 500–524.

[4] Blennerhassett, P.J., Lin, F. & Stiles, P.J. Heat transfer through strongly magnetized ferrofluids. *Proc. Roy. Soc. London A* **433** (1991), 165–177.

[5] Castellanos, A., Atten, P. & Velarde, M.G. Oscillatory and steady convection in dielectric liquid layers subjected to unipolar injection and temperature gradient. *Phys. Fluids* **27** (1984), 1607–1615.

[6] Castellanos, A., Atten, P. & Velarde, M.G. Electrothermal convection: Felici's hydraulic model and the Landau picture of non-equilibrium phase transitions. *J. Non-Equilib. Thermodyn.* **9** (1984), 235–244.

[7] Chandrasekhar, S. Axisymmetric fields and fluid motions. *Astrophys. J.* **124** (1956), 232–243.

[8] Chandrasekhar, S. *Hydrodynamic and hydromagnetic stability.* Dover publications. New York (1981).

[9] Curtis, R.A. Flows and wave propagation in ferrofluids. *Phys. Fluids* **14** (1971), 2096–2102.

[10] Deo, B.J.S. & Richardson, A.T. Generalized energy methods in electrohydrodynamic stability theory. *J. Fluid Mech.* **137** (1983), 131–151.

[11] Eringen, A.C. Theory of thermomicrofluids. *J. Math. Anal. Appl.* **38** (1972), 480–496.

[12] Eringen, A.C. Theory of thermo-microstretch fluids and bubbly liquids. *Int. J. Engng. Sci.* **28** (1990), 133–143.

[13] Finlayson, B.A. Convective instability of ferromagnetic fluids. *J. Fluid Mech.* **40** (1970), 753–767.

[14] González, H. & Castellanos, A. The effect of an axial electric field on the stability of a rotating dielectric cylindrical liquid bridge. *Phys. Fluids A* **2** (1990), 2069–2071.

[15] González, H., McCluskey, F.M.J., Castellanos, A. & Barrero, A. Stabilization of dielectric liquid bridges by electric fields in the absence of gravity. *J. Fluid Mech.* **206** (1989), 545–561.

[16] Hoburg, J.F. Temperature-gradient driven electro - hydrodynamic instability with unipolar injection in air. *J. Fluid Mech.* **132** (1983), 231–245.

[17] Khonsari, M.M. Whirl orbits of a journal lubricated with micropolar fluids. *Acta Mech.* **81** (1990), 235–244.

[18] Khonsari, M.M. & Brewe, D. On the performance of finite journal bearings lubricated with micropolar fluids. *STLE Tribology Trans.* **32** (1989), 155–160.

[19] Lacroix, J.C., Atten, P. & Hopfinger, E.J. Electro-convection in a dielectric liquid layer subjected to unipolar injection. *J. Fluid Mech.* **69** (1975), 539–563.

[20] Lalas, D.P. & Carmi, S. Thermoconvective stability of ferrofluids. *Phys. Fluids* **14** (1971), 436–437.

[21] Landau, L.D., Lifshitz, E.M. & Pitaevskii, L.P. *Electrodynamics of continuous media.* (2nd. Edition), Pergamon Press, Oxford, (1984).

[22] Lindsay, K.A. & Straughan, B. Penetrative convection in a micropolar fluid. *Int. J. Engng. Sci.* To appear.

[23] McCluskey, F.M.J. & Atten, P. Modifications to the wake of a wire across Poiseuille flow due to a unipolar space charge. *J. Fluid Mech.* **197** (1988), 81–104.

[24] Martin, P.J. & Richardson, A.T. Conductivity models of electrothermal convection in a plane layer of dielectric liquid. *J. Heat Transfer* **106** (1984), 131–136.

[25] Mohamed, A.A. & El Shehawey, E.F. Nonlinear electrohydrodynamic Rayleigh-Taylor instability. Part 1. A perpendicular field in the absence of surface charges. *J. Fluid Mech.* **129** (1983), 473–494.

[26] Morro, A. Variational methods for nonlinear dielectrics. *Atti Sem. Mat. Fis. Univ. Modena* **36** (1988), 339–353.

[27] Morro, A., Drouot, R. & Maugin, G.A. Thermodynamics of polyelectrolyte solutions in an electric field. *J. Non Equilib. Thermodyn.* **10** (1985), 131–144.

[28] Morro, A., & Parodi, M. A variational approach to nonlinear dielectrics: application to polyelectrolytes. *J. Electrostatics* **20** (1987), 219–232.

[29] Morro, A., & Parodi, M. Lagrangian and energy for nonlinear dielectrics. *Ferroelectrics* **89** (1989), 211–217.

[30] Müller, I. *Thermodynamics.* Pitman Press, Boston - London - Melbourne (1985).

[31] Muzikar, P. & Pethick, C.J. Flux bunching in type-II superconductors. *Phys. Rev. B* **24** (1981), 2533–2539.

[32] Prakash, J. & Sinha, P. Lubrication theory for micropolar fluids and its application to a journal bearing. *Int. J. Engng. Sci.* **18** (1975), 217–232.

[33] Rionero, S. Sulla stabilità asintotica in media in magnetoidrodinamica. *Ann. Matem. Pura Appl.* **76** (1967), 75–92.

[34] Rionero, S. Sulla stabilità asintotica in media in magnetoidrodinamica non isoterma. *Ricerche Matem.* **16** (1967), 250–263.

[35] Rionero, S. Metodi variazionali per la stabilità asintotica in media in magnetoidrodinamica. *Ann. Matem. Pura Appl.* **78** (1968), 339–364.

[36] Rionero, S. Sulla stabilità magnetoidrodinamica in media con vari tipi di condizioni al contorno. *Ricerche Matem.* **17** (1968), 64–78.

[37] Rionero, S. Sulla stabilità non lineare asintotica in media in magnetofluidodinamica. *Ricerche Matem.* **19** (1970), 269–285.

[38] Rionero, S. Sulla stabilità magnetofluidodinamica nonlineare asintotica in media in presenza di effetto Hall. *Ricerche Matem.* **20** (1971), 285–296.

[39] Roberts, P.H. *An introduction to magnetohydrodynamics.* Longmans, London (1967).

[40] Roberts, P.H. Electrohydrodynamic convection. *Q. Jl. Mech. Appl. Math.* **22** (1969), 211–220.

[41] Roberts, P.H. Equilibria and stability of a fluid type II superconductor. *Q. Jl. Mech. Appl. Math.* **34** (1981), 327–343.

[42] Rodriguez-Luis, A., Castellanos, A. & Richardson, A.T. Stationary instabilities in a dielectric liquid layer subjected to an arbitrary unipolar injection and an adverse thermal gradient. *J. Phys. D: Appli. Phys.* **19** (1986), 2115–2122.

[43] Rosensweig, R.E. *Ferrohydrodynamics.* Cambridge University Press, (1985).

[44] Rosensweig, R.E. Magnetic fluids. *Ann. Review Fluid Mech.* **19** (1987), 437–463.

[45] Rosensweig, R.E., Zahn, M. & Shumovich, R. Labyrinthine instability in magnetic and dielectric fluids. *J. Magnetism & Magnetic Materials* **39** (1983), 127–132.

[46] Shercliff, J.A. Thermoelectric magnetohydrodynamics. *J. Fluid Mech.* **91** (1979), 231–251.

[47] Shliomis, M.I. Magnetic fluids. *Sov. Phys. Usp.* **17** (1974), 153–169.

[48] Sneyd, A.D. Stability of fluid layers carrying a normal electric current. *J. Fluid Mech.* **156** (1985), 223–236.

[49] Straughan, B. *The energy method, stability, and nonlinear convection.* Springer Ser. in Appl. Math. Sci. (1992), vol. **91**.

[**50**] Tipei, N. Lubrication with a micropolar fluid and its application to short journal bearings. *J. Lubrication Tech.* **101** (1979), 356–363.

[**51**] Turnbull, R.J. Electroconvective instability with a stabilizing temperature gradient. I. Theory. *Phys. Fluids* **11** (1968), 2588–2596.

[**52**] Vaidyanathan, G., Sekar, R. & Balasubramanian, R. Ferroconvective instability of fluids saturating a porous medium. *Int. J. Engng. Sci.* **29** (1991), 1259–1267.

[**53**] Worraker, W.J. & Richardson, A.T. The effect of temperature-induced variations in charge carrier mobility on a stationary electrohydrodynamnic instability. *J. Fluid Mech.* **93** (1979), 29–45.

Brian Straughan,
Department of Mathematics, University of Glasgow,
Glasgow, G12 8QW - SCOTLAND, U.K.

Mathematical Results for Compressible Flows

ALBERTO VALLI

1 – Mathematical formulation

In this Section we present the mathematical formulation of the initial boundary-value problems we are interested in.

1.A. Conservation laws and constitutive equations

Let us begin by writing the equations which describe the motion of a fluid in Eulerian coordinates (see for instance Serrin [Sr1], pag. 135, 132, 177)

$$\rho[v_t + (v \cdot \nabla)v - f] = \operatorname{div} T \quad \text{(conservation of momentum)} \tag{1.1}$$

$$\rho_t + \operatorname{div}(\rho v) = 0 \quad \text{(conservation of mass)} \tag{1.2}$$

$$\rho[e_t + v \cdot \nabla e - r] = T : D - \operatorname{div} q \quad \text{(conservation of energy)}, \tag{1.3}$$

where

$$(\operatorname{div} T)_i := \sum_j D_j T_{ji}$$

$$(T : D) := \sum_{i,j} T_{ij} D_{ij} J \ ,$$

$\rho > 0$ is the density of the fluid, v the velocity and e the internal energy per unit mass; $T = T(v,p)$ is the stress tensor and $D = D(v)$ is the deformation tensor

$$D_{ij} := (1/2)(D_j v_i + D_i v_j);$$

q is the heat flux; f and r are the (assigned) external force field per unit mass and the (assigned) heat supply per unit mass per unit time, respectively.

In the sequel of this paper, we will limit our study to the physical situations in which the following constitutive equations hold:

$$T_{ij} = [-p + (\varsigma - 2\mu/d) \operatorname{div}\ v]\delta_{ij} + 2\mu D_{ij} \ , \tag{1.4}$$

Stokesian fluid linearly dependent on D_{ij};

$$q = -\chi \nabla \theta \quad \text{(Fourier's law)}, \tag{1.5}$$

where d is the physical dimension, p is the pressure, μ and ς are the shear and bulk viscosity coefficients, respectively, $\theta > 0$ is the (absolute) temperature and χ is heat conductivity coefficient.

1.B. Thermodynamic state equations

Choosing as thermodynamic unknowns the density ρ and the temperature θ, to close the system we must add the following state equations

$$p = P(\rho, \theta) \tag{1.6}$$

$$e = E(\rho, \theta) \tag{1.7}$$

$$\mu = \mu^*(\rho, \theta), \ \varsigma = \varsigma^*(\rho, \theta), \ \chi = \chi^*(\rho, \theta) \ , \tag{1.8}$$

where P, E, μ^*, ς^* and χ^* are known functions subjected to the thermodynamic restrictions (Clausius-Duhem inequalities)

$$\mu^* \geq 0 \ , \ \varsigma^* \geq 0 \ , \ \chi^* \geq 0 \ . \tag{1.9}$$

For physical reasons, it will be always supposed that $E_\theta > 0$ (this latter condition is a general attribute of all real materials), and in the assumptions of some of the following theorems it will be also required that $P_\rho > 0$ (this condition is generally true for one-phase fluids, while van der Waals flows don't satisfy it).

Moreover, from the well-known relation

$$dE = \theta dS - Pd(\rho^{-1}), \quad S \text{ specific entropy,} \tag{1.10}$$

E and P must satisfy the compatibility condition

$$E_\rho = \rho^{-2}(P - \theta P_\theta). \tag{1.11}$$

Hence we can finally rewrite the equations for the unknowns ρ, v, θ in the following way:

$$\rho[v_t + (v \cdot \nabla)v - f] = -\nabla P + \sum_j D_j(\mu^* D_j v + \mu^* \nabla v_j) + \nabla[(\varsigma^* - 2\mu^*/d) \operatorname{div} v] \tag{1.12}$$

$$\rho_t + \operatorname{div}(\rho v) = 0 \tag{1.13}$$

$$\begin{aligned} \rho E_\theta(\theta_t + v \cdot \nabla \theta) &= -\theta P_\theta \operatorname{div} v + \rho r + \operatorname{div}(\chi^* \nabla \theta) \\ &\quad + (\mu^*/2) \sum_{i,j} (D_i v_j + D_j v_i)^2 + (\varsigma^* - 2\mu^*/d)(\operatorname{div} v)^2. \end{aligned} \tag{1.14}$$

It must be noticed that if P, μ^* and ς^* do not depend on θ, then equation (1.14) can be separated from (1.12), (1.13). We will refer to this situation as the *barotropic* case.

1.C. Boundary conditions

Several boundary conditions could be considered with respect to different physical situations. In the sequel we will consider the most frequently used, describing the motion of a fluid in a rigid container Ω, a bounded connected open subset of $\mathbb{R}^d$, $d \geq 1$. However, some remarks on different kinds of boundary conditions can be found at the end of this Section.

The different structure of the equations leads to the necessity of distinguishing between viscous and inviscid fluids.

(i) Case $\mu^* > 0$, $\varsigma^* \geq 0$ (viscous fluids).

In this case, the physical effects due to the presence of the shear viscosity coefficient yield the validity of the no-slip condition:

$$v = 0 \text{ on } \partial\Omega \ . \tag{1.15}$$

(ii) Case $\mu^* = 0$, $\varsigma^* > 0$ (bulk-viscous fluids).

Since only the bulk viscosity coefficient is different from zero, in this situation the slip boundary condition

$$v \cdot n = 0 \text{ on } \partial\Omega \tag{1.16}$$

is assumed (here and in the sequel $n = n(x)$ denotes the unit outward normal vector to $\partial\Omega$).

(iii) Case $\mu^* = 0$, $\varsigma^* = 0$ (inviscid fluids).

Also in this case the slip boundary condition (1.16) is assumed.

For what is concerned with the absolute temperature, the boundary condition takes different form in the two alternative cases $\chi^* > 0$ and $\chi^* = 0$.

(iv) Case $\chi^* > 0$ (conductive fluids).

Several boundary conditions have a physical meaning. Just to limit ourselves to the most common cases, we can require

$$\theta = \theta^* \text{ on } \partial\Omega \text{ (Dirichlet), or} \tag{1.17}$$

$$\chi^* \frac{\partial\theta}{\partial n} = q^* \text{ on } \partial\Omega \text{ (Neumann), or} \tag{1.18}$$

$$\chi^* \frac{\partial\theta}{\partial n} + k\theta = k\theta^* \text{ on } \partial\Omega \text{ (third type),} \tag{1.19}$$

where $\theta^* > 0$ and q^* are known functions and $k > 0$ is a given constant.

(v) Case $\chi^* = 0$ (non conductive fluids).

No boundary condition has to be imposed on θ if (1.15) or (1.16) are satisfied, since in these cases the temperature is not subjected to transport phenomena through the boundary.

Let us present here shortly other types of boundary conditions which are physically interesting. We will mainly focus on the velocity field and the density, since conditions (1.17)-(1.19) are general enough for the absolute temperature. First of all, in many situations (inflow-outflow problems) the velocity cannot be assumed to vanish on $\partial\Omega$. This is the case, for instance, for the flow around an airfoil, where an inflow region is naturally present upstream (and an outflow region appears in the wake), or the flow near a rigid body, where the velocity can be assumed to vanish only on the boundary of the body.

In these cases, several different boundary conditions may be prescribed. Let us start by considering the viscous case. For what is concerned with the velocity field, a (non-zero) Dirichlet boundary condition can be imposed everywhere, or, alternatively, only in the inflow region (i.e., the subset of $\partial\Omega$ where $v \cdot n < 0$), whereas in remaining part of the boundary the conditions $v \cdot n = \psi^+ \geq 0$ and $[n \cdot D(v)] \cdot \tau = 0$ have to be prescribed (here τ is a unit tangent vector on $\partial\Omega$, and $D(v)$ is the deformation tensor). Let us moreover remark that the conditions $v \cdot n = 0$ and $[n \cdot D(v)] \cdot \tau = 0$ could also be considered on the whole $\partial\Omega$: in this case, however, no inflow or outflow regions would be present.

More important is to analyze the boundary condition for the density, since now it turns out that it is necessary to prescribe it on the inflow region. In fact, the first order hyperbolic equation (1.2) can be solved by means of the theory of characteristics (see also Section 2 below), and the boundary datum for ρ on the inflow region is indeed a (necessary) Cauchy datum for the density on a non-characteristic surface. Let us notice that if the heat conductivity coefficient χ is vanishing, the same type of Dirichlet-inflow boundary condition has to be imposed on the temperature θ, since in that case also equation (1.14) is of hyperbolic type.

More complicated is the situation when the inviscid case is considered. In fact, in this case system (1.12)-(1.14) is a first order hyperbolic one, and the number of boundary conditions is different if the flow is subsonic or supersonic. A simple computation shows that the local sound speed is given by $c = [P^*_\rho(\rho, S)]^{1/2}$, where the pressure p has been expressed as a function P^* of ρ and the specific entropy S.

Take for example $d = 3$; an analysis of the sign of the eigenvalues of the associated characteristic matrix yields the conclusion that the number of boundary conditions must be five or four on a inflow boundary, depending if the flow is supersonic or subsonic, and zero or one on an outflow boundary, again depending if the flow is supersonic or subsonic. We will not enter more deeply in this argument, and in Section 3 we will only treat the inviscid case subjected to the slip boundary condition (1.16), for which the boundary is a characteristic surface.

Further informations on inflow-outflow boundary-value problems for compressible Navier-Stokes and Euler equations can be found in two interesting papers due to Gustafsson-Sundström [GSu] and Oliger-Sundström [OlSu].

Another interesting set of boundary conditions appears when we consider the free-boundary problem, i.e., a problem for which the fluid is not contained in a given domain but can move freely. In this case the vector $n \cdot T(v,p)$ is prescribed on $\partial\Omega$, where moreover $v \cdot n$ is required to be zero (stationary case) or equal to the normal velocity of the boundary itself (non-stationary case). The value of $n \cdot T(v,p)$ can be zero (free expansion of a fluid in the vacuum), or to $-P_e n + \sigma_0 \mathcal{H} n$, where P_e is the external pressure, $\sigma_0 \geq 0$ is the surface tension and $\mathcal{H}/2$ is the mean curvature of the boundary. It has to be noticed that we are now imposing one more condition on $\partial\Omega$, since it is an unknown of the problem. (In the non-stationary case, an initial condition for the boundary has to be added too).

In the sequel we will comment again on some of these alternative sets of boundary conditions (see Section 5, 7 and 8).

1.D. Initial conditions

If we are concerned with non-stationary problems, suitable initial conditions have to be added. Looking at the preceding evolution equations (1.12)-(1.14), we see at once that it is necessary to assign

$$v_{|t=0} = v_0(x),\ \rho_{|t=0} = \rho_0(x) > 0,\ \theta_{|t=0} = \theta_0(x) > 0. \tag{1.20}$$

Moreover, as we already noticed, when considering the free-boundary problem an initial condition for the boundary $\partial\Omega(t)$ has to be imposed.

2 – Some remarks about the mathematical structure of the problem

Let us make now some remarks on the problems we are going to study in the sequel. First of all, for simplicity, we will always assume that (1.14) can be separated from the other equations (i.e. P, μ^* and ς^* do not depend on θ, or equivalently, as we said before, we are dealing with a barotropic fluid). However, some remarks concerning the non-barotropic case will be added in each Section. Furthermore, we are only going to consider problems in a *bounded* domain (i.e., a bounded connected open set) Ω, and we will give a few comments for problems in more general domains at the end of the following Sections.

Before starting our analysis, let us underline that the mathematical theories for viscous and inviscid fluids and for compressible and incompressible fluids are strongly different.

Roughly speaking, one can say that the equations for viscous fluids (the so-called *Navier-Stokes equations*) are parabolic and that the equations for inviscid fluids (the *Euler equations*) are hyperbolic. But, looking more deeply to the structure of the equations, it is easy to see that this is not completly true. In fact, equation (1.13) concerning the density ρ is of hyperbolic type, regardless of the viscosity.

Hence one can say more precisely that Navier-Stokes equations are hyperbolic-parabolic (or incompletely parabolic, following the definition proposed in Belov-Yanenko [BeY] and Strikwerda [Sk], where one can find interesting considerations concerning the mathematical properties of these equations).

Let us spend now some words about the main difference between the structure of the equations for compressible and incompressible fluids. In this latter case, the pressure is no more related to thermodynamic unknowns (see Serrin [Sr1], pag. 234, for a discussion of this important point), and instead of the state equation (1.6) one has to impose that any given amount of fluid cannot change its volume along the motion. As is well known, this is done by requiring

$$\operatorname{div} v = 0\ . \tag{2.1}$$

As a consequence, equations (1.12) and (1.13) become

$$\rho[v_t + (v\cdot\nabla)v - f] = -\nabla p + \sum_j D_j(\mu^* D_j v + \mu^* \nabla v_j) \tag{2.2}$$

$$\rho_t + v\cdot\nabla\rho = 0, \tag{2.3}$$

where now p is an unknown (in some sense, it is a Lagrangian multiplier associated to the constraint (2.1)).

We can remark that the solution ρ of (2.3), $(1.20)_2$ (where for a moment v has to be considered as an assigned vector satisfying (1.15) or (1.16)) is given by the well-known formula

$$(2.4) \qquad \rho(t,x) = \rho_0(U(0,t,x)),$$

where the characteristic direction $U(t,s,x)$ is the solution of the system of ordinary differential equations

$$(2.5) \qquad \begin{cases} U(t,s,x)_t = v(t,U(t,s,x)) \\ U(s,s,x) = x. \end{cases}$$

($U(t,s,x)$ is also called the flow of the vector field $v(t,x)$).

Hence, directly from (2.4) one easily obtains that $0 < \inf_\Omega \rho_0 \leq \rho(t,x) \leq \sup_\Omega \rho_0$. This fact has several consequences: the most important one is that the coefficient of the time derivative of v in (2.2) is not vanishing, i.e., (2.2) cannot degenerate; secondly, if $\rho_0(x) = \rho^*$, a positive constant, then the solution to (2.3), $(1.20)_2$ is given by $\rho(t,x) = \rho^*$ for each (t,x), and (2.2) thus becomes

$$(2.6) \qquad v_t + v \cdot \nabla v - f = -\nabla p^* + \nu^* \Delta v,$$

where $p^* := p/\rho^*$ and $\nu^* := \mu^*(\rho^*)/\rho^*$.

Equations (2.1) and (2.6) are usually called Navier-Stokes (if $\mu^* > 0$) or Euler (if $\mu^* = 0$) equations for *homogeneous* incompressible fluids, whereas (2.1)-(2.3) are the Navier-Stokes or Euler equations for *non-homogeneous* incompressible fluids. We will not treat here the mathematical theory of these equations, which have been intensively studied during the last sixty years; let us just refer the interested reader to the books of Ladyzhenskaya [Ld], Shinbrot [Si], Temam [Te], Antoncev-Kazhikov-Monakhov [AKMo] (especially for the non-homogeneous case), von Wahl [W], Constantin-Foias [CnF] for the Navier-Stokes equations, and Majda [Ma2] for the Euler equations (for the non-homogeneous case see also Valli [V5] and the references therein).

When we consider the compressible case, from (1.13), $(1.20)_2$ we find

$$(2.7) \qquad \rho(t,x) = \rho_0(U(0,t,x)) \exp\left[-\int_0^t (\text{div } v)(s,U(s,t,x))ds \right];$$

hence ρ can degenerate at the finite time t^* in the point x if

$$\left| \int_0^{t^*} (\text{div } v)(s,U(s,t^*,x))ds \right| = +\infty.$$

We can affirm that one of the main mathematical problems concerning compressible fluids is to find a-priori estimates assuring the non-degeneration of the density ρ. This is obviously easier locally in time, or for small data, as we will see in the sequel.

In the following Sections, we are going to present some results concerning the existence of a (unique "regular") solution for stationary and non-stationary problems (let us underline here that we will not consider to case of weak solutions). We will try to explain the main ideas of the proofs through an informal presentation, without entering in all the details, for these referring the interested reader to the papers where these problems have been studied in a complete way. Moreover, let us repeat that for the sake of simplicity we will mainly consider the barotropic case in a bounded domain, leaving the extensio9wlore general cases to some short comments.

Let us conclude this Section by mentioning a review paper due to Solonnikov-Kazhikhov [SoK], where, for viscous fluids, the existence theory available at the beginning of the 80's is presented, together with a detailed bibliography.

3 – Compressible inviscid fluids ($p = P(\rho)$, $\mu^* = 0$, $\varsigma^* = 0$)

We want to present here a result due to Beirão da Veiga [B1], [B10]. (A similar result has been obtained independently by Agemi [Ag]; the subsonic case was firstly solved by Ebin [E]). We are concerned with the local existence and uniqueness of strong solutions: we will not even touch the case of weak solutions and shock waves, whose existence can be proven by completely different methods (see, e.g., the books of Courant-Friedrichs [CFr], Roždestvenskiĭ-Yanenko [RY], Lax [Lx], Smoller [Sm], Majda [Ma1]).

Theorem A. *Let $\Omega \subset \mathbb{R}^d$ $(d = 2,3)$ be a bounded domain with $\partial\Omega \in C^5$, $v_0 \in H^3(\Omega)$, $\rho_0 \in H^3(\Omega)$, $\inf_\Omega \rho_0(x) > 0$, $f \in L^2(0,T_0;H^3(\Omega))$, $f_t \in L^2(0,T_0;H^2(\Omega))$, $f_{tt} \in L^2(0,T_0;H^1(\Omega))$, $f_{ttt} \in L^2(0,T_0;L^2(\Omega))$, $P \in C^4(\mathbb{R}^+)$, $P_\rho(\xi) > 0$ for $\xi > 0$. Set $\dot{v}_0 := [f(0) - (v_0 \cdot \nabla)v_0 - (\rho_0)^{-1}\nabla P(\rho_0)]$, and assume that the (necessary) compatibility conditions*

$$v_0 \cdot n = 0J \ , \ \dot{v}_0 \cdot n = 0 \ ,$$
$$\{f_t(0) - (\dot{v}_0 \cdot \nabla)v_0 - (v_0 \cdot \nabla)\dot{v}_0 + \nabla[(\rho_0)^{-1}P_\rho(\rho_0)\operatorname{div}(\rho_0 v_0)]\} \cdot n = 0$$

are satisfied on $\partial\Omega$. Then there exist $T^* \in]0,T_0]$,

$$v \in C^0([0,T^*];H^3(\Omega)),\ v_t \in C^0([0,T^*];H^2(\Omega)),\ v_{tt} \in C^0([0,T^*];H^1(\Omega)),$$
$$\rho \in C^0([0,T^*];H^3(\Omega)),\ \rho_t \in C^0([0,T^*];H^2(\Omega)),\ \rho_{tt} \in C^0([0,T^*];H^1(\Omega))$$

such that (v,ρ) is the (unique) solution of (1.12) (with $\mu^* = \varsigma^* = 0$), (1.13), (1.16), $(1.20)_{1,2}$. Moreover, $\inf_{]0,T^*[\times\Omega} \rho(t,x) > 0$.

The assumptions on f could be substituted, as in [B1], by $f \in L^\infty(0,T_0;H^3(\Omega))$, $f_t \in L^\infty(0,T_0;H^2(\Omega))$, $f_{tt} \in L^\infty(0,T_0;H^1(\Omega))$.

This result has been extended to the complete system (1.12) (with $\mu^* = \varsigma^* = 0$), (1.13), (1.14) (with $\chi^* = 0$) by Schochet [Sc], assuming $f = 0$ and $r = 0$. More recent and general results, especially concerned with the existence of more regular solutions and with the difficult problem of well-posedness, can be found in Beirão da Veiga [B8], [B9], [B10].

To start with, let us assume for simplicity that Ω is a simply-connected domain, $\partial\Omega$ is connected, $f = 0$ and $P(\xi) = \xi$. We are going to consider only the three-dimensional case, the two-dimensional one being simpler. First of all, introduce the unknown

$$g(t,x) := \log \rho(t,x) \tag{3.1}$$

and set $g_0(x) := \log \rho_0(x)$.

The problem can be thus rewritten in the form

$$v_t + (v \cdot \nabla)v + \nabla g = 0 \quad \text{in } Q_T :=]0,T[\times\Omega \tag{3.2}$$

$$g_t + v \cdot \nabla g + \operatorname{div} v = 0 \quad \text{in } Q_T \tag{3.3}$$

$$(v \cdot n)_{|\partial\Omega} = 0 \quad \text{on } \Sigma_T := (0,T) \times \partial\Omega \tag{3.4}$$

$$v_{|t=0} = v_0(x),\ g_{|t=0} = g_0(x) \quad \text{in } \Omega, \tag{3.5}$$

which is a first order symmetric hyperbolic system.

The idea of the proof consists in transforming problem (3.2)-(3.5) in an equivalent one for the vorticity and divergence of v. Let us at first apply the curl operator to (3.2). We obtain

$$(\operatorname{curl} v)_t + (v \cdot \nabla) \operatorname{curl} v - (\operatorname{curl} v \cdot \nabla)v + (\operatorname{div} v) \operatorname{curl} v = 0 \quad \text{in } Q_T, \tag{3.6}$$

plus the initial condition

$$(\operatorname{curl} v)_{|t=0} = \operatorname{curl} v_0 \quad \text{in } \Omega. \tag{3.7}$$

Applying the div operator to (3.2) we find

$$(\operatorname{div} v)_t + (v \cdot \nabla) \operatorname{div} v + \sum_{i,j} D_i v_j D_j v_i + \Delta g = 0 \quad \text{in } Q_T,$$

i.e., using (3.3),

$$(\partial_t + v \cdot \nabla)^2 g - \Delta g = \sum_{i,j} D_i v_j D_j v_i \quad \text{in } Q_T\ . \tag{3.8}$$

Moreover, by taking the scalar product of (3.2) by n on the boundary, we have

$$(\frac{\partial g}{\partial n})_{|\partial\Omega} = \sum_{i,j} v_i v_j D_i n_j \quad \text{on } \Sigma_T\ . \tag{3.9}$$

Here n is a suitable extension of the unit normal vector to $\bar{\Omega}$, and we have made use of the fact that v and $\nabla(v \cdot n)$ are orthogonal on $\partial\Omega$ (being v tangential, $\nabla(v \cdot n)$ is parallel to n), hence

$$0 = \sum_{i,j} v_i D_i(v_j n_j) = \sum_{i,j} v_i (D_i v_j) n_j + \sum_{i,j} v_i v_j D_i n_j\ .$$

From (3.2) evaluated at $t = 0$ we finally obtain

$$(g_t)_{|t=0} = -(\operatorname{div} v_0 + v_0 \nabla g_0) \quad \text{in } \Omega. \tag{3.10}$$

We claim that the system given by (3.3), (3.4), $(3.5)_2$-(3.10) is equivalent to (3.2)-(3.5). In fact, it is well known that in a simply-connected domain one has

$$\begin{cases} \operatorname{curl} V = 0 & \text{in } \Omega \\ \operatorname{div} V = 0 & \text{in } \Omega \\ (V \cdot n)_{|\partial\Omega} = 0 & \text{on } \partial\Omega \end{cases}$$

if and only if $V = 0$. At first we apply this remark (for each $t \in [0, T]$) to the left hand side of (3.2), i.e., we choose $V = v_t + (v \cdot \nabla)v + \nabla g$. A straightforward computation thus shows that (3.2) is satisfied. Moreover, one repeats the same argument choosing now $V = v_{|t=0} - v_0$, and thus also $(3.5)_1$ is easily seen to hold.

To prove the existence of a solution to (3.3), (3.4), $(3.5)_2$-(3.10) we intend to apply a fixed point argument of Schauder type. The choice of the convex, compact set $\mathbf{K}$ which will be mapped into itself by a suitable continuous transformation is suggested by the a-priori estimate that we are going to obtain in the sequel. However, to simplify the exposition let us define at once this set.

Introduce at first the Banach space

$$\mathcal{X}_T := L^\infty(0, T; H^2(\Omega)) \cap H^{1,\infty}(0, T; H^1(\Omega)) \cap H^{2,\infty}(0, T; L^2(\Omega)) \, , \tag{3.11}$$

endowed with the norm

$$\|\phi\|_{\mathcal{X}_T} := \sup_{]0,T[} \|\phi(t)\|_2 + \sup_{]0,T[} \|\phi_t(t)\|_1 + \sup_{]0,T[} \|\phi_{tt}(t)\|_0 \, . \tag{3.12}$$

A couple $(\omega, \vartheta) \in \mathcal{X}_T \times \mathcal{X}_T$ will be said to satisfy the compatibility conditions up to the order one if

$$\begin{aligned} \omega_{|t=0} &= \operatorname{curl} v_0 \, , \quad (\omega_t)_{|t=0} = -\operatorname{curl}[(v_0 \cdot \nabla)v_0] \\ \vartheta_{|t=0} &= \operatorname{div} v_0 \, , \quad (\vartheta_t)_{|t=0} = -\operatorname{div}[(v_0 \cdot \nabla)v_0] - \Delta g_0 \, . \end{aligned} \tag{3.13}$$

Now, given $A \geq 1$, define

$$\begin{aligned} \mathbf{K}_{A,T} := \{(\omega, \vartheta) \in \mathcal{X}_T \times \mathcal{X}_T \mid \|\omega\|_{\mathcal{X}_T} + \|\vartheta\|_{\mathcal{X}_T} \leq A \, , \; (\omega, \vartheta) \text{ satisfy } (3.13) \, , \\ \operatorname{div} \omega(t, x) = 0 \quad \forall (t, x) \in Q_T \, , \; \int_\Omega \vartheta(t) = 0 \quad \forall t \in [0, T] \, \} \, , \end{aligned} \tag{3.14}$$

where A and T will be chosen in the sequel. $\mathbf{K}_{A,T}$ is a convex set, and by means of the Ascoli's theorem it is not difficult to show that it is compact in $C^0([0, T]; H^1(\Omega)) \times C^0([0, T]; H^1(\Omega))$.

We are in a position now to construct the solution to (3.3), (3.4), $(3.5)_2$-(3.10).

Step 1. Given $(\omega, \vartheta) \in \mathbf{K}_{A,T}$, solve

$$\begin{cases} \operatorname{curl} v = \omega & \text{in } Q_T \\ \operatorname{div} v = \vartheta & \text{in } Q_T \\ (v \cdot n)_{|\partial\Omega} = 0 & \text{on } \Sigma_T \end{cases} \, . \tag{3.15}$$

The necessary and sufficient conditions for the solvability of this first order elliptic system are satisfied due to our definition of $\mathbf{K}_{A,T}$. Moreover, from well known elliptic estimates we have

$$\|v(t)\|_3 + \|v_t(t)\|_2 + \|v_{tt}(t)\|_1 \leq c\,A \ , \tag{3.16}$$

while, integrating in $]0,T[$ the time derivative of v and v_t,

$$\|v(t)\|_2 + \|v_t(t)\|_1 \leq c(1+AT) \ , \tag{3.17}$$

where here and in the sequel each constant c only depends on the data of the problem.

Step 2. Solve now

$$\begin{cases} \omega_t^* + (v\cdot\nabla)\omega^* - (\omega^*\cdot\nabla)v + (\operatorname{div} v)\omega^* = 0 & \text{in } Q_T \\ \omega^*_{|t=0} = \operatorname{curl} v_0 & \text{in } \Omega \ . \end{cases} \tag{3.18}$$

The existence of the solution is a consequence of the theory of characteristics. Moreover, multiplying by ω^* and integrating by parts the term coming from $(v\cdot\nabla)\omega^*$ one easily obtains

$$\frac{d}{dt}\|\omega^*(t)\|_0^2 \leq c\|v(t)\|_3\|\omega^*(t)\|_0^2 \ .$$

Repeating the same argument for $D\omega^*$ and $D^2\omega^*$ and applying Gronwall's lemma we find

$$\|\omega^*(t)\|_2 \leq c\exp(cAT) \ .$$

Applying D_t and D_{tt} to $(3.18)_1$ and employing a similar procedure to that described above, one finally has

$$\|\omega^*(t)\|_2 + \|\omega_t^*(t)\|_1 + \|\omega_{tt}^*(t)\|_0 \leq c_1\exp(cAT) \ . \tag{3.19}$$

Step 3. Being known v by Step 1, it is possible to find the solution of

$$\begin{cases} (\partial_t + v\cdot\nabla)^2 g - \Delta g = \sum_{i,j} D_i v_j D_j v_i & \text{in } Q_T \\ \left(\frac{\partial g}{\partial n}\right)_{|\partial\Omega} = \sum_{i,j} v_i v_j D_i n_j & \text{on } \Sigma_T \\ g_{|t=0} = g_0 & \text{in } \Omega \\ (g_t)_{|t=0} = -v_0\cdot\nabla g_0 - \operatorname{div} v_0 & \text{in } \Omega \ . \end{cases} \tag{3.20}$$

The existence and uniqueness of a solution g to this "wave-like" hyperbolic problem, and especially the proof of suitable a-priori estimates for g is probably the most difficult point of the overall procedure. In fact, the non-homogeneous Neumann boundary condition and the "non-regularity" of the coefficients depending on v makes this problem an unusual one. Another point which is not trivial is concerned with the verification of the compatibility conditions necessary for finding the solution. We will not give here the details of the proof, which can be found in the papers of Beirão da Veiga [B1], [B10]; there it is obtained

$$\|g(t)\|_3 + \|g_t(t)\|_2 + \|g_{tt}(t)\|_1 \leq c(1+A^kT)\exp(cA^kT) \ , \tag{3.21}$$

for a suitable integer $k \geq 1$.

Step 4. Let us simply define now

$$\delta := -g_t - v \cdot \nabla g \ . \tag{3.22}$$

Notice that from $(3.30)_1$ it follows

$$\delta_t + v \cdot \nabla \delta = -\Delta g - \sum_{i,j} D_i v_j D_j v_i \quad \text{in } Q_T \ . \tag{3.23}$$

Let us estimate this function δ. From (3.22) we have at once

$$\begin{aligned}
&\|\delta(t)\|_2 \leq c(\|g_t(t)\|_2 + \|v(t)\|_2\|g(t)\|_3) \\
&\|\delta_t(t)\|_1 \leq c(\|g_{tt}(t)\|_1 + \|v_t(t)\|_1\|g(t)\|_3 + \|v(t)\|_2\|g_t(t)\|_2) \ ,
\end{aligned}$$

hence from (3.21) and (3.17)

$$\|\delta(t)\|_2 + \|\delta_t(t)\|_1 \leq c(1 + A^{k+1}T)\exp(cA^kT) \ . \tag{3.24}$$

To estimate δ_{tt} we differentiate (3.23) with respect to t. We thus find

$$\|\delta_{tt}(t)\|_0 \leq c(\|v_t(t)\|_1\|\delta(t)\|_2 + \|v(t)\|_2\|\delta_t(t)\|_1 + \|g_t(t)\|_2 + \|Dv_t(t)Dv(t)\|_0) \ .$$

Moreover, since by the Sobolev embedding theorem $H^{7/4}(\Omega) \subset C^0(\bar{\Omega})$, we have

$$\|Dv_t(t)Dv(t)\|_0 \leq c\|v_t(t)\|_1\|v(t)\|_{11/4} \ .$$

By interpolation $\|v(t)\|_{11/4} \leq c\|v(t)\|_2^{1/4}\|v(t)\|_3^{3/4}$, consequently

$$\|\delta_{tt}(t)\|_0 \leq c[(1 + A^{5/4}T)A^{3/4} + (1 + A^{k+2}T)\exp(cA^kT)] \ . \tag{3.25}$$

From (3.24) and (3.25) it follows

$$\|\delta(t)\|_2 + \|\delta_t(t)\|_1 + \|\delta_{tt}(t)\|_0 \leq c_2(1 + A^{k+2}T)[A^{3/4} + \exp(cA^kT)] \ . \tag{3.26}$$

Step 5. We have thus constructed the map $\Phi : (\omega, \vartheta) \to (\omega^*, \delta)$, mapping $\mathcal{X}_T \times \mathcal{X}_T$ into itself. Moreover, it is not difficult to show that Φ is continuous in the topology of $C^0([0,T]; H^1(\Omega)) \times C^0([0,T]; H^1(\Omega))$. We only need to prove that $\Phi(\mathbf{K}_{A,T}) \subset \mathbf{K}_{A,T}$ for some A and T.

First of all, (ω^*, δ) obviously satisfy the compatibility conditions (3.13).

Moreover, recalling that equation $(3.18)_1$ can be written as

$$\omega_t^* + v \operatorname{div} \omega^* - \operatorname{curl}(v \wedge \omega^*) = 0 \ ,$$

we see that

$$\begin{cases} (\operatorname{div} \omega^*)_t + v \cdot \nabla \operatorname{div} \omega^* + \operatorname{div} v \operatorname{div} \omega^* = 0 & \text{in } Q_T \\ (\operatorname{div} \omega^*)_{|t=0} = 0 & \text{in } \Omega \ , \end{cases}$$

hence $\operatorname{div} \omega^* = 0$ in Q_T.

Finally, choosing $A > c_0(1+A^{3/4})$, $c_0 := \max(1, c_1, c_2)$ and T small enough (depending on A), say $T = T^*$, from (3.19) and (3.26) it is readily seen that $\|\omega^*\|_{X_{T^*}} + \|\delta\|_{X_{T^*}} \le A$.

Now we just need to show that $\int_\Omega \delta(t) = 0$ for each $t \in [0, T^*]$. Unfortunately, in general this is not true.

Step 6. For each $t \in [0, T^*]$, let us project δ over the subspace of functions having vanishing mean value, i.e., define

$$\vartheta^*(t,x) = \delta(t,x) - |\Omega|^{-1} \int_\Omega \delta(t) \qquad (|\Omega| := \mathrm{meas}(\Omega)) \ . \tag{3.27}$$

All the estimates concerning δ are still valid for ϑ^*; moreover, this last function obviously satisfies the compatibility conditions $(3.13)_2$, since $\vartheta^*(0,x) = \delta(0,x)$ in Ω. Hence, by applying the Schauder fixed point theorem, we find a fixed point of the map $\Psi : (\omega, \vartheta) \to (\omega^*, \vartheta^*)$.

In principle, this fixed point is not the solution we are looking for, unless we can verify, *for the function* δ *corresponding to the fixed point*, that $\int_\Omega \delta(t) = 0$ for each $tJ \in [0, T^*]$, i.e., that, though in general the maps Φ and Ψ are different, for the fixed point (ω, ϑ) we have $\Phi(\omega, \vartheta) = \Psi(\omega, \vartheta)$.

Step 7. The final step is concerned with the proof of the following result: ϑ and δ satisfy

$$\begin{cases} \frac{d}{dt} \int_\Omega \delta(t) = \int_\Omega \vartheta(t)[\delta(t) - \vartheta(t)] & \text{in }]0, T^*[\\ \int_\Omega \delta(0) = 0 \ . \end{cases} \tag{3.28}$$

Equation $(3.28)_2$ is trivial. Let us prove $(3.28)_1$: using the divergence theorem and $(3.20)_2$ we find

$$\int_\Omega \Delta g = \int_{\partial\Omega} \sum_{i,j} v_i v_j D_i n_j = -\int_\Omega \mathrm{div}[(v \cdot \nabla) v] = -\int_\Omega (\sum_{i,j} D_i v_j D_j v_i + v \cdot \nabla \vartheta) \ .$$

Hence, integrating (3.23) in Ω

$$\frac{d}{dt} \int_\Omega \delta = -\int_\Omega (v \cdot \nabla \delta + \Delta g + \sum_{i,j} D_i v_j D_j v_i) = \int_\Omega v \cdot \nabla(\vartheta - \delta) \ .$$

Applying once again the divergence theorem, $(3.28)_1$ follows at once.

If $\vartheta = \vartheta^*$ is the fixed point of the map Ψ, the corresponding δ satisfies

$$\frac{d}{dt} \int_\Omega \delta(t) = \int_\Omega \vartheta^*(t)[\delta(t) - \vartheta^*(t)] = 0 \ ,$$

since ϑ^* is the $L^2(\Omega)$-projection of δ over the space of functions having vanishing mean value. In conclusion, for each $t \in [0, T^*]$ we have $\int_\Omega \delta(t) = \int_\Omega \delta(0) = 0$, and the proof is finally complete. (Indeed, we have proven the existence of a solution in $L^\infty(0, T^*; H^3(\Omega)) \cap H^{1,\infty}(0, T^*; H^2(\Omega)) \cap H^{2,\infty}(0, T^*; H^1(\Omega))$; to show that it is in fact continuous with respect to t requires some additional remarks, and for them we refer to [B10]).

The uniqueness of the solution has been proven by Graffi [Gr], Serrin [Sr3].

Let us finish by mentioning some additional results: the barotropic case in a time-dependent domain has been considered by Secchi [Se5]; the problem modelling the evolution of a gaseous star in the presence of self-gravitation has been studied by Makino [M] (barotropic case) and Makino-Ukai [MU]; the existence of a solution for the barotropic magnetohydrodynamic case has been proven by Yanagisawa [Ya] and Yanagisawa-Matsumura [YaMt].

The existence for the Cauchy problem (i.e., $\Omega = \mathbb{R}^3$) is a consequence of the general theory concerning quasi-linear symmetric hyperbolic systems (see, e.g., Kato [Kt]). The case of a solution with compact support has been considered by Makino-Ukai-Kawashima [MUKw]. An existence theorem for the barotropic case in the half-space has been given by Beirão da Veiga [B7]. Finally, let us mention that a counter-example to global existence of a regular solution for the Cauchy problem has been given by Sideris [Sd].

4 – Compressible bulk-viscous fluids ($p = P(\rho)$, $\mu^* = 0$, $\varsigma^* > 0$)

We just want to mention an interesting result due to Secchi [Se1], which, under suitable assumptions, can be also extended to the complete system (1.12) (with $\mu^* = 0$), (1.13), (1.14) (with $\chi^* > 0$ or $\chi^* = 0$).

Theorem B. *Let $\Omega \subset \mathbb{R}^d$ ($d = 2, 3$) be a bounded domain with $\partial\Omega \in C^4$, $v_0 \in H^3(\Omega)$, $\rho_0 \in H^3(\Omega)$, $\inf_\Omega \rho_0(x) > 0$, $f \in L^2(0, T_0; H^2(\Omega))$, $\operatorname{curl} f \in L^2(0, T_0; H^2(\Omega))$, $f_t \in L^2(0, T_0; L^2(\Omega))$, $P \in C^3(\mathbb{R}^+)$, $\varsigma^* \in C^3(\mathbf{R}^+)$, $\varsigma^*(\xi) > 0$ for $\xi > 0$. Assume that the (necessary) compatibility conditions $v_0 \cdot n = 0$, $\{f(0) - (v_0 \cdot \nabla)v_0 - (\rho_0)^{-1}\nabla[P(\rho_0) - \varsigma^*(\rho_0) \operatorname{div} v_0]\} \cdot n = 0$ are satisfied on $\partial\Omega$. Then there exist $T^* \in]0, T_0]$,*

$$v \in C^0([0, T^*]; H^3(\Omega)),\ v_t \in L^2(0, T^*; H^2(\Omega)),\ v_{tt} \in L^2(0, T^*; L^2(\Omega)),$$
$$\operatorname{div} v \in L^2(0, T^*; H^3(\Omega))$$
$$\rho \in C^0([0, T^*]; H^3(\Omega)),\ \rho_t \in L^\infty(0, T^*; H^2(\Omega)),\ \rho_{tt} \in L^2(0, T^*; H^1(\Omega))$$

such that (v, ρ) is the (unique) solution of (1.12) (with $\mu^* = 0$), (1.13), (1.16), $(1.20)_{1,2}$. Moreover, $\inf_{]0,T^*[\times\Omega} \rho(t, x) > 0$.

To better understand this result, one has to notice that equation (1.12) (with $\mu^* = 0$) is a second order equation which is *not* parabolic in the usual sense. On the other hand, $\operatorname{div} v$ essentially satisfies the heat equation with Neumann boundary condition, and this is a crucial remark in order to find the solution.

The existence of a global solution (even assuming small data) is an open problem, excepting in spatial dimension equal to one (in this last case there is no distinction between viscous and bulk-viscous fluids: see Section 9). In this respect, from the mathematical point of view we can say that, if the spatial dimension d is greater than 1, bulk-viscous fluids are more similar to inviscid fluids than to viscous ones.

5 – Compressible viscous fluids ($p = P(\rho)$, $\mu^* > 0$, $\varsigma^* \geq 0$)

We want to present a global in time existence theorem which has been proven by Valli [V3], extending some ideas of Matsumura [Mt] and Matsumura-Nishida [MtN3], [MtN4].

It is useful to introduce the mean density

$$\rho^* := |\Omega|^{-1} \int_\Omega \rho(t,x), \qquad (|\Omega| := \text{meas}(\Omega)),$$

which is a positive constant as a consequence of (1.13), (1.15) and $(1.20)_2$. (More precisely, $\rho^* = |\Omega|^{-1} \int_\Omega \rho_0(x)$).

Theorem C. *Let $\Omega \subset \mathbb{R}^d$ $(d \leq 3)$ be a bounded domain with $\partial\Omega \in C^3$, $v_0 \in H^2(\Omega) \cap H_0^1(\Omega)$, $\rho_0 \in H^2(\Omega)$, $\inf_\Omega \rho_0(x) > 0$, $f \in L^\infty(\mathbb{R}^+; H^1(\Omega))$, $f_t \in L^\infty(\mathbb{R}^+; H^{-1}(\Omega))$, $P \in C^2(\mathbb{R}^+)$, $\mu^* \in C^2(\mathbb{R}^+)$, $\varsigma^* \in C^2(\mathbb{R}^+)$, $P_\rho(\xi) > 0$, $\mu^*(\xi) > 0$ and $\varsigma^*(\xi) \geq 0$ for $\xi > 0$. Assume that $\|v_0\|_2$, $\|\rho_0 - \rho^*\|_2$ and $\sup_{\mathbb{R}^+}(\|f(t)\|_1 + \|f_t(t)\|_{-1})$ are small enough. Then there exist*

$$v \in L^2_{loc}(\mathbb{R}^+; H^3(\Omega)) \cap C^0_B(\mathbb{R}_+; H^2(\Omega)), \; v_t \in L^2_{loc}(\mathbb{R}^+; H^1(\Omega)) \cap C^0_B(\mathbb{R}_+; L^2(\Omega))$$
$$\rho \in C^0_B(\mathbb{R}_+; H^2(\Omega)), \; \rho_t \in C^0_B(\mathbb{R}_+; H^1(\Omega))$$

such that (v, ρ) is the (unique) solution of (1.12), (1.13), (1.15), $(1.20)_{1,2}$. Moreover, $\inf_{\mathbb{R}^+\times\Omega} \rho(t,x) > 0$.

Here $C^0_B(\mathbb{R}_+; X)$ means the space of continuous and bounded functions on $\mathbb{R}_+ := [0, +\infty[$ valued in the Banach space X, and $\mathbb{R}^+ :=]0, +\infty[$.

These results have been extended to the system (1.12)-(1.14) by Valli-Zajączkowski [VZ], where some results on *inflow-outflow* problems (i.e. $v \cdot n_{|\partial\Omega} \neq 0$) can also be found.

Before starting the proof, let us assume for simplicity that $\mu^* = 1$, $\varsigma^* = 2/d$, $\rho^* = 1$, $P(\xi) = \xi$, and consider the new unknown

$$\sigma(t,x) := \rho(t,x) - 1, \tag{5.1}$$

which satisfies $\int_\Omega \sigma = 0$.

We can rewrite the equations in this form

$$v_t + (v \cdot \nabla)v - f = (\sigma + 1)^{-1}(-\nabla\sigma + \Delta v + \nabla \operatorname{div} v) \quad \text{in } Q_\infty :=]0, +\infty[\times\Omega \tag{5.2}$$

$$\sigma_t + \operatorname{div}(\sigma v) + \operatorname{div} v = 0 \quad \text{in } Q_\infty \tag{5.3}$$

$$v_{|\partial\Omega} = 0 \quad \text{on } \Sigma_\infty := (0, +\infty) \times \partial\Omega \tag{5.4}$$

$$v_{|t=0} = v_0(x), \; \sigma_{|t=0} = \rho_0(x) - 1 > -1 \quad \text{in } \Omega. \tag{5.5}$$

Let us just show how to find some global a-priori estimates in $L^\infty(\mathbb{R}^+; H^2(\Omega))$, since the proof of the existence of a local solution can be obtained by linearization plus a fixed point argument, and is not difficult in concept, though precise estimates on the linear problem and several calculations are needed.

Rewrite the problem in the following form:

$$v_t - \Delta v - \nabla \operatorname{div} v + \nabla \sigma = f + H \quad \text{in } Q_\infty \tag{5.6}$$

$$\sigma_t + \operatorname{div} v = L \quad \text{in } Q_\infty, \tag{5.7}$$

where

$$H := -(v \cdot \nabla)v - (\sigma + 1)^{-1}\sigma(-\nabla\sigma + \Delta v + \nabla \operatorname{div} v) \tag{5.8}$$

$$L := -\sigma \operatorname{div} v - v \cdot \nabla\sigma. \tag{5.9}$$

In this and in the following Sections we will indicate by $\|\cdot\|_k$ each *equivalent* norm in $H^k(\Omega)$.

Step 1. It is possible to eliminate the contributions of the terms $\nabla\sigma$ and $\operatorname{div} v$ multiplying (5.6) by v and (5.7) by σ and integrating in Ω. In fact, adding the two equations, from

$$\int_\Omega \nabla\sigma \cdot v = -\int_\Omega \sigma \operatorname{div} v \tag{5.10}$$

one obtains

$$\frac{d}{dt}(\|v\|_0^2 + \|\sigma\|_0^2) + \|\nabla v\|_0^2 \le c\|f\|_{-1}^2 + N.L., \tag{5.11}$$

where here and in the sequel N.L. indicates some norms related to nonlinear terms (which we expect to be "good", since they will be controlled by the linear terms because we are assuming small initial data).

This procedure can be repeated for v_t (and for "tangential" and "interior" derivatives of v up to order two), since in all these cases the boundary conditions permit integration by parts as in (5.10). Remark that the term $v \cdot \nabla\sigma$ contained in L (and the others similar to it which will appear in the sequel) must be integrated by parts in this way

$$\int_\Omega (v \cdot \nabla D^\alpha\sigma)D^\alpha\sigma = -\frac{1}{2}\int_\Omega \operatorname{div} v (D^\alpha\sigma)^2, \ |\alpha| \le 2.$$

Moreover, considering equations (5.6) and (5.7) by themselves, one obtains

$$\frac{d}{dt}\|v\|_1^2 + \|v\|_2^2 \le c(\|\sigma\|_1^2 + \|f\|_0^2) + N.L. \tag{5.12}$$

$$\frac{d}{dt}\|\sigma\|_2^2 \le c\|v\|_3^2 + N.L.. \tag{5.13}$$

Finally, $\|\sigma_t\|_1$ can be estimated directly from (5.7).

Adding all these estimates we (essentially) find

$$(5.14)\qquad \frac{d}{dt}(\|v\|_1^2 + \|\sigma\|_2^2 + \|v_t\|_0^2 + \|\sigma_t\|_0^2) + \|v\|_3^2 + \|\sigma\|_2^2 + \|v_t\|_1^2 + \|\sigma_t\|_1^2 \leq$$
$$\leq c(\|v\|_3^2 + \|\sigma\|_2^2) + c(\|f\|_0^2 + \|f_t\|_{-1}^2) + N.L..$$

Step 2. We need to estimate only one of the norms $\|v\|_3^2$ and $\|\sigma\|_2^2$ (since they are related to each other by (5.6)). However, as we said, we can control "interior" and "tangential" derivatives of v, but the normal derivatives give some difficulties. On the other hand, we shall see that the normal derivatives of σ can be estimated, whereas it is not clear how to do the same for tangential derivatives. The trick to overcome this difficulty is to consider (5.6), (5.7) as a Stokes problem, i.e. to utilize the a-priori estimate

$$(5.15)\quad \|v\|_3^2 + \|\sigma\|_2^2 \leq c(\| - \Delta v + \nabla\sigma\|_1^2 + \|\operatorname{div} v\|_2^2) \leq c(\|v_t\|_1^2 + \|\operatorname{div} v\|_2^2 + \|f\|_1^2) + N.L..$$

Since v_t is already controlled in terms of f_t and nonlinear terms(see (5.11) for v_t and σ_t), the crucial point is to estimate

$$\|\operatorname{div} v\|_2^2.$$

Step 3. Since, as we said before, we have a control on the "interior" and "tangential" derivatives of v, it is only necessary to estimate the first and second order normal derivatives of $\operatorname{div} v$. Let us begin observing that on $\partial\Omega$ we have

$$(5.16)\qquad \Delta v \cdot n \cong (\nabla J \operatorname{div} v) \cdot n$$

(more precisely, the equality holds up to first order terms and second order terms containing only one normal derivative of v at most).

Hence, by adding to (5.6) the gradient of (5.7) multiplied by $2n$, we obtain

$$2(\nabla\sigma \cdot n)_t + (\nabla\sigma \cdot n) \cong -v_t \cdot n + f \cdot n + N.L..$$

In this way we obtain an estimate for $\nabla\sigma \cdot n$ and we can repeat the same argument for the second order normal derivative of σ. (For doing that, one has indeed to obtain also a suitable estimate for the tangential derivatives of the second order normal derivative of v. This can be done by means of a Stokes problem for the tangential derivatives of v in local coordinates near the boundary). On the other hand, by (5.6) and (5.16),

$$2(\nabla \operatorname{div} v) \cdot n \cong \nabla\sigma \cdot n + v_t \cdot n - f \cdot n + N.L.,$$

hence the estimates just proven for $\nabla\sigma \cdot n$ furnish corresponding estimates for the normal derivatives of $\operatorname{div} v$.

By considering (5.14), (5.15) and these estimates on $\operatorname{div} v$, we have

$$(5.17)\qquad \mathcal{E}_t + \mathcal{D} \leq c(\|f\|_1^2 + \|f_t\|_{-1}^2) + N.L.,$$

where

$$\mathcal{E} := \|v\|_1^2 + \|\sigma\|_2^2 + \|v_t\|_0^2 + \|\sigma_t\|_0^2$$
$$\mathcal{D} := \|v\|_3^2 + \|\sigma\|_2^2 + \|v_t\|_1^2 + \|\sigma_t\|_1^2.$$

(Remark moreover that $\mathcal{D} \geq \mathcal{E}$).

Step 4. It is necessary now to estimate in a "good" way the nonlinear terms. We need to obtain something like

$$N.L. \leq c\mathcal{D}(\mathcal{E}^{\alpha} + \mathcal{E}^{\beta}), \quad 0 < \alpha \leq \beta \ . \tag{5.18}$$

Let us underline that here it is essential that the "dissipative" term $\mathcal{D}$ has an exponent less or equal to 1. In fact, from (5.17) and (5.18) it follows

$$\mathcal{E}_t \leq -\mathcal{D}[1 - c_0(\mathcal{E}^{\alpha} + \mathcal{E}^{\beta})] + c_0(\|f\|_1^2 + \|f_t\|_{-1}^2) \ .$$

One easily sees now that, if $\mathcal{E}(0) \leq \epsilon_0$ and $\|f\|_1^2 + \|f_t\|_{-1}^2 \leq (2c_0)^{-1}\epsilon_0$, ϵ_0 small enough, then $\mathcal{E}(t) \leq \epsilon_0$ and $\int_0^t \mathcal{D} \leq c[\mathcal{E}(0) + \int_0^t(\|f\|_1^2 + \|f_t\|_{-1}^2)]$ in $\mathbb{R}^+$ (in particular, $\int_0^t \mathcal{D} \leq c_1(1+t)$ in $\mathbb{R}^+$, c_1 a small positive constant).

Estimate (5.18) heavily depends on the structure of the nonlinear terms. We have indeed to admit at this point that initially we were optimist assuming that these terms *had* to be "good": for instance, we would not be able to find (5.18) if the nonlinear terms were quadratic in Dv (which is the case of non-barotropic compressible flows!). However, in the present situation, looking back carefully at the proof of the preceding estimates, one can see that (5.18) holds with $\alpha = 1$ and $\beta = 2$. Some sharp estimates concerning multiplication in Sobolev spaces must be repeatedly used, and, for the sake of shortness, it is not possible to present all the details here.

The case of non-barotropic compressible flows require some additional estimates which permit to "enrich" the "energy" term $\mathcal{E}$ and the "dissipative" term $\mathcal{D}$ in (5.17), yielding thus the proof of (5.18) (let us refer to Valli-Zajączkowski [VZ] for the details).

Step 5. Having found a global a-priori estimate on $\mathcal{E}$, it is easy to prove that

$$\|v\|_2^2 + \|\sigma\|_2^2 \leq c(\mathcal{E} + \mathcal{E}^3 + \|f\|_0^2),$$

which is the a-priori estimate we need to extend the local solution on the whole real half-line $\mathbb{R}_+$.

We want to give now some informations about previous and related results. First of all, the uniqueness theorem has been proven by GraffiJ[Gr], Serrin [Se3], Itaya [I2] and Valli [V1], and other local existence theorems were obtained by Solonnikov [So], Tani [T1], [T4] (for the slip boundary condition $v \cdot n = 0$ and $[n \cdot D(v)] \cdot \tau = 0$), [T6] (for more general slip boundary conditions), Valli [V2], Łukaszewicz [Łu2] (this last one for the inflow-outflow case), Ströhmer [St]. An a-priori estimate for the minimum of the temperature θ is shown in Łukaszewicz [Łu1]. The existence of global in time and periodic solutions for the barotropic magnetohydrodynamic case has been proven by Štědrý-Vejvoda [ŠýVe]. The free boundary problem (local in time) was solved by Tani [T2], [T3] and Secchi-Valli [SeV]; more recently, the free boundary problem modelling the motion of a star subjected to self-gravitation (and eventually in the presence of radiation and surface tension) has been considered by Solonnikov-Tani [SoT1], [SoT2], Tani [T5] and Secchi [Se2], [Se3], [Se4].

Let us add now some comments concerning problems in different domains. The Cauchy problem ($\Omega = \mathbb{R}^d$) has been studied by Nash [Ns], Itaya [I1], Vol'pert-Hudjaev

[VoH], Matsumura-Nishida [MtN1], [MtN2], Ponce [Po]; problems in exterior domains or in the half-space have been considered by Matsumura-Nishida [MtN5], Deckelnick [D1], [D2] (this last author is specially interested in the time-asymptotic behaviour of the solution).

We conclude this Section by mentioning some open problems:

(i) global existence (for large data) if $\Omega \subset \mathbb{R}^2$, finding some new a-priori estimates on $\inf \rho$ and $\sup \rho$. As we already said, this is related to good estimates for the L^∞-norm of $\operatorname{div} v$.

(ii) global existence for the free-boundary problem, even for small data. No result of this type is known in spatial dimension larger than one.

6 – A stability result for compressible viscous flows

In Valli [V3] the stability of the global solution to (1.12), (1.13), (1.15), $(1.20)_{1,2}$ has been also studied. More precisely, the following Theorem was proven:

Theorem D. *Assume that Ω, P, μ^* and ς^* are as in Theorem C. Let (v_1, ρ_1) and (v_2, ρ_2) be solutions of (1.12), (1.13), (1.15) such that $\inf_{\mathbb{R}_+ \times \Omega} \rho_i(t,x) > 0$ $(i = 1,2)$ and*

$$|\Omega|^{-1} \int_\Omega \rho_1(0) = |\Omega|^{-1} \int_\Omega \rho_2(0) := \rho^* > 0 \tag{6.1}$$

$$\|v_i(t)\|_2^2 + \|\rho_i(t) - \rho^*\|_2^2 \le B_1 \ , \quad t \in \mathbb{R}_+ \ , \quad i = 1,2 \tag{6.2}$$

$$\int_0^t \|v_1\|_3^2 \le B_2 + B_3 t \ , \quad t \in \mathbb{R}_+ \ , \tag{6.3}$$

where $B_1 > 0$ and $B_3 > 0$ are small enough. Then there exists $\lambda > 0$ such that the difference (w, η) between (v_1, ρ_1) and (v_2, ρ_2) satisfies

$$\|w(t)\|_0^2 + \|\eta(t)\|_0^2 \le c(\|w(0)\|_0^2 + \|\eta(0)\|_0^2) e^{-\lambda t} \ , \quad t \in \mathbb{R}_+ \ .$$

Let us recall that, due to (1.13), (1.15), conditions (6.1) imply $|\Omega|^{-1} \int_\Omega \rho_1(t) = |\Omega|^{-1} \int_\Omega \rho_2(t) = \rho^*$ for each $t \in \mathbb{R}_+$.

Choose now for simplicity $\mu^* = 1$, $\varsigma^* = 2/d$, $\rho^* = 1$ and $P(\xi) = \xi$, and introduce the unknowns $\sigma_i(t,x) = \rho_i(t,x) - 1$ as in (5.1). We start the proof by writing in a convenient form the equations for w and η:

$$w_t - \Delta w - \nabla \operatorname{div} w + \nabla \eta = H_1 - H_2 \quad \text{in } Q_\infty \tag{6.4}$$

$$\eta_t + \operatorname{div} w = L_1 - L_2 = -\operatorname{div}(\eta v_1 + \sigma_2 w) \quad \text{in } Q_\infty, \tag{6.5}$$

plus the homogeneous Dirichlet boundary condition (5.4) for w. (We are not reporting here the precise definitions of H_i and L_i, $i = 1,2$, which can be easily deduced by (5.8) and (5.9)).

Step 1. As in the first step in Section 5, we can eliminate the terms $\nabla\eta$ and $\operatorname{div} w$ multiplying (6.4) by w and (6.5) by η, integrating in Ω and adding the two relations thus obtained. A simple computation of the right hand sides yields

$$\frac{d}{dt}(\|w\|_0^2 + \|\eta\|_0^2) + \|w\|_1^2 \leq c(\|\sigma_2\|_2^2\|\eta\|_0^2 + \|v_1\|_3\|\eta\|_0^2 + \|H_1 - H_2\|_{-1}^2) \ . \tag{6.6}$$

The problem now is to find a suitable estimate for $\|\eta\|_0^2$.

Step 2. The idea is to take advantage from the presence of the term $\nabla\eta$ in (6.4). Thus multiply this equation by z, where z is the solution of the Stokes problem

$$\begin{cases} -\Delta z + \nabla\pi = 0 \\ \operatorname{div} z = \eta \\ z_{|\partial\Omega} = 0 \ , \end{cases} \tag{6.7}$$

which satisfies

$$\|z\|_1^2 \leq c\|\eta\|_0^2 \ . \tag{6.8}$$

(Let us notice here that, as a consequence of (6.1), we have $\int_\Omega \eta = 0$; hence, the necessary condition for solving (6.7) is satisfied).

After integration by parts, one obtains

$$\|\eta\|_0^2 \leq \int_\Omega w_t \cdot z + c(\|w\|_1 + \|H_1 - H_2\|_{-1})\|z\|_1 \ ,$$

hence

$$\|\eta\|_0^2 \leq c\int_\Omega w_t \cdot z + c(\|w\|_1^2 + \|H_1 - H_2\|_{-1}^2) \ . \tag{6.9}$$

From (6.6) and (6.9) it follows at once

$$\begin{aligned} &\frac{d}{dt}(\|w\|_0^2 + \|\eta\|_0^2) + \|w\|_1^2 + \delta\|\eta\|_0^2 \leq \\ &\qquad \leq c\delta\int_\Omega w_t \cdot z + c(\|\sigma_2\|_2^2\|\eta\|_0^2 + \|v_1\|_3^2\|\eta\|_0^2 + \|H_1 - H_2\|_{-1}^2) \ , \end{aligned} \tag{6.10}$$

where $\delta > 0$ can be chosen arbitrarily small.

Step 3. Rewrite the term $\int_\Omega w_t \cdot z$ as

$$\int_\Omega w_t \cdot z = (\int_\Omega w \cdot z)_t - \int_\Omega w \cdot z_t \ ;$$

differentiating (6.7) with respect to t, one sees that $\operatorname{div}(z_t) = \eta_t = -\operatorname{div}(\eta v_1 + \sigma_2 w + w)$.

Introduce now the solution (V, Q) to the Stokes problem

$$\begin{cases} -\Delta V + \nabla Q = w \\ \operatorname{div} V = 0 \\ V_{|\partial\Omega} = 0 \ , \end{cases} \tag{6.11}$$

which satisfies

$$\|V\|_2^2 + \|\nabla Q\|_0^2 \le c\|w\|_0^2 \, .$$

By simple integrations by parts one finds

$$\left|\int_\Omega w \cdot z_t\right| = \left|\int_\Omega \nabla Q \cdot (\eta v_1 + \sigma_2 w + w)\right| \le c\|w\|_0 \|\eta v_1 + \sigma_2 w + w\|_0 \, .$$

Now, from estimate (6.8), choosing δ small enough and recalling the smallness assumption (6.2), by integrating on $]0,t[$ we obtain

$$\begin{aligned} &(\|w(t)\|_0^2 + \|\eta(t)\|_0^2) + \int_0^t (\|w\|_1^2 + \|\eta\|_0^2) \le \\ &\le c(\|w(0)\|_0^2 + \|\eta(0)\|_0^2) + c\int_0^t (\|v_1\|_3^2 \|\eta\|_0^2 + \|H_1 - H_2\|_{-1}^2) \, . \end{aligned} \tag{6.12}$$

Step 4. An easy computation shows that

$$\|H_1 - H_2\|_{-1}^2 \le cB_1(\|w\|_1^2 + \|\eta\|_0^2) + c\|v_1\|_3^2 \|\eta\|_0^2 \, ,$$

where B_1 is assumed to be sufficiently small (see (6.2)). One can thus apply Gronwall lemma, and, recalling (6.3), it yields

$$\begin{aligned} &(\|w(t)\|_0^2 + \|\eta(t)\|_0^2) + \int_0^t (\|w\|_1^2 + \|\eta\|_0^2) \le \\ &\le c(\|w(0)\|_0^2 + \|\eta(0)\|_0^2) \exp[c^*(B_2 + B_3 t)] \, . \end{aligned} \tag{6.13}$$

Step 5. The final step consists in repeating the same argument for

$$w_\alpha(t,x) := \exp(\alpha t)\; w(t,x)\; \eta_\alpha(t,x) := \exp(\alpha t)\; \eta(t,x) \, ,$$

which satisfy a problem similar to (6.4), (6.5). Taking into account the few necessary modifications, and choosing α small enough (but independent of B_3), one easily finds

$$\begin{aligned} &(\|w(t)\|_0^2 + \|\eta(t)\|_0^2) + \int_0^t (\|w\|_1^2 + \|\eta\|_0^2) \le \\ &\le c(\|w(0)\|_0^2 + \|\eta(0)\|_0^2) \exp[-(2\alpha - c^* B_3)t] \, . \end{aligned} \tag{6.14}$$

The resulHrhus follows by assuming that $c^* B_3 < 2\alpha$ and $\lambda := 2\alpha - c^* B_3$.

Let us finish by recalling that the stability for the compressible Bénard problem has been investigated by Padula [P1] and Coscia-Padula [CsP1], [CsP2].

7 – Time-periodic compressible viscous flows

By following an approach proposed by Serrin [Se2] for the incompressible case, the results contained in the previous Sections yield the existence of a time-periodic solution.

Theorem E. *Let $\Omega \subset \mathbb{R}^d$ $(d \leq 3)$ be a bounded domain with $\partial\Omega \in C^3$, $f \in L^\infty(\mathbb{R}^+; H^1(\Omega))$, $f_t \in L^\infty(\mathbb{R}^+; H^{-1}(\Omega))$, $P \in C^2(\mathbb{R}^+)$, $\mu^* \in C^2(\mathbb{R}^+)$, $\varsigma^* \in C^2(\mathbb{R}^+)$, $P_\rho(\xi) > 0$, $\mu^*(\xi) > 0$ and $\varsigma^*(\xi) \geq 0$ for $\xi > 0$. Assume that $\sup_{\mathbb{R}^+}(\|f(t)\|_1 + \|f_t(t)\|_{-1})$ is small enough and f is a time-periodic function of period $T > 0$. Then for each $\rho^* > 0$ there exist*

$$v \in L^2_{loc}(\mathbb{R}^+; H^3(\Omega)) \cap C^0_B(\mathbb{R}_+; H^2(\Omega)),\ v_t \in L^2_{loc}(\mathbb{R}^+; H^1(\Omega)) \cap C^0_B(\mathbb{R}_+; L^2(\Omega)$$
$$\rho \in C^0_B(\mathbb{R}_+; H^2(\Omega)),\ \rho_t \in C^0_B(\mathbb{R}_+; H^1(\Omega))$$

such that (v, ρ) is a time-periodic solution of (1.12), (1.13), (1.15) of period T, satisfying $|\Omega|^{-1} \int_\Omega \rho(0) = \rho^*$ and $\inf_{\mathbb{R}^+ \times \Omega} \rho(t, x) > 0$. Moreover, this solution is unique among any other solution (v_1, ρ_1) satisfying (6.2) and $|\Omega|^{-1} \int_\Omega \rho_1(0) = \rho^*$.

The proof is based on the remark that, if a time-periodic solution satisfying the assumptions (6.2), (6.3) of Theorem D does exist, then it is (locally) asymptotically stable. Hence, we can find its initial value by taking the limit of $(\bar{v}(nT), \bar{\rho}(nT))$, where $(\bar{v}, \bar{\rho})$ is any global solution constructed by means of Theorem C.

Let us present this argument in more detail. Take an initial data satisfying the assumptions of Theorem C, and denote by $\bar{U}(t, x) := (\bar{v}(t, x), \bar{\rho}(t, x))$ the corresponding global solution. It is easily seen that $\bar{U}(nT)$ is convergent in $L^2(\Omega)$. In fact, define for $m > n$

$$V_{m,n}(t) := \bar{U}(t + (m - n)T) \ ;$$

since f is periodic, $V_{m,n}$ is a solution to (1.12), (1.13), (1.15) with initial datum $\bar{U}((m - n)T)$. By Theorem D, we have

$$\|\bar{U}(t) - V_{m,n}(t)\|_0^2 \leq c(\|\bar{U}(0) - \bar{U}((m - n)T)\|_0^2) e^{-\lambda t} \ , \tag{7.1}$$

and taking $t = nT$

$$\|\bar{U}(nT) - \bar{U}(mT)\|_0^2 \leq c(\|\bar{U}(0) - \bar{U}((m - n)T)\|_0^2) e^{-\lambda nT} \ . \tag{7.2}$$

Since $\bar{U}(t)$ is uniformly bounded in $L^2(\Omega)$, $\bar{U}(nT)$ is a Cauchy sequence, and we denote by U_* the limit value.

Take now U_* as initial datum, and let (v, ρ) be the corresponding global solution of (1.12), (1.13), (1.15). We claim that it is a T-periodic function. In fact, set

$$W_n(t) := \bar{U}(t + nT) \ ,$$

which is a solution to (1.12), (1.13), (1.15) with initial datum $\bar{U}(nT)$. Hence, by Theorem D we have

$$\|(v, \rho)(t) - W_n(t)\|_0^2 \leq c(\|U_* - \bar{U}(nT)\|_0^2 e^{-\lambda t} \ , \tag{7.3}$$

and for $t = T$

$$\|(v, \rho)(T) - \bar{U}((n + 1)T)\|_0^2 \leq c(\|U_* - \bar{U}(nT)\|_0^2 e^{-\lambda T} \ . \tag{7.4}$$

Taking the limit for n going to infinity, we find $(v, \rho)(T) = U_* = (v, \rho)(0)$.

Through the results proven in Theorems C and D it is also possible to show the existence of almost-periodic solutions. Moreover, these solutions are (locally) asymptotically stable (see Marcati-Valli [MrV]).

Furthermore, it is easily proven that, if f is independent of t, the solution obtained in Theorem E is periodic of any rational period, hence independent of t by continuity. The existence of stationary solutions has been proven also by other approaches, the most interesting one due to Beirão da Veiga [B2], [B3] (see also Padula [P2], Valli [V4]). We are going to present two of these approaches in the next Section.

8 – Steady compressible viscous flows

In this Section we want to present two different methods for proving the existence of a stationary solution for compressible viscous fluids.

The equations describing the steady motion are simply obtained by (1.12)-(1.14) by dropping out the terms v_t, ρ_t and θ_t. One has also to impose boundary conditions, for instance we will consider the Dirichlet boundary conditions (1.15), (1.17), and moreover the total amount of fluid $\int_\Omega \rho$ must be assigned, say equal to $|\Omega|\rho^*$, $\rho^* > 0$ a given constant. (This constraint is in fact related to the boundary conditions (1.15) or (1.16), and it is not necessary for inflow-outflow problems).

As in the previous Sections, we only consider in detail the barotropic case.

Theorem F. *Let $\Omega \subset \mathbb{R}^d$ $(d \leq 3)$ be a bounded domain with $\partial\Omega \in C^3$, $f \in H^1(\Omega)$, $P \in C^2(\mathbb{R}^+)$, $\mu^* \in C^2(\mathbb{R}^+)$, $\varsigma^* \in C^2(\mathbb{R}^+)$, $P_\rho(\xi) > 0$, $\mu^*(\xi) > 0$ and $\varsigma^*(\xi) \geq 0$ for $\xi > 0$. Assume that $\|f\|_1$ is small enough. Then for each $\rho^* > 0$ there exist*

$$v \in H^3(\Omega), \ \rho \in H^2(\Omega)$$

such that (v, ρ) is a solution to the stationary problem associated to (1.12), (1.13), (1.15), satisfying $|\Omega|^{-1} \int_\Omega \rho = \rho^*$ and $\inf_\Omega \rho(x) > 0$. This solution is unique among any other solution (v_1, ρ_1) close enough to $(0, \rho^*)$ and such that $|\Omega|^{-1} \int_\Omega \rho_1 = \rho^*$.

(Let us point out that for a misprint in [V4] it is only required $\mu^* \in C^1(\mathbb{R}^+)$ and $\varsigma^* \in C^1(\mathbb{R}^+)$).

Assume for simplicity that μ^* and ς^* are constant, $\rho^* = 1$, $P(\xi) = \xi$. Introducing the unknown σ as in (5.1), the steady motion can be rewritten in the following way:

$$\text{(8.1)} \quad \begin{cases} -\mu^* \Delta v - [\varsigma^* + (d-2)\mu^*/d] \nabla \operatorname{div} v + \nabla \sigma = (\sigma + 1)[f - (v \cdot \nabla) v] & \text{in } \Omega \\ \operatorname{div} v + \operatorname{div}(\sigma v) = 0 & \text{in } \Omega \\ v_{|\partial\Omega} = 0 & \text{on } \partial\Omega \\ \int_\Omega \sigma = 0 \,. \end{cases}$$

The two methods we are going to present here make use of a fixed point argument which is based upon an existence theorem for the following linear problem:

$$\text{(8.2)} \quad \begin{cases} -\mu^* \Delta w - [\varsigma^* + (d-2)\mu^*/d] \nabla \operatorname{div} w + \nabla \eta = F & \text{in } \Omega \\ \operatorname{div} w + \operatorname{div}(\eta v) = 0 & \text{in } \Omega \\ w_{|\partial\Omega} = 0 & \text{on } \partial\Omega \\ \int_\Omega \eta = 0, \end{cases}$$

where v is a given vector field, satisfying $v_{|\partial\Omega} = 0$. Assigned v and σ in a suitable convex compact set $\mathbf{K}$ (see (8.18)) and choosing $F = F(v, \sigma, f)$ as in (8.19), the Schauder theorem will give a fixed point for the map $\Phi : (v, \sigma) \to (w, \eta)$.

8.A. The linear problem (8.2)

We want to prove that there exists a unique solution to (8.2) satisfying

$$\|w\|_3 + \|\eta\|_2 \leq c\|F\|_1 \tag{8.3}$$

for each $v \in H^3(\Omega) \cap H^1_0(\Omega)$, $\|v\|_3 \leq B$, B small enough, and $F \in H^1(\Omega)$. It is imortant to remark that the constant c depends in a continuous way on μ^*,ς^* and B (but does not depend on v).

Remark that system (8.2) (which is a generalization of Stokes system) is *not* elliptic in the sense of Agmon-Douglis-Nirenberg (excepting for $v = 0$: in this case it is equivalent to the Stokes system). Adopting their notations, after a simple computation one finds in fact that, if v is not identically vanishing,

$$\det\{l'_{ij}(x, \xi)\} = (-1)^d (\mu^*)^{d-1} [\varsigma^* + 2(d-1)\mu^*/d] |\xi|^{2d} v(x) \cdot \xi \, .$$

(i) **First method** (see Valli [V4]).

Let us remark at first that, though estimate (8.3) is clearly crucial, the existence of a solution does not seem to follow at once from it. (By elliptic approximation one can only show the existence of a solution in $H^2(\Omega) \times H^1(\Omega)$, since no conditions are imposed to η on $\partial\Omega$. Moreover, the usual regularization procedures cannot be applied since system (8.2) is not elliptic). Indeed, to prove the existence one can use a continuity method (see [V4] for the details), showing at first the existence of a solution for $\varsigma^* >> \mu^*$ as in Padula [P2], and then proving the same result for general ς^* and μ^* by means of the a-priori estimate (8.3). In order to make clear how to apply the continuity method, this time we have not chosen particular values of μ^* and ς^*, and we have underlined that the constant c in (8.3) depends in a continuous way on μ^* and ς^*. This last assertion can be easily checked by following the proof of the estimates (8.4)-(8.6).

Let us show now that (8.3) holds. The procedure is strictly related to the one presented in Section 5. In fact, proceding as in Step 1 there, one finds

$$\|w\|_1^2 \leq c(\|F\|_{-1}^2 + \|v\|_3 \|\eta\|_0^2) \, . \tag{8.4}$$

(As usual, the term coming from $vJ \cdot \nabla\eta$ must be integrated by part in a suitable way). The same argument can be repeated for the "interior" and "tangential" derivatives of first and second order.

The a-priori estimate for the Stokes system gives

$$\|w\|_3^2 + \|\eta\|_2^2 \leq c(\|F\|_1^2 + \|\operatorname{div} w\|_2^2) \, . \tag{8.5}$$

Hence, the crucial point is to estimate $\|\operatorname{div} w\|_2^2$, and this can be done as in Section 5, Step 3, yielding

$$\|\operatorname{div} w\|_2^2 \leq c(\|F\|_1^2 + \|v\|_3 \|\eta\|_2^2) + \epsilon\|\eta\|_2^2 \, , \tag{8.6}$$

where ϵ is arbitrarily small. (In the proof of this inequality, estimate (8.4) for w and its "interior" ad "tangential" derivatives of first and second order have to be used).

From (8.5) and (8.6), choosing ϵ and B small enough, one obtains at once (8.3).

To complete the proof, it is now necessary to show that we are in a condition to apply a continuity argument, i.e., we have to prove that, at least for one choice of the parameters μ^* and ς^*, there exists a solution of the linear problem (8.2). The second method we are going to present in the sequel is indeed an existence theorem for any choice of μ^* and ς^*; let us present, however, an alternative approach, which, as we already recalled, is essentially due to Padula [P2].

Denoting

$$\pi := \eta/\mu^* - [\varsigma^*/\mu^* + (d-2)/d] \operatorname{div} w \ ,$$

problem (8.2) can be transformed into

$$(8.7) \qquad \begin{cases} -\Delta w + \nabla \pi = F/\mu^* & \text{in } \Omega \\ \operatorname{div} w = [\varsigma^*/\mu^* + (d-2)/d]^{-1}(\eta/\mu^* - \pi) & \text{in } \Omega \\ w_{|\partial\Omega} = 0 & \text{on } \partial\Omega \\ \int_\Omega \pi = 0 \end{cases}$$

$$(8.8) \qquad \begin{cases} (\mu^*)^{-1}[\varsigma^*/\mu^* + (d-2)/d]^{-1}\eta + \operatorname{div}(v\eta) = [\varsigma^*/\mu^* + (d-2)/d]^{-1}\pi & \text{in } \Omega \\ \int_\Omega \eta = 0 \ . \end{cases}$$

(This second problem is solvable if $\|v\|_3 \leq c(\mu^*)^{-1}[\varsigma^*/\mu^* + (d-2)/d]^{-1}$).

Writing (η^*, π^*) instead of (η, π) in the right hand side of (8.7), it is not difficult to see that, if $\varsigma^* >> \mu^*$, the affine map $(\eta^*, \pi^*) \to (\eta, \pi)$ is a contraction in $H^2(\Omega) \times H^2(\Omega)$ (see [V4] for the details). Hence, the existence of a solution to (8.2) is proven for suitable parameters μ^* and ς^*.

(ii) Second method (see Beirão da Veiga [B2], [B3]).

This method directly gives the existence of a solution to (8.2) satisfying (8.3). Since a continuity argument concerning μ^* and ς^* is no more introduced, we can for simplicity assume $\mu^* = 1$ and $\varsigma^* = 2/d$.

The main idea of the proof is based on this interesting remark: a solution (v, σ) of (8.1) with $\rho = (\sigma + 1) > 0$ must satisfy $\operatorname{div} v = 0$ on $\partial\Omega$. In fact one has

$$(8.9) \qquad \operatorname{div} v = -(\sigma + 1)^{-1} v \cdot \nabla\sigma,$$

hence from $v_{|\partial\Omega} = 0$ it follows $(\operatorname{div} v)_{|\partial\Omega} = 0$.

Thinking (8.2) as a useful linearization of (8.1), we can thus assume that v in (8.2) also satisfies the condition $\operatorname{div} v \in H_0^1(\Omega)$.

By taking the divergence of $(8.2)_1$ one obtains

$$(8.10) \qquad \begin{cases} -2\Delta \operatorname{div} w = \operatorname{div} F - \Delta\eta & \text{in } \Omega \\ \operatorname{div} w = (-v \cdot \nabla\eta - \eta \operatorname{div} v) = 0 & \text{on } \partial\Omega \ . \end{cases}$$

By applying Δ to $(8.2)_2$ and by using $(8.10)_1$ one has

$$(8.11) \qquad \begin{aligned} &(1/2)\Delta\eta + v \cdot \nabla(\Delta\eta) = \\ &\qquad = (1/2) \operatorname{div} F - [2Dv : D^2\eta + \Delta v \cdot \nabla\eta + \Delta(\eta \operatorname{div} v)] \quad \text{in } \Omega. \end{aligned}$$

We shall construct a fixed point for the following map: given $\tau \in H^2(\Omega)$, find the solution $\lambda \in L^2(\Omega)$ of

$$(8.12) \qquad (1/2)\lambda + v \cdot \nabla\lambda = (1/2)\,\mathrm{div}\,F - [2Dv : D^2\tau + \Delta v \cdot \nabla\tau + \Delta(\tau\,\mathrm{div}\,v)] \quad \text{in } \Omega.$$

(Let us remark that this solution exists if $\|v\|_3$ is small enough; moreover, the estimate

$$(8.13) \qquad \|\lambda\|_0 \leq c(\|F\|_1 + \|v\|_3\|\tau\|_2)$$

holds true).

Hence find the solution $\theta \in H^2(\Omega) \cap H^1_0(\Omega)$ of

$$(8.14) \qquad \begin{cases} -2\Delta\theta = \mathrm{div}\,F - \lambda & \text{in } \Omega \\ \theta_{|\partial\Omega} = 0 & \text{on } \partial\Omega \ . \end{cases}$$

Finally solve the Stokes problem

$$(8.15) \qquad \begin{cases} -\Delta w + \nabla\eta = F + \nabla\theta & \text{in } \Omega \\ \mathrm{div}\,w = \theta - \theta_\Omega & \text{inJ } \Omega \\ w_{|\partial\Omega} = 0 & \text{on } \partial\Omega \\ \int_\Omega \eta = 0, & \end{cases}$$

where $\theta_\Omega := |\Omega|^{-1}\int_\Omega \theta$. We claim that a fixed point $\tau = \eta$ gives the solution (w, η) of (8.2).

In fact, one has only to verify that $(8.2)_2$ holds. From $(8.15)_1$, $(8.15)_2$ (by taking the divergence)

$$\Delta\eta = \mathrm{div}\,F + 2\Delta\theta,$$

i.e., from $(8.14)_1$ $\Delta\eta = \lambda$. Moreover from (8.12), $(8.14)_2$

$$(8.16) \qquad \begin{cases} \Delta[\mathrm{div}\,w + \mathrm{div}(v\eta)] = 0 & \text{in } \Omega \\ [\mathrm{div}\,w + \mathrm{div}(v\eta)]_{|\partial\Omega} = -\theta_\Omega & \text{on } \partial\Omega \ , \end{cases}$$

hence $\mathrm{div}\,w + \mathrm{div}(v\eta) = -\theta_\Omega$ in $\bar{\Omega}$. Finally, the mean value in Ω of $\mathrm{div}\,w + \mathrm{div}(v\eta)$ is zero, hence $\theta_\Omega = 0$ and $(8.2)_2$ holds.

To prove the existence of a fixed point of the map $\tau \to \eta$, from (8.12)-(8.15) one easily verifies that

$$(8.17) \qquad \|w\|_3 + \|\eta\|_2 \leq c(\|F\|_1 + \|v\|_3\|\tau\|_2),$$

and consequently, if $\|v\|_3$ is small enough, the affine map $\tau \to \eta$ is a contraction in $H^2(\Omega)$.

We want to remark that the method just presented can be extended without particular difficulties to the case: d arbitrary, $F \in H^{k,p}(\Omega)$, $k \geq 0$, $1 < p < \infty$, $k + 1 > n/p$ (see [B3] and Defranceschi [De]).

On the other hand, for what is concerned with the solvability of the linear system (8.2), it requires that the given vector field v satisfies the additional condition $(\mathrm{div}\,v)_{|\partial\Omega} = 0$.

8.B. The non-linear problem

The solution of (8.1) is obtained by a fixed point argument. Taking

$$
\begin{aligned}
\mathbf{K} := \Big\{(v,\sigma) \in H^3(\Omega) \times H^2(\Omega) | \\
v_{|\partial\Omega} = 0,\ (\operatorname{div} v)_{|\partial\Omega} = 0,\ \int_\Omega \sigma = 0, \|v\|_3 + \|\sigma\|_2 \leq B\Big\}
\end{aligned}
\tag{8.18}
$$

$$
F := (\sigma + 1)[f - (v \cdot \nabla)v] , \tag{8.19}
$$

estimate (8.3) gives

$$
\|w\|_3 + \|\eta\|_2 \leq c\|F\|_1 \leq c(B+1)(\|f\|_1 + B^2).
$$

Choosing $B := \|f\|_1^{1/2}$ and small enough, the map $\Phi : (v,\sigma) \to (w,\eta)$ defined before satisfies $\Phi(\mathbf{K}) \subset \mathbf{K}$, and Schauder theorem easily shows the existence of a fixed point, hence a solution of (8.1).

An analogous result can be obtained for non-barotropic compressible viscous fluids (see [B2], [B3], [V4]).

Let us add now some comments. Another method to prove the existence of a stationary solution in the non-barotropic case has been proposed by Farwig [Fa1], who in [Fa2] has extended his approach to the case of the slip boundary condition $v \cdot n = 0$ and $[n \cdot D(v)] \cdot \tau = 0$ on $\partial\Omega$. This last boundary-value problem has also been studied by Secchi [Se6], [Se7] for the more general case of a compressible fluid in the presence of self-gravitation. Using a perturbation argument, the existence of a spatially periodic solution at high Mach number has been proven by Boldrini [Bo]. Following the approach of Farwig [Fa1], Novotný-Padula [NoP] have shown the existence of a stationary solution for a polytropic gas (i.e., $E(\rho,\theta) = c_V\theta$, c_V a positive constant, $P(\rho,\theta) = \theta G(\rho)$) when the external force field $f = \nabla\phi + f_0$, f_0 small enough. (In this respect, it has to be noticed that the necessary and sufficient conditions for the existence of an equilibrium solution, i.e., a stationary solution with vanishing velocity, under the action of a potential force field have been given by Beirão da Veiga [B3]; see also the discussion at the end of Section 9). The free-boundary problem for the barotropic case has been studied by Pilackas-Zajączkowski [PiZ]. Finally, Matsumura-Nishida [MtN6] have proven the existence theorem for the flow in an unbounded exterior domain.

The inflow-outflow case is still an open problem; for some remarks about it, see Valli-Zajączkowski [VZ].

9 – One-dimensional compressible viscous flows ($p = P(\rho)$, $\varsigma^* > 0$)

Let us remark that in the one-dimensional case (i.e., $d = 1$ in (1.4)) the stress tensor becomes $T = -p + \varsigma Dv$, hence the shear viscosity coefficient μ does not appear anymore.

Moreover, a suitable change of variable yields a simpler formulation which is an useful tool for showing the existence of a global in time solution for large data. We want to present a result due to Kazhikhov [K1], [K3] (for $f = 0$).

Theorem G. *Let* $\Omega =]a,b[$, $v_0 \in H_0^1(\Omega)$, $\rho_0 \in H^1(\Omega)$, $\inf_\Omega \rho_0(x) > 0$, $f \in L^1(\mathbb{R}^+;$-$L^\infty(\Omega)) \cap L^2(\mathbb{R}^+; L^\infty(\Omega))$, $P \in C^1(\mathbb{R}^+)$, $P_\rho(\xi) > 0$ *for* $\xi > 0$, $\varsigma^* > 0$ *a constant. Then there exist*

$$v \in L^2(\mathbb{R}^+; H^2(\Omega)) \cap C_B^0(\mathbb{R}_+; H^1(\Omega)), \; v_t \in L^2(\mathbb{R}^+; L^2(\Omega))$$
$$\rho \in C_B^0(\mathbb{R}_+; H^1(\Omega)), \; \rho_t \in C_B^0(\mathbb{R}_+; L^2(\Omega))$$

such that (v,ρ) *is the (unique) solution of (1.12), (1.13), (1.15)* $(1.20)_{1,2}$. *Moreover,* $\inf_{\mathbb{R}^+\times\Omega} \rho(t,x) > 0$.

Define
$$M := \int_a^b \rho(t,x) = \int_a^b \rho_0(x) \;,$$
and set
$$y = y(t,x) := \int_a^{U(0,t,x)} \rho_0(\lambda) d\lambda = \int_a^x \rho(t,\lambda) d\lambda, \quad x = x(t,y) = a + \int_0^y \eta(t,\xi) d\xi$$
$$u(t,y) := v(t,x), \quad \eta(t,y) := 1/\rho(t,x),$$
where $U(t,s,x)$ is defined in (2.5). Since $\inf_\Omega \rho_0(x) > 0$, the map $x \to y(0,x)$ has an inverse $\Lambda : y \to x$.

It is easy to see that equations (1.12), (1.13), (1.15), $(1.20)_{1,2}$ are transformed into

$$u_t + P(\eta^{-1})_y - (\varsigma^* \eta^{-1} u_y)_y = f(t, a + \int_0^y \eta(t,\xi) d\xi) \quad \text{in } M_\infty :=]0,+\infty[\times]0,M[\tag{9.1}$$
$$\eta_t = u_y \qquad \text{in } M_\infty \tag{9.2}$$
$$u(t,0) = u(t,M) = 0 \qquad \text{in } \mathbb{R}^+ \tag{9.3}$$
$$u_{|t=0} = u_0(y), \quad \eta_{|t=0} = \eta_0(y) > 0 \quad \text{in }]0,M[, \tag{9.4}$$

where $u_0(y) := v_0(\Lambda(y))$, $\eta_0(y) := (1/\rho_0)(\Lambda(y))$.

For simplicity, choose $a = 0$, $b = \rho^* = 1$ (hence $M = 1$), $\varsigma^* = 1$. We just want to present some a-priori estimates for u and η in $L^\infty(\mathbb{R}^+; H^1(\Omega))$, which permit to find a global in time solution for large data.

Step 1. Set
$$R(\gamma) := \int_1^\gamma [P(1) - P(\xi^{-1})] d\xi$$
(which satisfies $R(\gamma) \geq 0$, since $P_\rho(\xi) > 0$), multiply (9.1) by u and integrate over $]0,1[$. Recalling that
$$\int_0^1 P(\eta^{-1})_y u = -\int_0^1 P(\eta^{-1}) u_y = -\int_0^1 P(\eta^{-1}) \eta_t \;, \quad \int_0^1 \eta_t = \int_0^1 u_y = 0,$$
one easily obtains
$$\frac{d}{dt} \int_0^1 [\frac{u^2}{2} + R(\eta)] + \int_0^1 \eta^{-1} u_y^2 = \int_0^1 u\, f(t, \int_0^y \eta) \leq (\sup_y |f(t, \int_0^y \eta)|)(\int_0^1 u^2)^{1/2}. \tag{9.5}$$

Hence by Gronwall lemma

$$\int_0^1 u^2 \le c(u_0,\eta_0)\exp\Big(\int_0^{+\infty} \|f(\tau)\|_\infty\Big) , \tag{9.6}$$

and

$$\int_0^t\int_0^1 \eta^{-1}u_y^2 \le c(u_0,\eta_0,f), \tag{9.7}$$

where $\|\cdot\|_\infty$ is the norm in $L^\infty(0,1)$.

Step 2. We need now to find an estimate for $\inf_{M_\infty}\eta(t,y)$ and $\sup_{M_\infty}\eta(t,y)$. Since

$$(\eta^{-1}u_y)_y = (\eta^{-1}\eta_t)_y = (\log\eta)_{yt}, \tag{9.8}$$

multiplying (9.1) by $(\log\eta)_y$ we have

$$\frac{1}{2}\frac{d}{dt}\int_0^1(\log\eta)_y^2 = \int_0^1 u_t(\log\eta)_y + \int_0^1 P(\eta^{-1})_y(\log\eta)_y - \int_0^1 f\Big(t,\int_0^\xi\eta\Big)(\log\eta)_y.$$

Moreover

$$\int_0^1 u_t(\log\eta)_y = \frac{d}{dt}\int_0^1 u(\log\eta)_y - \int_0^1 u(\log\eta)_{yt} = \frac{d}{dt}\int_0^1 u(\log\eta)_y + \int_0^1\eta^{-1}u_y^2,$$
$$\int_0^1 P(\eta^{-1})_y(\log\eta)y = -\int_0^1 P_\rho(\eta^{-1})\eta^{-3}\eta_y^2 \le 0,$$

hence integrating over $]0,t[$

$$\frac{1}{2}\int_0^1(\log\eta)_y^2 + \int_0^t\int_0^1 P_\rho(\eta^{-1})\eta^{-3}\eta_y^2 = \frac{1}{2}\int_0^1(\log\eta_0)_y^2 + \int_0^1 u(\log\eta)_y -$$
$$-\int_0^1 u_0(\log\eta_0)_y + \int_0^t\int_0^1\eta^{-1}u_y^2 - \int_0^t\int_0^1 f\Big(\tau,\int_0^\xi\eta\Big)(\log\eta)_y.$$

Using now (9.6) and (9.7) we find

$$\int_0^1(\log\eta)_y^2 + \int_0^t\int_0^1 P'(\eta^{-1})\eta^{-3}\eta_y^2 \le c(u_0,\eta_0,f) + c\int_0^t \|f(\tau)\|_\infty\Big(\int_0^1(\log\eta)_y^2\Big)^{1/2},$$

and from Gronwall lemma

$$\int_0^1(\log\eta)_y^2 + \int_0^t\int_0^1 P_\rho(\eta^{-1})\eta^{-3}\eta_y^2 \le c(u_0,\eta_0,f). \tag{9.9}$$

On the other hand, from (9.2) and (9.3) we know that

$$\int_0^1\eta = \int_0^1\eta_0 = 1,$$

hence for each $t\in\mathbb{R}_+$ there exists a point $y_1(t)\in]0,1[$ such that

$$\eta(t,y_1(t)) = 1.$$

As a consequence, Poincaré inequality shows that for each $t \in \mathbb{R}_+$

$$\|\log \eta(t)\|_1 \le c(u_0, \eta_0, f). \tag{9.10}$$

By using Sobolev embedding theorem $H^1(0,1) \subset C^0([0,1])$ we finally obtain

$$0 < \inf_{M_\infty} \eta(t,y) \le \sup_{M_\infty} \eta(t,y) < +\infty. \tag{9.11}$$

Step 3. From (9.7), (9.9)-(9.11) one sees at once that for each $t \in \mathbb{R}_+$

$$\|\eta(t)\|_1 + \int_0^t \int_0^1 \eta_y^2 + \int_0^t \int_0^1 u_y^2 \le c(u_0, \eta_0, f), \tag{9.12}$$

which is one of the global estimates we need.

Moreover, multiplying (9.1) by u_{yy} and integrating by parts we find

$$\begin{aligned} \frac{1}{2}\frac{d}{dt}\int_0^1 u_y^2 + \int_0^1 \eta^{-1} u_{yy}^{-2} = \int_0^1 P(\eta^{-1})_y u_{yy} + \\ + \int_0^1 \eta^2 \eta_y u_y u_{yy} - \int_0^1 f(t, \int_0^y \eta) u_{yy}. \end{aligned} \tag{9.13}$$

By (9.12) and interpolation

$$\int_0^1 \eta_y^2 u_y^2 \le (\sup_y u_y^2) \int_0^1 \eta_y^2 \le c(u_0, \eta_0, f)\|u_y\|_0 \|u_{yy}\|_0,$$

hence a standard argument gives

$$\int_0^1 u_y^2 + \int_0^t \int_0^1 u_{yy}^2 \le c(u_0, \eta_0, f), \tag{9.14}$$

which is the last estimate we need.

Some interesting questions arise if we assume that f is independent of t, i.e., $f = f(x) \in L^\infty(\Omega)$. We find again a global in time solution, since we can repeat the arguments we used in the proof of Theorem G considering the problem in the set Q_T for each $T \in \mathbb{R}^+$. But in general we are not able to show that (9.11) holds, since the estimate on $\log \eta$ depends on T. Can we say anything about the asymptotic behaviour of η (i.e. of the density ρ)? Is it possible to obtain that (9.11) is valid also in this case?

It is not difficult to realize that an answer to these questions has to be related to the existence of a *stationary* solution (u, η) satisfying (9.11). A result due to Beirão da Veiga [B3] shows that such a stationary solution does exist if and only if the pressure P and the external force field f satisfy a suitable compatibility condition. For instance, if $P(\rho) = A\rho^\gamma$ and $f(x) = \lambda$ ($A > 0$, $\gamma > 1$ and λ given constants), then the stationary solution satisfying (9.11) exists if and only if

$$|\lambda| < A(\gamma/\gamma - 1)^\gamma. \tag{9.15}$$

(In case the compatibility condition is not assumed, some results concerning the existence of cavitation solutions to the stationary problem are contained in the paper of Lovicar-Straškraba [LS]).

Starting from this result, a first answer concerning the asymptotic behaviour of the density is given by Straškraba-Valli [SV], where it is proved that a (regular enough) solution to (9.1)-(9.4) satisfying (9.11) exists only if the same compatibility condition on P and f discovered in [B3] for the stationary solution holds. Hence, when such a condition does not hold, the global solution must asymptotically develop either vacuum or infinite density.

As a consequence, it is not possible to extend Theorem F to the case $f = f(x)$ if this compatibility condition is not satisfied. An interesting open problem is *to prove* this existence result when such a compatibility condition holds (for instance, choose $P(\rho) = A\rho^{\gamma}$ and $f(x) = \lambda$ satisfying (9.15)). A particular case was solved by Shelukhin [Sh2], assuming that

$$\exists\, c \geq 1: \quad c^{-1}\xi^{-1} \leq P_{\rho}(\xi) \leq c\,\xi^{-1}, \quad \forall \xi > 0.$$

Under this hypothesis the compatibility condition established in [B3] is satisfied for each $f = f(x)$.

A more precise answer has been given by Beirão da Veiga [B4], where the existence of a global solution satisfying (9.11) has been proven under fairly general assumptions. Moreover, the same author has shown the stability of the stationary solutions (see [B5]).

Another open problem is concerned with the existence of an attractor for this system of equations. The presence of the (hyperbolic) equation (9.2) makes difficult to apply the known methods, since they usually take advantage of the presence of dissipative terms. A partial result has been obtained in Straškraba-Valli [SV], where it is shown that if a (regular enough) solution to (9.1)-(9.4) satisfying (9.11) does exist, it has to converge to the stationary solution (remark that under these assumptions the compatibility condition between P and f must be satisfied). Another interesting result is contained in a paper of Beirão da Veiga [B6], where the attracting properties of one dimensional barotropic viscous flows are studied in more detail, and as a consequence the existence of periodic solutions is also proven. (Periodic solutions for the particular case $P(\rho) = A\rho$ have been obtained by Shelukhin [Sh1] and Matsumura-Nishida [MtN7]). The behaviour of a solution defined on the whole real line $\mathbb{R}$ and satisfying (9.11) on $\mathbb{R}\times]0,M[$ has been studied by Lovicar-Straškraba-Valli [LSV].

The extension of the existence result given by Theorem G to the non-barotropic case requires some effort. First of all, it has been always assumed that $f = 0$. Secondly, the structure of the thermodynamic state equations plays in general a relevant role. The most considered case is that of ideal polytropic gases, i.e., $P(\rho,\theta) = R\rho\theta$, $E(\rho,\theta) = c_V\theta$, R, c_V, ς^* and χ^* positive constants. For these state equations, the theorem has been proven by Kazhikhov-Shelukhin [KSh] and Kazhikhov [K5]. More general boundary conditions have been considered by Nagasawa [Na1]. The general case (P and E satisfying $P_\rho > 0$ and $E_\theta > 0$, and viscosity and heat conductivity coefficients eventually dependent on ρ and θ) was firstly solved for small initial data by Okada-Kawashima [OKw], and without smallness restriction (but for constant viscosity coefficient ς^*) by Kawhol [Kh] (where some non-monotone dependences of P on ρ are also considered; other results in this direction are contained in Kazhikhov [K4] and Kazhikhov-Nikolaev [KNi], but the van der Waals case $P(\rho,\theta) = (1 - b\rho)^{-1}R\rho\theta - a\rho^2$, R, a and b positive constants, is still an open problem).

The existence of a solution to the free-boundary problem was established for $P(\rho) = A\rho^\gamma$ by Kazhikhov [Ka2] and Okada [O] (this last one assuming small data, but especially studying the asymptotic behaviour as t goes to infinity); for the ideal polytropic case by Kazhikhov [Ka2], Okada [O] (this second one for small data) and Nagasawa [Na2], [Na3], [Na4], [Na5]. A review paper on several of these results (for fixed or free-boundary) has been written by Nishida [N].

For what is concerned to the Cauchy problem, in the barotropic case let us mention Kanel' [Ka1], Itaya [I3], [I4] (this last author for $P(\rho) = A\rho$), and for ideal polytropic gases Kanel' [Ka2] and Kawashima-Nishida [KwN] for small data, and Kazhikhov [K5], [K6] without smallness assumptions. The general case (assuming small initial data) has been considered by Okada-Kawashima [OKw], Kawashima-Okada [KwO] (in this last paper, the results also hold true for magnetohydrodynamics).

Let us conclude by mentioning some open problems:

(i) in the barotropic case, global existence of a solution satisfying (9.11) when $f = f(x)$ and under the only assumption that a stationary solution having the same property does exist, i.e., that the compatibility condition established in [B3] is satisfied;

(ii) for the general case, global existence with non-constant viscosity, in particular $\varsigma^* = \varsigma^*(\theta)$;

(iii) global existence for the van der Waals case ($P(\rho,\theta) = (1 - b\rho)^{-1}R\rho\theta - a\rho^2$).

REFERENCES

[Ag] R. Agemi, *The initial boundary value problem for inviscid barotropic fluid motion*, Hokkaido Math. J., **10** (1981), 156-182.

[AKMo] S.N. Antontsev, A.V. Kazhikhov, V.N. Monakhov, *Boundary value problems in mechanics of inhomogeneous fluids*, Nauka, Novosibirsk, 1983 [Russian] = North Holland, Amsterdam, 1990.

[B1] H. Beirão da Veiga, *On the barotropic motion of compressible perfect fluids*, Ann. Scuola Norm. Sup. Pisa, (4) **8** (1981), 317-351.

[B2] H. Beirão da Veiga, *Stationary motions and the incompressible limit for compressible viscous fluids*, Houston J. Math., **13** (1987), 527-544.

[B3] H. Beirão da Veiga, *An L^p-theory for the n-dimensional, stationary, compressible Navier-Stokes equations, and the incompressible limit for compressible fluids. The equilibrium solutions*, Comm. Math. Phys., **109** (1987), 229-248.

[B4] H. Beirão da Veiga, *Long time behavior for one-dimensional motion of a general barotropic viscous fluid*, Arch. Rational Mech Anal., **108** (1989), 141-160.

[B5] H. Beirão da Veiga, *The stability of one dimensional stationary flows of compressible viscous fluids*, Ann. Inst. Henri Poincaré. Anal. Non Linéaire, **7** (1990), 259-268.

[B6] H. Beirão da Veiga, *Attracting properties for one dimensional flows of a general barotropic viscous fluid. Periodic flows*, Università di Pisa, Istituto di Matematiche Applicate "Ulisse Dini", preprint **14**, 1988; Ann. Mat. Pura Appl., to appear.

[B7] H. Beirão da Veiga, *On the existence theorem for the barotropic motion of a compressible inviscid fluid in the half-space*, Ann. Mat. Pura Appl., to appear.

[B8] H. Beirão da Veiga, *Data dependence in the mathematical theory of compressible inviscid fluids*, Arch. Rational Mech. Anal., to appear.

[B9] H. Beirão da Veiga, *Perturbation theory and well-posedness in Hadamard's sense of hyperbolic initial-boundary value problems*, Università di Pisa, Dipartimento di Matematica, preprint **2.71 (582)**, 1991.

[B10] H. Beirão da Veiga, *Perturbation theorems for linear hyperbolic mixed problems and applications to the compressible Euler equations*, Università di Pisa, Dipartimento di Matematica, preprint **2.80 (600)**, 1991.

[BeY] Yu.Ya. Belov, N.N. Yanenko, *Influence of viscosity on the smoothness of solutions of incompletely parabolic systems*, Mat. Zametki, **10** (1971), 93-99 [Russian] = Math. Notes, **10** (1971), 480-483.

[Bo] J.L. Boldrini, *Stationary spatially periodic compressible flows at high Mach number*, Rend. Sem. Mat. Univ. Padova, **84** (1990), 201-215.

[CnF] P. Constantin, C. Foias, *Navier-Stokes equations*, University of Chicago Press, Chicago, 1988.

[CsP1] V. Coscia, M. Padula, *Quantitative bounds for nonlinear stability in compressible Bénard problem*, Hydrosoft, **2** (1989), 182-185.

[CsP2] V. Coscia, M. Padula, *Nonlinear convective stability in a compressible atmosphere*, Geophys. Astrophys. Fluid Dyn., **54** (1990), 49-83.

[CFr] R. Courant, K.O. Friedrichs, *Supersonic flow and shock waves*, Wiley-Interscience, New York, 1948.

[D1] K. Deckelnick, *Decay estimates for the compressible Navier-Stokes equations in the half-space*, Bonn Universität, Sonderforschungsbereich 256, preprint **46**, 1988.

[D2] K. Deckelnick, *Das zeitasymptotische Verhalten von Lösungen der kompressiblen Navier-Stokes Gleichungen in unbeschräntken Gebieten*, Bonn Universität, Sonderforschungsbereich 256, preprint **105**, 1990.

[De] A. Defranceschi, *On the stationary, compressible and incompressible Navier-Stokes equations*, Ann. Mat. Pura Appl., (4) **149** (1987), 217-236.

[E] D.G. Ebin, *The initial boundary value problem for sub-sonic fluid motion*, Comm. Pure Appl. Math., **32** (1979), 1-19.

[Fa1] R. Farwig, *Stationary solutions of compressible Navier-Stokes equations with slip boundary condition*, Bonn Universität, Sonderforschungsbereich 256, preprint **22**, 1988.

[Fa2] R. Farwig, *Stationary solutions of the Navier-Stokes equations for a compressible, viscous and heat-conductive fluid*, Comm. Partial Differ. Equations, **14** (1989), 1579-1606.

[Gr] D. Graffi, *Il teorema di unicità nella dinamica dei fluidi compressibili*, J. Rational Mech. Anal., **2** (1953), 99-106.

[GSu] B. Gustafsson, A. Sundström, *Incompletely parabolic problems in fluid dynamics*, SIAM J. Appl. Math., **35** (1978), 343-357.

[I1] N. Itaya, *On the Cauchy problem for the system of fundamental equations describing the movement of compressible viscous fluid*, Kodai Math. Sem. Rep., **23** (1971), 60-120.

[I2] N. Itaya, *On the initial value problem of the motion of compressible viscous fluid, especially on the problem of uniqueness*, J. Math. Kyoto Univ., **16** (1976), 413-427.

[I3] N. Itaya, *A survey on the generalized Burgers' equation with a pressure model term*, J. Math. Kyoto Univ., **16** (1976), 223-240.

[I4] N. Itaya, *A survey on two model equations for compressible viscous fluid*, J. Math. Kyoto Univ., **19** (1979), 293-300.

[Ka1] Ya.I. Kanel', *On a model system of equations of one-dimensional gas motion*, Differ. Uravn., **4** (1968), 721-734 [Russian] = Differ. Equations, **4** (1968), 374-380.

[Ka2] Ya.I. Kanel', *Cauchy problem for the equations of gasdynamics with viscosity*, Sib. Mat. Zh., **20** (1979), 293-306 [Russian] = Sib. Math. J., **20** (1979), 208-218.

[Kt] T. Kato, *Quasi-linear equations of evolution, with applications to partial differential equations*, in "Spectral theory and differential equations", W.N. Everitt ed., Springer-Verlag, Berlin, 1975, 25-70.

[KwN] S. Kawashima, T. Nishida, *Global solutions to the initial value problem for the equations of one-dimensional motion of viscous polytropic gases*, J. Math. Kyoto Univ., **21**, (1981), 825-837.

[KwO] S. Kawashima, M. Okada, *Smooth global solutions for the one-dimensional equations in magnetohydrodynamics*, Proc. Japan Acad. Ser. A, **58** (1982), 384-387.

[Kh] B. Kawhol, *Global existence of large solutions to initial boundary value problems for a viscous, heat-conducting, one dimensional real gas*, J. Differ. Equations, **58** (1985), 76-103.

[K1] A.V. Kazhikhov, *Correctness "in the large" of mixed boundary value problems for a model system of equations of a viscous gas*, Din. Sploshnoj Sredy, **21** (1975), 18-47 [Russian].

[K2] A.V. Kazhikhov, *On global solvability of one-dimensional boundary value problems for equations of motion of a viscous heat-conducting gas*, Din. Sploshnoj Sredy, **24** (1976), 45-61 [Russian].

[K3] A.V. Kazhikhov, *Stabilization of solutions of an initial-boundary-value problem for the equations of motion of a barotropic viscous fluid*, Diff. Uravn., **15** (1979), 662-667 [Russian] = Differ. Equations, **15** (1979), 463-467.

[K4] A.V. Kazhikhov, *Some problems of the theory of the Navier-Stokes equations for a compressible fluid*, Din. Sploshnoj Sredy, **38** (1979), 33-47 [Russian].

[K5] A.V. Kazhikhov, *On the theory of boundary value problems for equations of the one-dimensional time dependent motion of a viscous heat-conducting gas*, Din. Sploshnoj Sredy, **50** (1981), 37-62 [Russian].

[K6] A.V. Kazhikhov, *Cauchy problem for viscous gas equations*, Sib. Mat. Zh., **23** (1982), 60-64 [Russian] = Sib. Math. J., **23** (1982), 44-49.

[KNi] A.V. Kazhikhov, V.B. Nikolaev, *On the correctness of boundary value problems for the equations of a viscous gas with nonmonotone state function*, Chisl. Metody Mekh. Sploshnoi Sredy, **10** (1979), 77-84 [Russian] = Amer. Math. Soc. Transl., (2) **125** (1985), 45-50.

[KSh] A.V. Kazhikhov, V.V. Shelukhin, *Unique global solution with respect to time of initial-boundary value problems for one-dimensional equations of a viscous gas*, Prikl. Math. Mekh., **41** (1977), 282-291 [Russian] = J. Appl. Math. Mech., **41** (1977), 273-282.

[Ld] O.A. Ladhyzhenskaya, *The mathematical theory of viscous incompressible flow*, Gosudarstv. Izdat. Fiz.-Mat. Lit., Moscow, 1961 [Russian] = Gordon and Breach, New York, 1963 (2nd ed. 1969).

[Lx] P.D. Lax, *Hyperbolic systems of conservation laws and the mathematical theory of shock waves*, SIAM, Philadelphia, 1973.

[LS] V. Lovicar, I. Straškraba, *Remark on cavitation solutions of stationary compressible Navier-Stokes equations in one dimension*, Czech. Math. J., (1991), to appear.

[LSV] V. Lovicar, I. Straškraba, A. Valli, *On bounded solutions of one-dimensional compressible Navier-Stokes equations*, Rend. Sem. Mat. Univ. Padova, **83** (1990), 81-95.

[Lu1] G. Lukaszewicz, *On an estimate of the temperature of a viscous compressible fluid*, Bull. Acad. Polon. Sci., **30** (1982), 31-37.

[Lu2] G. Lukaszewicz, *An existence theorem for compressible viscous and heat conducting fluids*, Math. Meth. Appl. Sci., **6** (1984), 234-247.

[Ma1] A. Majda, *Compressible fluid flow and systems of conservation laws in several space variables*, Springer-Verlag, New York, 1984.

[Ma2] A. Majda, *Mathematical foundations of incompressible fluid flow*, Princeton University, Department of Mathematics, Lecture Note, 1985.

[M] T. Makino, *On a local existence theorem for the evolution equation of gaseous stars*, in "Pattern and waves. Qualitative analysis of nonlinear differential equations", T. Nishida & al. eds., Kinokuniya, Tokyo & North-Holland, Amsterdam, 1986, 459-479.

[MU] T. Makino, S. Ukai, *Sur l'existence des solutions locales de l'équation d'Euler-Poisson pour l'évolution d'étoiles gazeuses*, J. Math. Kyoto Univ., **27** (1987), 387-399.

[MUKw] T. Makino, S. Ukai, S. Kawashima *Sur la solution à support compact de l'équation d'Euler compressible*, Japan J. Appl. Math., **3** (1986), 249-257.

[MrV] P. Marcati, A. Valli, *Almost-periodic solutions to the Navier-Stokes equations for compressible fluids*, Boll. Un. Mat. Ital. B, (6) **4** (1985), 969-986.

[Mt] A. Matsumura, *An energy method for the equations of motion of compressible viscous and heat-conductive fluids*, University of Wisconsin - Madison, MRC Technical Summary Report **2194**, 1981.

[MtN1] A. Matsumura, T. Nishida, *The initial value problem for the equations of motion of compressible viscous and heat-conductive fluids*, Proc. Japan Acad. Ser. A, **55** (1979), 337-342.

[MtN2] A. Matsumura, T. Nishida, *The initial value problem for the equations of motion of viscous and heat-conductive gases*, J. Math. Kyoto Univ., **20** (1980), 67-104.

[MtN3] A. Matsumura, T. Nishida, *The initial boundary value problem for the equations of motion of compressible viscous and heat-conductive fluid*, University of Wisconsin - Madison, MRC Technical Summary Report **2237**, 1981.

[MtN4] A. Matsumura, T. Nishida, *Initial boundary value problems for the equations of motion of general fluids*, in "Computing methods in applied sciences and engineering, V", R. Glowinski & J.L. Lions eds., North-Holland, Amsterdam, 1982, 389-406.

[MtN5] A. Matsumura, T. Nishida, *Initial boundary value problems for the equations of motion of compressible viscous and heat-conductive fluids*, Comm. Math. Phys., **89** (1983), 445-464.

[MtN6] A. Matsumura, T. Nishida, *Exterior stationary problems for the equations of motion of compressible viscous and heat-conductive fluids*, in "Differential equations", C.M. Dafermos & al. eds., Marcel Dekker, New York, 1989, 473-479.

[MtN7] A. Matsumura, T. Nishida, *Periodic solutions of a viscous gas equation*, in "Recent topics in nonlinear PDE, IV", M. Mimura & T. Nishida eds., Kinikuniya, Tokyo & North-Holland, Amsterdam, 1989, 49-82.

[Na1] T. Nagasawa, *On the one-dimensional motion of the polytropic ideal gas non-fixed on the boundary*, J. Differ. Equations, **65** (1986), 49-67.

[Na2] T. Nagasawa, *On the asymptotic behavior of the one-dimensional motion of the polytropic ideal gas with stress-free condition*, Quart. Appl. Math., **46** (1988), 665-679.

[Na3] T. Nagasawa, *On the outer pressure problem of the one-dimensional polytropic ideal gas*, Japan J. Appl. Math., **5** (1988), 53-85.

[Na4] T. Nagasawa, *Global asymptotics of the outer pressure problem with free boundary*, Japan J. Appl. Math., **5** (1988), 205-224.

[Na5] T. Nagasawa, *On the one-dimensional free boundary problem for the heat-conductive compressible viscous gas*, in "Recent topics in nonlinear PDE, IV", M. Mimura & T. Nishida eds., Kinikuniya, Tokyo & North-Holland, Amsterdam, 1989, 83-99.

[Ns] J. Nash, *Le problème de Cauchy pour les équations différentielles d'un fluide général*, Bull. Soc. Math. France, **90** (1962), 487-497.

[N] T. Nishida, *Equations of motion of compressible viscous fluids*, in "Pattern and waves. Qualitative analysis of nonlinear differential equations", T. Nishida & al. eds., Kinokuniya, Tokyo & North-Holland, Amsterdam, 1986, 97-128.

[NoP] A. Novotný, M. Padula, *Existence and uniqueness of stationary solutions for viscous compressible heat-conductive fluid with large potential and small nonpotential external forces*, Università di Ferrara, Dipartimento di Matematica, preprint **164**, 1991.

[OKw] M. Okada, S. Kawashima, *On the equations of one-dimensional motion of compressible viscous fluids*, J. Math. Kyoto Univ., **23** (1983), 55-71.

[O] M. Okada, *Free boundary value problems for equations of one-dimensional motion of compressible viscous fluids*, Japan J. Appl. Math., **4** (1987), 219-235.

[OlSu] J. Oliger, A. Sundström, *Theoretical and practical aspects of some initial boundary value problems in fluid dynamics*, SIAM J. Appl. Math., **35** (1978), 419-446.

[P1] M. Padula, *Nonlinear energy stability for the compressible Bénard problem*, Boll.Un. Mat. It., (6) **5-B** (1986), 581-602.

[P2] M. Padula, *Existence and uniqueness for viscous steady compressible motions*, Arch. Rat. Mech. Anal., **97** (1987), 89-102.

[PiZ] K. Pileckas, W.M. Zajączkowski, *On the free boundary problem for stationary compressible Navier-Stokes equations*, Comm. Math. Phys., **129** (1990), 169-204.

[Po] G. Ponce, Global existence of small solutions to a class of nonlinear evolution equations, Nonlinear Anal., **9** (1985), 399-418.

[RY] B.L. Roždestvenskiĭ, N.N. Janenko [N.N. Yanenko], *Systems of quasilinear equations and their applications to gas dynamics*, Nauka, Moscow, 1968 (2nd ed. 1978) [Russian] = American Mathematical Society, Providence, 1983.

[Sc] S. Schochet, *The compressible Euler equations in a bounded domain: existence of solutions and the incompressible limit*, Comm. Math. Phys., **104** (1986), 49-75.

[Se1] P. Secchi, *Existence theorems for compressible viscous fluids having zero shear viscosity*, Rend. Sem. Mat. Univ. Padova, **70** (1983), 73-102.

[Se2] P. Secchi, *On the evolution equations of viscous gaseous stars*, Ann. Scuola Norm. Sup. Pisa, (4) **18** (1991), 295-318.

[Se3] P. Secchi, *On the motion of gaseous stars in the presence of radiation*, Comm. Partial Differ. Equations, **15** (1990), 185-204.

[Se4] P. Secchi, *On the uniqueness of motion of viscous gaseous stars*, Math. Meth. Appl. Sci., **13** (1990), 391-404.

[Se5] P. Secchi, On nonviscous compressible fluids in a time-dependent domain, Ann. Inst. Henri Poincaré. Anal. Non Linéaire, (1992), to appear.

[Se6] P. Secchi, *On a stationary problem for compressible Navier-Stokes equations*, Università di Trento, Dipartimento di Matematica, preprint **353**, 1991.

[Se7] P. Secchi, *On the stationary motion of compressible viscous fluids*, Università di Trento, Dipartimento di Matematica, preprint **354**, 1991.

[SeV] P. Secchi, A. Valli, *A free boundary problem for compressible viscous fluids*, J. Reine Angew. Math., **341** (1983), 1-31.

[Sr1] J. Serrin, *Mathematical principles of classical fluid mechanics*, Handbuch der Physik VIII/1, Springer-Verlag, Berlin, 1959.

[Sr2] J. Serrin, *A note on the existence of periodic solutions of the Navier-Stokes equations*, Arch. Rational Mech. Anal., **3** (1959), 120-122.

[Sr3] J. Serrin, *On the uniqueness of compressible fluid motion*, Arch. Rational Mech. Anal., **3** (1959), 271-288.

[Sh1] V.V. Shelukhin, *Periodic flows of a viscous gas*, Din. Sploshnoj Sredy, **42** (1979), 80-102 [Russian].

[Sh2] V.V. Shelukhin, *Bounded, almost-periodic solutions of a viscous gas equation*, Din. Sploshnoj Sredy, **44** (1980), 147-163 [Russian].

[Si] M. Shinbrot, *Lectures on fluid mechanics*, Gordon and Breach, New York, 1973.

[Sd] T.C. Sideris, *Formation of singularities in three-dimensional compressible fluids*, Comm. Math. Phys., **101** (1985), 475-485.

[Sm] J. Smoller, *Shock waves and reaction-diffusion equations*, Springer-Verlag, New York, 1983.

[So] V.A. Solonnikov, *Solvability of the initial-boundary value problem for the equations of motion of a viscous compressible fluid*, Zap. Naucn. Sem. Leningrad. Otdel. Mat. Inst. Steklov. (LOMI), **56** (1976), 128-142 [Russian] = J. Soviet Math., **14** (1980), 1120-1133.

[SoK] V.A. Solonnikov, A.V. Kazhikhov, *Existence theorems for the equations of motion of a compressible viscous fluid*, Ann. Rev. Fluid Mech., **13** (1981), 79-95.

[SoT1] V.A. Solonnikov, A. Tani, *A problem with a free-boundary for Navier-stokes equations for a compressible fluid in the presence of surface tension*, Zap. Naucn. Sem. Leningrad. Otdel. Mat. Inst. Steklov. (LOMI), **182** (1990), 142-148 [Russian].

[SoT2] V.A. Solonnikov, A. Tani, *Free boundary problem for a viscous compressible flow with a surface tension*, in "Constantin Carathéodory: an international tribute", Th.M. Rassias ed., World Scientific, Singapore, 1991, 1270-1303.

[ŠýVe] M. Štědrý, O. Vejvoda, *Equations of magnetohydrodynamics of compressible fluid: periodic solutions*, Apl. Mat., **30** (1985), 77-91.

[SV] I. Straškraba, A. Valli, *Asymptotic behaviour of the density for one-dimensional Navier-Stokes equations*, Manuscripta Math., **62** (1988), 401-416.

[Sk] J.C. Strikwerda, *Initial boundary value problems for incompletely parabolic systems*, Comm. Pure Appl. Math., **30** (1977), 797-822.

[St] G. Strömer, *About a certain class of parabolic-hyperbolic systems of differential equations*, Analysis, **9** (1989), 1-39.

[T1] A. Tani, *On the first initial-boundary value problem of compressible viscous fluid motion*, Publ. Res. Inst. Math. Sci. Kyoto Univ., **13** (1977), 193-253.

[T2] A. Tani, *On the free boundary value problem for compressible viscous fluid motion*, J. Math. Kyoto Univ., **21** (1981), 839-859.

[T3] A. Tani, *Two phase free boundary problem for compressible viscous fluid motion*, J. Math. Kyoto Univ., **24** (1984), 243-267.

[T4] A. Tani, *The initial value problem for the equations of the motion of compressible viscous fluid with some slip boundary condition*, in "Pattern and waves. Qualitative analysis of nonlinear differential equations", T. Nishida & al. eds., Kinokuniya, Tokyo & North-Holland, Amsterdam, 1986, 675-684.

[T5] A. Tani, *Free boundary problems for general fluids*, Kyoto University, RIMS-Kōkyuroku, **698** (1989), 146-170.

[T6] A. Tani, *The initial value problem for the equations of the motion of general fluids with general slip boundary condition*, Kyoto University, RIMS-Kōkyuroku, **734** (1990), 123-142.

[Te] R. Temam, *Navier Stokes equations. Theory and numerical analysis*, North-Holland, Amsterdam, 1977 (3rd ed. 1984).

[V1] A. Valli, *Uniqueness theorems for compressible viscous fluids, especially when the Stokes relation holds*, Boll. Un. Mat. It., Anal. Funz. Appl., (5) **18-C** (1981), 317-325.

[V2] A. Valli, *An existence theorem for compressible viscous fluids*, Ann. Mat. Pura Appl., (4) **130** (1982), 197-213; Ann. Mat. Pura Appl., (4) **132** (1982), 399-400.

[V3] A. Valli, *Periodic and stationary solutions for compressible Navier-Stokes equations via a stability method*, Ann. Scuola Norm. Sup. Pisa, (4) **10** (1983), 607-647.

[V4] A. Valli, *On the existence of stationary solutions to compressible Navier-Stokes equations*, Ann. Inst. Henri Poincaré. Anal. Non Linéaire, **4** (1987), 99-113.

[V5] A. Valli, *An existence theorem for non-homogeneous inviscid incompressible fluids*, in "Differential equations", C.M. Dafermos & al. eds., Marcel Dekker, New York, 1989, 691-698.

[VZ] A. Valli, W.M. Zajączkowski, *Navier-Stokes equations for compressible fluids: global existence and qualitative properties of the solutions in the general case*, Comm. Math. Phys., **103** (1986), 259-296.

[VoH] A.I. Vol'pert, S.I. Hudjaev, *On the Cauchy problem for composite systems of nonlinear differential equations*, Mat. Sb., **87** (1972), 504-528 [Russian] ≐ Math. USSR-Sb., **16** (1972), 517-544.

[W] W. von Wahl, *The equations of Navier-Stokes and abstract parabolic equations*, Friedr. Vieweg & Sohn, Braunschweig, 1985.

[Ya] T. Yanagisawa, *The initial boundary value problem for the equations of ideal magneto-hydrodynamics*, Hokkaido Math. J., **16** (1987), 295-314.

[YaMt] T. Yanagisawa, A. Matsumura, *The fixed boundary value problems for the equations of ideal magneto-hydrodynamics with a perfectly conducting wall condition*, Comm. Math. Phys., **136** (1991), 119-140.

Alberto Valli,
Dipartimento di Matematica, Università di Trento,
38050 Povo (Trento) - ITALY

Communications

Une Solution Numérique des Équations de Navier–Stokes Stationnaires

OCTAVIAN BAN

Introduction

Le sujet de cet exposé fait partie d'un travail de recherche complet concernant l'étude des équations de Navier–Stokes d'évolution par schémas aux différences finies. Le présent exposé concerne le début de ce travail, plus précisément, on trouve une solution numérique des équations de Navier-Stokes stationnaires pour le nombre de Reynolds $Re = 1$. Le travail prochain va concerner sur les solutions stationnaires pour différents Re, en précisant le nombre de Reynolds critique, où il y a perte de la stabilité de la solution stationnaire et puis l'étude de l'évolution des équations de Navier–Stokes, en précisant la transition vers la turbulence et l'écoulement caotique.

Un travail simillaire a été fait pour un domaine semblable à celui du présent exposé, mais avec des obstacles cylindriques, par éléments finis, en utilisant la méthode de lagrangien augmenté et il appartient à André Fortin de l'École Polytechnique de Montréal. Pour $Re = 1$, il a obtenu la même solution.

1 – Le problème posé

On se propose de trouver les solutions stationnaires et d'essayer de déterminer le nombre de Reynolds critique pour lequel il y a perte de stabilité de la solution stationnaire, dans un torre avec des obstacles. Plus précisément, du point de vue de la géométrie du domaine, on considère un torre à l'intérieur du quel il y a des obstacles de forme parallélipipédique, équidistantes. On considère la section plane du domaine qui contient un seul obstacle et qui se répète l'une après l'autre pour former le torre. Elle se présente de la manière suivante:

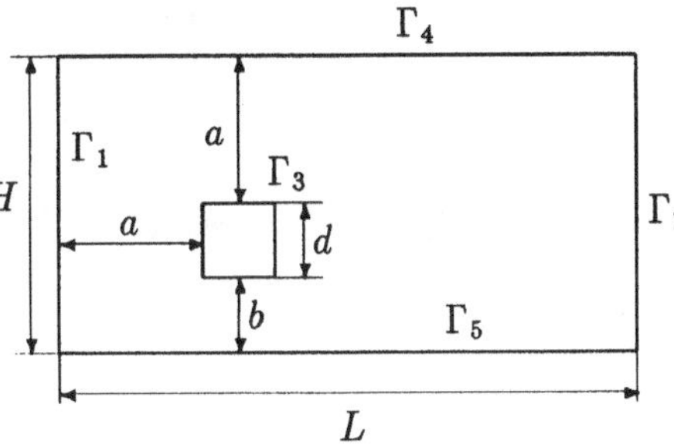

Fig 1: La déscription du problème

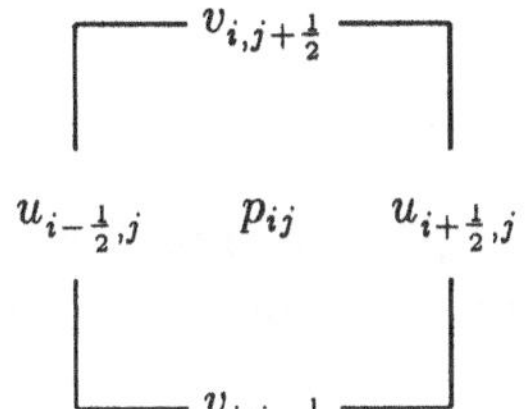

Fig 2: La pression et la vitesse sur la maille (i, j)

En considérant la section de l'obstacle comme un carré, d est le côté du carré; a est la distance de l'obstacle par rapport à l'entrée du domaine et par rapport au mur supérieur; b est la distance de l'obstacle par rapport au mur inférieur; L est la longueur du domaine et H est la hauteur du domaine.

2 – Conditions aux limites et périodicité

On va étudier l'écoulement d'un fluide incompressible dans les conditions suivantes:

- On ne connait pas la vitesse à l'entrée, mais on se donne le débit Q à l'entrée et on impose qu'il soit le même à la sortie du domaine, i.e. que l'écoulement soit périodique. Donc la vitesse à l'entrée doit avoir le même profile qu'à la sortie, à cause de la périodicité imposée.
- Le gradient de la pression est périodique, c'est-à-dire le gradient de la pression à l'entrée est le même qu'à la sortie.
- Le fluide adhère aux murs (supérieur et inférieur) du domaine et il adhère aussi aux murs de l'obstacle.

Donc on a les conditions aux limites suivantes:

$$Q|_{\Gamma_1} = Q|_{\Gamma_2}\,; \quad \mathbf{u}|_{\Gamma_1} = \mathbf{u}|_{\Gamma_2}\,; \quad \mathbf{u}|_{\Gamma_3} = \mathbf{u}|_{\Gamma_4} = \mathbf{u}_{\Gamma_5} = 0\,.$$
$$\nabla p|_{\Gamma_1} = \nabla p|_{\Gamma_2}\,.$$

Dans ces conditions, en utilisant une méthode multigrille, à l'aide des schémas aux différences finies, on va trouver pour le moment pour $Re = 1$, une solution numérique des équations de Navier–Stokes stationnaires.

3 – Le problème mathématique et l'approximation

Les équations de Navier–Stokes stationnaires, adimensionnelles d'un fluide visqueux incompressible en deux dimensions sont écrites de la manière suivante en variables primitives:

$$\text{(3.1)} \qquad \begin{cases} -\dfrac{1}{Re}\Delta u + u\dfrac{\partial u}{\partial x} + v\dfrac{\partial u}{\partial y} + \dfrac{\partial p}{\partial x} = f_1; \quad -\dfrac{1}{Re}\Delta v + u\dfrac{\partial v}{\partial x} + v\dfrac{\partial v}{\partial y} + \dfrac{\partial p}{\partial y} = f_2\,; \\ \dfrac{\partial u}{\partial x} + \dfrac{\partial v}{\partial y} = 0\,, \end{cases}$$

où $\mathbf{q} = (u, v)$ et p représente la vitesse et la pression, $\mathbf{f} = (f_1, f_2)$ représente les forces extérieurs et Re est le nombre de Reynolds.

On reécrit les équations (3.1) sous la forme:

$$\text{(3.2)} \qquad \begin{cases} N(u,v) + \nabla p = \mathbf{f} \\ \operatorname{div}\mathbf{q} = 0\,, \end{cases}$$

où $N(u, v)$ est la somme du terme linéaire de diffusion $D(u, v)$ et du terme non linéaire de convection $C(u, v)$. On va approcher ces opérateurs par différences finies sur des grilles décalées, en considérant la pression au centre de la maille et les composantes de la vitesse au milieu des côtés de la maille (voir fig. 2).

Pour approcher le terme linéaire, on a utilisé des différences finies centrées. Donc sur la maille (i,j) on a pour les variables $u_{i-\frac{1}{2},j}$, $v_{i,j-\frac{1}{2}}$ et p_{ij}, les formules suivantes:

(3.3)

$$[D(u,v)]_{(i,j)} = \begin{cases} \dfrac{1}{Re(\Delta x)^2}(2u_{i-\frac{1}{2},j} - u_{i+\frac{1}{2},j} - u_{i-\frac{3}{2},j}) + \frac{1}{Re(\Delta y)^2}(2u_{i-\frac{1}{2},j} - u_{i-\frac{1}{2},j+1} - u_{i-\frac{1}{2},j-1}), \\ \dfrac{1}{Re(\Delta x)^2}(2v_{i,j-\frac{1}{2}} - v_{i+1,j-\frac{1}{2}} - v_{i-1,j-\frac{1}{2}}) + \frac{1}{Re(\Delta y)^2}(2v_{i,j-\frac{1}{2}} - v_{i,j+\frac{1}{2}} - v_{i,j-\frac{3}{2}}) \ , \end{cases}$$

$$[\nabla p]_{(i,j)} = \begin{cases} \dfrac{1}{\Delta x}(p_{i,j} - p_{i-1,j}), \\ \frac{1}{\Delta y}(p_{i,j} - p_{i,j-1}) \ , \end{cases}$$

$$[\mathrm{div}\,\mathbf{q}]_{(i,j)} = \frac{1}{\Delta x}(u_{i+\frac{1}{2},j} - u_{i-\frac{1}{2},j}) + \frac{1}{\Delta y}(v_{i,j+\frac{1}{2}} - v_{i,j-\frac{1}{2}}) \ ,$$

où u, v, p sont les variables discrétisées sur toutes les grilles G_ℓ et Δx et Δy représentent les pas uniformes des grilles dans les directions x et y, respectivement.

Pour approcher le terme non linéaire de convection, on a utilisé des différences finies descentrées du second ordre:

(3.4)

$$[C(u,v)]_{(i,j)} = \begin{cases} \left[\begin{array}{ll} \dfrac{u^{n-1}_{i-1,j}}{3\Delta x}(4u_{i-\frac{1}{2},j} - 5u^{n-1}_{i-\frac{3}{2},j} + u^{n-1}_{i-\frac{5}{2},j}), & \text{si } u^{n-1}_{i-1,j} > 0 \\ -\dfrac{u^{n-1}_{i,j}}{3\Delta x}(4u_{i-\frac{1}{2},j} - 5u^{n-1}_{i+\frac{1}{2},j} + u^{n-1}_{i+\frac{3}{2},j}), & \text{si } u^{n-1}_{i-1,j} < 0 \end{array}\right. \\ + \left[\begin{array}{ll} \dfrac{v^{n-1}_{i-\frac{1}{2},j-\frac{1}{2}}}{3\Delta y}(4u_{i-\frac{1}{2},j} - 5u^{n-1}_{i-\frac{1}{2},j-1} + u^{n-1}_{i-\frac{1}{2},j-2}), & \text{si } v^{n-1}_{i-\frac{1}{2},j-\frac{1}{2}} > 0 \\ -\dfrac{v^{n-1}_{i-\frac{1}{2},j+\frac{1}{2}}}{3\Delta y}(4u_{i-\frac{1}{2},j} - 5u^{n-1}_{i-\frac{1}{2},j+1} + u^{n-1}_{i-\frac{1}{2},j+2}), & \text{si } v^{n-1}_{i-\frac{1}{2},j+\frac{1}{2}} < 0 \end{array}\right. \\ \left[\begin{array}{ll} \dfrac{u^{n-1}_{i-\frac{1}{2},j-\frac{1}{2}}}{3\Delta x}(4v_{i,j-\frac{1}{2}} - 5v^{n-1}_{i-1,j-\frac{1}{2}} + v^{n-1}_{i-2,j-\frac{1}{2}}), & \text{si } u^{n-1}_{i-\frac{1}{2},j-\frac{1}{2}} > 0 \\ -\dfrac{u^{n-1}_{i+\frac{1}{2},j-\frac{1}{2}}}{3\Delta x}(4v_{i,j-\frac{1}{2}} - 5v^{n-1}_{i+1,j-\frac{1}{2}} + v^{n-1}_{i+2,j-\frac{1}{2}}), & \text{si } u^{n-1}_{i+\frac{1}{2},j-\frac{1}{2}} < 0 \end{array}\right. \\ + \left[\begin{array}{ll} \dfrac{v^{n-1}_{i,j-1}}{3\Delta y}(4v_{i,j-\frac{1}{2}} - 5v^{n-1}_{i,j-\frac{3}{2}} + v^{n-1}_{i,j-\frac{5}{2}}), & \text{si } v^{n-1}_{i,j-1} > 0 \\ -\dfrac{v^{n-1}_{i,j}}{3\Delta y}(4v_{i,j-\frac{1}{2}} - 5v^{n-1}_{i,j+\frac{1}{2}} + v^{n-1}_{i,j+\frac{3}{2}}), & \text{si } u^{n-1}_{i,j} < 0 \ . \end{array}\right. \end{cases}$$

Finalement, on obtient un opérateur non linéaire, de dimension finie, L_ℓ (sur la grille G_ℓ) et on a à résoudre:

$$L_\ell(u,v,p) = f_\ell \ . \tag{3.5}$$

Dans la relation (3.4), $u^{n-1}_{i-1,j}$, $u^{n-1}_{i,j}$, $v^{n-1}_{i-\frac{1}{2},j-\frac{1}{2}}$ et $v^{n+1}_{i-\frac{1}{2},j+\frac{1}{2}}$ peuvent être interprétées comme les composantes de la vitesse sur les côtés de la maille du point $(i-\frac{1}{2},j)$. Simillairement,

$u^{n-1}_{i-\frac{1}{2},j-\frac{1}{2}}$, $u^{n-1}_{i+\frac{1}{2},j-\frac{1}{2}}$, $v^{n-1}_{i,j-1}$ et $v^{n-1}_{i,j}$ représentent la même chose pour le point courrant $(i, j-\frac{1}{2})$.

4 – La méthode multigrille et la procédure de relaxation

Pour résoudre (3.5), on utilise une procédure FMG-FAS ("Full MultiGrid"-"Full Approximation Storage"). L'algorithme FAS consiste à evaluer le résidu, à projeter ce residu sur la grille grossière, à calculer une correction sur cette grille grossière, puis à faire l'extension de cette correction sur la grille fine et finalement à corriger la solution précédente.

L'algorithme FMG consiste à résoudre le problème, avec l'algorithme FAS sur une grille de départ G_k ($k \geq 1$), avec un test $r_k \leq \varepsilon_k$ (où r_k est le résidu sur la grille G_k et ε_k est la précision demandée sur la grille G_k) avant de passer à une grille plus fine et puis à faire l'extension de cette solution sur la grille G_{k+1} et ainsi de suite sur les grilles de plus en plus fines. Le seul controle sur les grilles successives est la valeur de ε_k et l'on a: $\varepsilon_k = \delta\, \varepsilon_{k+1}$, $\delta > 1$. D'habitude, on prend $\delta = 4$.

Dans la technique multigrille, le choix de la procédure de relaxation joue un rôle très important sur l'éficacité de la méthode.

On utilise un opérateur de lissage connu sous le nom "Symetrical Coupled Gauss-Seidel" (SCGS) qui consiste à faire une sous-relaxation maille par maille. Ça signifie que, avec les grilles décallées, les cinq inconnues de la maille (i, j) sont calculées simultanément, par la résolution d'un système linéaire 5×5, de la forme suivante:

$$(4.1) \qquad \begin{pmatrix} \text{diago}\,1 & 0 & 0 & 0 & \text{derco}\,1 \\ 0 & \text{diago}\,2 & 0 & 0 & \text{derco}\,2 \\ 0 & 0 & \text{diago}\,3 & 0 & \text{derco}\,3 \\ 0 & 0 & 0 & \text{diago}\,4 & \text{derco}\,4 \\ -1/\Delta x & 1/\Delta x & -1/\Delta y & 1/\Delta y & 0 \end{pmatrix} \times \begin{pmatrix} u_{i-\frac{1}{2},j} \\ u_{i+\frac{1}{2},j} \\ v_{i,j-\frac{1}{2}} \\ v_{i,j+\frac{1}{2}} \\ p_{i,j} \end{pmatrix} = \begin{pmatrix} b_1 \\ b_2 \\ b_3 \\ b_4 \\ b_5 \end{pmatrix},$$

où les termes qui se trouvent sur la diagonale de la matrice contiennent les contributions diagonaux de $D(u, v)$ et $C(u, v)$ et les seconds membres contiennent tous les termes de L_k qui restent et le second membre f_k. Après avoir calculé les cinq inconnues sur une maille, on fait la sous-relaxation des 5 variables correspondantes, par exemple:

$$(4.2) \qquad u^n_{i-\frac{1}{2},j} = \omega\, u_{i-\frac{1}{2},j} + (1-\omega)\, u^{n-1}_{i-\frac{1}{2},j} \qquad (0 < \omega \leq 1)\ ,$$

où n est l'indice des ittérations de relaxation (comme d'ailleurs partout dans l'exposé). Après, on passe à résoudre le problème sur la maille suivante et ainsi de suite. De cette manière, la pression est calculée une seule fois et la vitesse deux fois car ses composantes sont données sur les côtés des mailles.

5 – L'évaluation des types des mailles

Pour simplifier au maximum les subroutines dont on a besoin pour le calcul de la solution stationnaire et surtout celle qui calcule la solution du système dans chaque maille et qui fait la relaxation, à chaque maille d'une grille quelquonque, on a associé un indice. Plus précisément, on a plusieurs types de mailles sur n'importe quelle grille qu'on utilise, chaque type (ou chaque indice de maille) étant fonction de la position de la maille par rapport au bord ou à l'obstacle. Les indices des types de mailles sont calculés de la manière suivante:

- Si l'un des bytes de 1 à 16 de l'indice de la maille est égal à 0, alors on n'extrapole pas la valeur respective de la composante de la vitesse et on la prend telle qu'elle est au pas précédent.
- Si l'un des bytes de 1 à 16 de l'indice de la maille est égal à 1, alors on extrapole la valeur respective de la composante de la vitesse.
- Si l'un des bytes de 17 à 20 de l'indice de la maille est égal à 0, alors on calcule la composante de la vitesse qui correspond à ce byte (pour 17 on calcule U du côté gauche de la maille, pour 18 on calcule U du côté droite, pour 19 la composante V du côté bas et pour 20, V du côté haut).
- Si l'un des bytes de 17 à 20 de l'indice de la maille est égale à 1, alors on ne calcule pas la composante de la vitesse qui correspond à ce byte.

6 – Résultats numériques

On a trouvé la solution numérique pour $Re = 1$ en prenant:

$$\begin{array}{ll} H = 2 & Q = 1.36 \\ L = 6 & \omega = 0.5 \\ b = 0.4 & N1 = 15 \\ d = 0.4 & N2 = 5 \\ a = 1.2 & (N1, N2 = \text{les dimensions de la grille grossière}). \end{array}$$

On a initialisé la vitesse par un Poiseuille à l'entrée et nulle ailleurs et la pression nulle partout et on n'a pas de forces extérieurs. Donc on a pris:

$$\begin{array}{ll} U = y(2-y) \ \text{ à l'entrée} & \\ U = 0 \ \text{ ailleurs} & f_1 = 0 \\ V = 0 \ \text{ partout} & f_2 = 0\ . \\ P = 0 \ \text{ partout} & \end{array}$$

REFERENCES

[1] *Charles-Henri Bruneau & Claude Jouron – An efficient scheme for solving steady incompressible Navier–Stokes Equations,* Rapport de recherche, Lab. Anal. Num. Orsay, Univ. Paris-Sud.

[2] *M. Fortin, R. Peyret & R. Temam* - *J. Mécanique*, **10** (1971), p. 357.
[3] *R. Peyret & T.D. Taylor* - *Computational methods for fluid flow*, Springer-Verlag, New York, Heidelberg, Berlin, 1983.
[4] *A. Fortin, M. Fortin & J.J. Gervais* - *J. Comput. Phys.*, **70** (1987), p. 295.
[5] *S.P. Vanka* - *J. Comput. Phys.*, **65** (1986), p. 138.
[6] *K. Gustafson & R. Leben* - *Applied Math. and Comput.*, **19** (1986), p. 89.
[7] *W. Hackbusch* - *Multigrid Methods and Applications*, Springer-Verlag, New York, Heidelberg, Berlin, 1985.

Octavian Ban,
Laboratoire d'Analyse Numérique d'Orsay, Bât. 425,
CNRS and Université Paris-Sud,
91405 ORSAY - FRANCE

Stationary Solutions for a Bingham Flow with Nonlocal Friction

LUISA CONSIGLIERI

1 – Introduction and variational formulation

Let Ω be a bounded open set of $\mathbb{R}^n$ $(n = 2,3)$ with sufficiently smooth boundary $\Gamma = \partial\Omega$, which is assumed to consist of two disjoint parts Γ_1 and Γ_2, with $\text{meas}(\Gamma_1) > 0$. The flow of an incompressible Bingham fluid in stationary case is described by

- the incompressibility equation

$$\operatorname{div} u = u_{i,i} = \sum_{i=1}^{n} \frac{\partial u_i}{\partial x_i} = 0 \quad \text{in} \quad \Omega; \tag{1}$$

- the equation of motion where, for simplicity, we take the constant density $\rho = 1$,

$$u_j u_{i,j} = f_i + \sigma_{ij,j} \quad \text{in} \quad \Omega \qquad (i = 1, ..., n) \ ; \tag{2}$$

- the constitutive law:

$$\begin{cases} \sigma_{ij} = -p\delta_{ij} + 2\mu D_{ij} + g\dfrac{D_{ij}}{D_{II}^{\frac{1}{2}}} & \Leftrightarrow \ D_{II} > 0 \\ \sigma_{ij} = -p\delta_{ij} \ \Leftrightarrow \ D_{II} = 0 \ ; \end{cases} \tag{3}$$

where $u = (u_i)$ represents the velocity field, $\sigma = (\sigma_{ij})$ the stress tensor, $f = (f_i)$ the body forces, $D = (D_{ij})$ the rate of strain tensor, given by $D_{ij}(u) = \frac{1}{2}(u_{i,j}+u_{j,i})$, $D_{II} = \dfrac{1}{2}D_{ij}D_{ij}$, p denotes the pressure, μ the viscosity and g the threshold of plasticity;

- and the boundary conditions

$$\text{on} \quad \Gamma_1 : \quad u = 0 \tag{4}$$

$$\text{on} \quad \Gamma_2 : \quad u_N = 0 \tag{5}$$

$$\text{and} \quad \begin{cases} |\sigma_T| < k\,|\sigma_N| \implies u_T = 0 \\ |\sigma_T| = k\,|\sigma_N| \implies \exists \lambda \geq 0 \quad u_T = -\lambda\sigma_T \end{cases} \tag{6}$$

where $n = (n_i)$ is the unit outward normal to Γ, and

$$u_T = u - u_N\, n \quad , \quad u_N = u_i n_i$$
$$\sigma_{Ti} = \sigma_{ij}n_j - \sigma_N n_i \quad , \quad \sigma_N = \sigma_{ij}n_i n_j = (\sigma.n).n$$

are, respectively, the tangential and normal velocity, the components of the tangential stress tensor and the normal stress. The boundary condition (4) corresponds to the adherence to the boundary Γ_1 and the condition (6) is the usual friction law of Coulomb (see [DL]), where k is a friction coefficient, in addition to the no flux condition (5) across Γ_2.

Using the standard notations for Sobolev spaces, like for instance in [BC], we consider the spaces:

$$V = \{u \in [H^1(\Omega)]^n : \operatorname{div} u = 0 \quad \text{in} \quad \Omega, \ u = 0 \quad \text{on} \quad \Gamma_1, \ u_N = 0 \quad \text{on} \quad \Gamma_2\}$$
$$W = \{u|_{\Gamma_2} : \ u \in V\}$$

endowed, respectively, with the norms

$$||u||_V = \left(||u||^2_{[L^2(\Omega)]^n} + \sum_{i,j=1}^{n} ||u_{i,j}||^2_{L^2(\Omega)}\right)^{\frac{1}{2}} \quad \text{and} \quad ||w||_W = \inf_{\substack{v \in V \\ w = v|_{\Gamma_2}}} ||v||_V ,$$

where $u|_{\Gamma_2}$ denotes the restriction to Γ_2 of the trace of a function of V. We also denote by $(u, v) = \int_\Omega u_i v_i$ the $[L^2(\Omega)]^n$ inner product.

A formal application of Green's formula, using (2) and (6), leads to the variational principle for $u \in V$

$$\int_\Omega u_j u_{i,j} \, (v-u)_i \geq - \int_{\Gamma_2} k \, |\sigma_N| \, (|v_T| - |u_T|) - \int_\Omega \sigma_{ij} D_{ij}(v-u) + \int_\Omega f_i \, (v-u)_i \qquad \forall v \in V \, .$$

The term $\int_{\Gamma_2} k \, |\sigma_N| (|v_T| - |u_T|) \, ds$, in general has no sense, because we only have, by Green's formula, $\sigma.n$ defined by duality in W'. Therefore $|\sigma_N|$ is not, in general, well defined. A possible way to remove this mathematical difficulty, following [D] and [P], is to made a regularization by convolution with mollifiers, i.e., to replace $|\sigma_N|$ by $|[(\sigma.n).n] * \rho_m|$ where

$$\tau * \rho_m(x) = \int_{\mathbb{R}^n} \tau(x-y) \rho_m(y) dy, \quad \rho_m \in \mathcal{D}(\Omega) \, , \quad \rho_m \geq 0 \, , \quad \int_\Omega \rho_m = 1 \, .$$

This corresponds to consider a non local friction law. More generaly, we shall replace (6) on Γ_2 by

$$\begin{cases} |\sigma_T| < k \, \Phi(\sigma.n) \implies u_T = 0 \\ |\sigma_T| = k \, \Phi(\sigma.n) \implies \exists \lambda \geq 0 \quad u_T = -\lambda \sigma_T \end{cases} \tag{6'}$$

where the nonlocal operator Φ is such that ($\rightharpoonup$ denoting weak convergence)

$$\begin{aligned} &\Phi : W' \longrightarrow L^2_+(\Gamma_2) = \{\psi \in L^2(\Gamma_2) : \quad \psi \geq 0\} \quad \text{is weakly continuous, i.e.,} \\ &g_k \rightharpoonup g \qquad \text{in} \quad W' \Longrightarrow \Phi(g_k) \rightharpoonup \Phi(g) \qquad \text{in} \quad L^2(\Gamma_2) \end{aligned} \tag{7}$$

Under the assumptions:

$$\mu, g \in \mathbb{R}, \quad \mu, g > 0, \quad k \in L^\infty(\Gamma_2), \quad k \geq 0, \quad f \in [L^2(\Omega)]^n;$$

and the usual definitions (see [DL], for instance):

$$
\begin{aligned}
a :& V \times V \longrightarrow \mathbb{R} \\
& (u,v) \longmapsto a(u,v) = 2\int_\Omega D_{ij}(u)D_{ij}(v)dx \\
b :& V \times V \times V \longrightarrow \mathbb{R} \\
& (u,v,w) \longmapsto b(u,v,w) = \int_\Omega u_i v_{j,i} w_j dx \\
j :& V \longrightarrow \mathbb{R} \\
& v \longmapsto j(v) = 2\int_\Omega D_{II}^{\frac{1}{2}}(v)dx
\end{aligned}
$$

we can state the weak formulation of the

Main Problem: Find a function $u \in V$, such that:

$$
\textbf{(Pb)} \quad \mu a(u,v-u) + b(u,u,v-u) + gj(v) - gj(u) + \int_{\Gamma_2} k\Phi(\sigma(u).n)\left(|v_T| - |u_T|\right) \geq
$$
$$
\geq (f, v-u), \qquad \forall v \in V.
$$

We recall the useful properties of the functionals:

i) a is a bilinear form:

- continuous: $|a(u,v)| \leq 2\,||u||_V\,||v||_V$ (8)
- coercive, because of the Korn's inequality $\exists \alpha > 0$:

$$
a(u,u) \geq \alpha ||u||_V^2 \qquad \forall u \in [H^1(\Omega)]^n \quad : \quad u = 0 \quad \text{in} \quad \Gamma_1 . \tag{9}
$$

ii) b is a trilinear form:

- continuous: $\exists c_1 > 0$: $|b(u,v,w)| \leq c_1\,||u||_V\,||v||_V\,||w||_V$ (10)
- anti-symmetrical: $b(u,v,w) + b(u,w,v) = 0$ (11)

iii) j is a convex, continuous but not differentiable functional in V.

2 – Existence of solutions

Main Theorem. *Under the preceding assumptions, there exists at least one solution $u \in V$ of the problem* **(Pb)**.

Proof: The proof consists of two parts:

Part A) The study of an auxiliary well-posed problem, obtained by freezing the essential nonlinearities in b and in Φ of the main problem. In particular, we shall use the continuous dependence of the associated Lagrange multiplyers.

Part B) The application of a general fixed point theorem of Tychonov-Kakutani-Glicksberg for the multivalued mappings from $V \times W'$ into $\mathcal{P}(V \times W')$, with the weak topologies.

A) Properties of the solutions to the auxiliary problem

Let $\xi \in V$ and $\psi \in L^2_+(\Gamma_2)$.

Auxiliary Problem: Find a function $u = u(\xi, \psi) \in V$, such that:

$$(\mathbf{Pb}\ \xi,\psi) \quad \mu a(u, v-u) + b(\xi, u, v-u) + gj(v) - gj(u) + \varphi(\psi, v) - \varphi(\psi, u) \geq (f, v-u), \quad \forall v \in V,$$

where, for each $\psi \in L^2_+(\Gamma_2)$, $\varphi(\psi, .) : v \longmapsto \varphi(\psi, v) = \int_{\Gamma_2} k\psi |v_T|\, ds$ is a convex, continuous but not differentiable functional defined in V.

Proposition 1. *For every $\xi \in V$ and $\psi \in L^2_+(\Gamma_2)$, there exists a unique solution $u \in V$ of the problem* ($\mathbf{Pb}$ ξ, ψ)*, which satisfies the following estimate, independently of ξ and ψ:*

$$||u||_V \leq \frac{||f||_{V'}}{\mu\alpha}. \tag{12}$$

Proof: The existence and uniqueness are consequences of compactness and monotonicity methods (see [L]) on variational inequalities with convex functionals.
Taking $v = 0$ in $(Pb\ \xi, \psi)$, the estimate (12) follows easily. ∎

Proposition 2. *Let $\{\xi_m\}$ and $\{\psi_m\}$ be sequences in V and in $L^2_+(\Gamma_2)$, respectively, such that $\xi_m \rightharpoonup \xi$ in V and $\psi_m \rightharpoonup \psi$ in $L^2(\Gamma_2)$. Let u_m be the solution of $(Pb\ \xi_m, \psi_m)$, $\forall m \in \mathbb{N}$. Then u_m converges strongly to u in V, where u is the solution of $(Pb\ \xi, \psi)$.*

Proof: The weak convergence $u_m \rightharpoonup u$ follows from (12) and standard lower semicontinuity properties. Then, the weak convergence implies the convergence in the V-norm, by the coerciveness of the bilinear form $a(.,.)$. ∎

Proposition 3. *Let u be the solution of $(Pb\ \xi, \psi)$.*
Let $Y := \{q = (q_{ij}) : q_{ij} = q_{ji},\ q_{ij} \in L^2(\Omega)\} \times [L^2(\Gamma_2)]^n$.
Then, there exists a Lagrange multiplyer

$$p^* = (p_1^*, p_2^*) \in Y' = Y$$

such that:

i) $|p_1^*| \leq \sqrt{2}g$

ii) $2gD_{II}^{\frac{1}{2}}(u) + (p_1^*)_{ij} D_{ij}(u) = 0$ *in* Ω

iii) $|p_2^*| \leq k\psi$

iv) $k\psi|u_T| + p_2^*.u_T = 0$ *in* Γ_2

v) $2\mu D_{ij,j}(u) - (p_1^*)_{ij,j} + f_i = \xi_j\, u_{i,j}$ *in* Ω

vi) $(p_1^*)_{ij} n_j + (p_2^*)_i = 2\mu D_{ij}(u) n_j$ *in* Γ_2.

Conversely, let $u \in V$ and $p^* = (p_1^*, p_2^*) \in Y$ such that satisfy i), ii), iii), iv), v), vi), then u is the solution of $(Pb\ \xi, \psi)$.

Proof: We apply some duality theorems of convex optimization (see Remark 4.2 in [ET], page 59) with the following definitions:

$$
\begin{aligned}
&V := \{u \in [H^1(\Omega)]^n : \operatorname{div} u = 0 \quad \text{in} \quad \Omega, \ u = 0 \quad \text{on} \quad \Gamma_1, \ u_N = 0 \quad \text{on} \quad \Gamma_2\}\\
&Y := \{q = (q_{ij}) : q_{ij} = q_{ji}, \ q_{ij} \in L^2(\Omega)\} \times [L^2(\Gamma_2)]^n\\
&F : V \longrightarrow \mathbb{R}\\
&\quad v \longmapsto F(v) = \mu a(u,v) + b(\xi, u, v) - (f, v)\\
&G : Y \longrightarrow \mathbb{R}\\
&\quad p = (p_1, p_2) \longmapsto G(p) = G_1(p_1) + G_2(p_2)\\
&\qquad \text{where}\\
&\qquad G_1 : \{q = (q_{ij}) : q_{ij} = q_{ji}, \ q_{ij} \in L^2(\Omega)\} \longrightarrow \mathbb{R}\\
&\qquad\quad p_1 \longmapsto \int_\Omega \sqrt{2} g \left((p_1)_{ij}(p_1)_{ij}\right)^{\frac{1}{2}}\\
&\qquad G_2 : [L^2(\Gamma_2)]^n \longrightarrow \mathbb{R}\\
&\qquad\quad p_2 \longmapsto \int_{\Gamma_2} k\psi\, |p_2|\\
&\Lambda : V \longrightarrow Y\\
&\quad u \longmapsto (D_{ij}(u), u|_{\Gamma_2}) \ . \blacksquare
\end{aligned}
$$

Proposition 4. *Let $\{\xi_m\}$ and $\{\psi_m\}$ be sequences in V and $L^2_+(\Gamma_2)$, respectively, such that $\xi_m \rightharpoonup \xi$ in V and $\psi_m \rightharpoonup \psi$ in $L^2(\Gamma_2)$. Let u_m be the solutions of $(Pb\ \xi_m, \psi_m)$, p_m^* their Lagrange multiplyers, $\forall m \in \mathbb{N}$; and u the solution of $(Pb\ \xi, \psi)$. Then there exists a subsequence p_m^* weakly converging in Y to p^*, which is a Lagrange multiplyer of the solution u.*

Proof: We apply proposition 1 and proposition 2. ∎

Definition. Let u be the solution of $(Pb\ \xi, \psi)$ and $p^* = (p_1^*, p_2^*)$ its Lagrange multiplyer. We define the stress tensor by

$$\sigma_{ij}(u) := 2\mu D_{ij}(u) - (p_1^*)_{ij} \ .$$

Proposition 5. *Under the above notations, $\sigma(u).n = p_2^*$ in Γ_2.*

Proof: It is a consequence of the Green's formula. ∎

B) Existence of fixed points

We apply a fixed point theorem proved by Glicksberg (1952), which is an extension of the Tychonov's theorem (for single-valued mappings) and of the Kakutani's theorem (for multi-valued mappings in $\mathbb{R}^n$). Its proof and references can be found in [BC], pages 218-220.

Theorem TKG. *Let E be a locally convex Hausdorff topological vector space and let K be a non-empty convex compact set in E. If $h : K \longrightarrow 2^K_{\blacksquare ck}$ is upper semi-continuous then h has at least one fixed point.* ∎

We also apply an extension of closed graph's theorem (see [BC], page 413):

Theorem gf. *Let X and Y be two compact Hausdorff topological spaces and $h : X \longrightarrow 2^Y_{\blacksquare c}$. Then h is upper semi-continuous iff*

$$G_{XY}(h) = \{(x, y) \in X \times Y : \quad x \in X \quad \text{and} \quad y \in h(x)\} \quad \textit{is closed in } X \times Y . \blacksquare$$

Thus, we consider

$$\begin{aligned} &E := V \times W' \\ &K := \{\xi \in V : ||\xi||_V \leq r_1\} \times \{\tau \in W' : ||\tau||_{W'} \leq r_2\} \\ &h : K \longrightarrow 2^K_{\blacksquare ck} = \{R \in \mathcal{P}(K) : \ R \neq \emptyset, \ R \quad \text{closed convex } \} \\ &\quad (\xi, \tau) \longmapsto S \times T \end{aligned}$$

where V is endowed with the weak topology, W' with the weak-$*$ topology, and, thus, E the with product topology. Then E is a locally convex Hausdorff topological vector space and K is a non-empty convex compact set in E. And:

$$\begin{aligned} &r_1 \geq \frac{||f||_{V'}}{\mu\alpha}; \\ &r_2 \geq (\frac{2\mu + r_1 c_1}{\mu\alpha} + 1) \ ||f||_{V'} + \sqrt{2} g[\text{meas}(\Omega)]^{\frac{1}{2}} \\ &S = \{u(\xi, \Phi(\tau)) : \ u(\xi, \Phi(\tau)) \quad \text{is the solution of} \quad (Pb \ \xi, \Phi(\tau))\} \\ &\quad = \{u(\xi, \Phi(\tau))\} \\ &T = \{\sigma(u).n : \ \sigma \quad \text{is a stress tensor of the solution} \quad u = u(\xi, \Phi(\tau))\} \ . \end{aligned}$$

Recalling the properties of the solution to the auxiliary problem, we have the proposition:

Proposition 6. a) *h is well defined;*

b) *$G_{KK}(h)$ is closed in $K \times K$ with the product topology.* ∎

Then, using the proposition 6 and the theorem gf, we can apply the theorem TKG to obtain:

$$\exists(\xi, \tau) \in K \qquad (\xi, \tau) \in h((\xi, \tau)) \ .$$

Therefore, $\xi \in V$ is the solution of $(Pb \ \xi, \Phi(\tau))$, and $\tau = \sigma(\xi).n \in W'$, where σ is a stress tensor of the solution ξ, i.e., ξ is a solution of (Pb).

REFERENCES

[BC] *C. Baiocchi* and *A. Capelo* – *Variational and quasivariational inequalities: Applications to free boundary problems*, Wiley-Interscience, Chichester–New York, 1984.

[D] *G. Duvaut* – *Équilibre d'un solide élastique avec contact unilatéral et frottement de Coulomb*, C.R. Acad. Sc. Paris, **290**, (1980), 263-265.

[DL] *G. Duvaut* and *J.L. Lions* – *Les inéquations en mécanique et en physique*, Dunod, Paris, 1972.

[ET] *I. Ekeland* and *R. Temam* – *Analyse convexe et problèmes variationnels*, Dunod et Gauthier-Villars, Paris, 1974.

[L] *J.L. Lions* – *Quelques méthodes de résolution des problèmes aux limites non linéaires*, Dunod et Gauthier-Villars, Paris, 1969.

[P] *E.B. Pires* – *Analysis of nonclassical friction laws for contact problems in elastostatics*, Ph.D. Dissertation, The University of Texas at Austin, 1982.

Luisa Consiglieri,
CMAF/University of Lisbon,
Av. Prof. Gama Pinto, 2
1699 LISBOA Codex - PORTUGAL

Étude de la Stabilité du Couplage des Équations d'Euler et Maxwell à une Dimension d'Espace

SYLVIE FABRE

Le système des équations d'Euler et de Maxwell permet de modéliser un plasma dans une approche "fluides". Rappelons qu'un plasma est un gaz ou un fluide constitué de particules chargées (voir [1]). Ici nous présentons une étude de stabilité de méthodes numériques utilisées pour résoudre de façon approchée ces équations.

1 – Modélisation et analyse de stabilité

Dans notre modèle le plasma est représenté par un bifluide où l'on distingue les ions et les électrons. Pour simplifier, nous supposerons que les ions sont fixes, les hypothèses physiques portant sur le fluide des électrons sont les suivantes : le fluide des électrons est parfait et en évolution isotherme. On suppose de plus que le plasma est animé d'un certain mouvement thermique ($KT \neq 0$). L'évolution du fluide des électrons est donc modélisée par le système des équations d'Euler et Maxwell (écrit ici sans dimensions)

$$\begin{cases} \partial_t n + \partial_x nu = 0 \\ \partial_t nu + \partial_x(nu^2 + p) = -nE_x - nvB_z - \nu nu \\ \partial_t nv + \partial_x(nuv) = -nE_y + nuB_z - \nu nv \\ \partial_t B_z + \partial_x E_y = 0 \\ \partial_t E_y + \partial_x B_z = nv \\ \partial_x E_x = n_i - n \, . \end{cases}$$

où n, nu, nv, p désignent respectivement la densité, la quantité de mouvement et la pression du fluide electronique avec $P = nv_T^2$, où v_T est la vitesse thermique et v est la fréquence de collision ion-électron.

Si l'on se plaçait dans un cadre purement électrostatique (par exemple en prenant v, B, $E_y = 0$ dans les équations précédentes) une analyse faite dans [2,3] montre que le système d'équations restant nécessite une condition de stabilité très forte. En effet le champ électrique E_x doit absolument être introduit en implicite dans la deuxième équation. Si on pense à un cadre bidimensionnel, on peut imaginer le coût que représente une telle condition; en effet il faudrait résoudre de façon implicite les équations d'Euler et Maxwell à chaque pas de temps. Pour éviter ce problème, nous avons mis au point une correction à apporter à la façon dont est calculé le flux numérique dans la partie Euler, qui permet de traiter le second membre nE de façon semi-implicite au temps $n + 1/2$.

Soit $(g^n_{j+1/2})_{j,n}$ le flux numérique permettant de résoudre les équations d'Euler par une méthode décentrée classique de type Gudunov. Dans une zone proche de l'état d'équilibre, zone de linéarisation, le système précédent est d'une certaine façon gouverné par l'équation des ondes qui lui est sous-jacente et on peut voir facilement qu'un schéma décentré classique n'approche pas de façon précise cette équation des ondes. Par contre le schéma saute-mouton est parfaitement adapté à cette équation. La correction consiste alors à modifier le flux $(g^n_{j+1/2})$ dans un voisinage de l'état d'équilibre de telle façon à retrouver le schéma saute mouton (pour plus d'explications voir [2]).

2 – Résultats numériques et conclusion

Nous avons fait des tests comparatifs entre les trois méthodes suivantes.

(**1**) Schéma splitting de Van Leer pour les équations d'Euler (voir [2,4]) et second membre au temps $n + 1/2$.

(**2**) Idem avec second membre au temps $n + 1/2$ et correction "saute-mouton".

(**3**) Idem avec second membre implicite.

Grâce à l'étude de stabilité théorique nous savons que (1) est instable et (3) stable. Pour (2), nous savons que le schéma saute mouton couplé avec un second membre en $n + 1/2$ donne une méthode stable sur le problème linéarisé (voir [2,3]).

Le shéma (2) donne exactement les mêmes résultats numériques que le schéma (3), ce qui concorde aussi avec l'analyse théorique de stabilité. De plus, le schéma (2) permet d'envisager un modèle à deux dimensions d'espace beaucoup moins complexe que la méthode (3).

REFERENCES

[**1**] *F.F. Chen – Introduction to plasma physics and controlled fusion*, Plenum, New-York/ London, 1984.

[**2**] *S. Fabre* – Thèse de l'Ecole Polytechnique, 1990.

[**3**] *S. Fabre – Stability analysis of the Euler-Poisson equations*, to appear in Journal of Computational Physics (1992).

[**4**] *B. Van Leer – Flux-vector splitting for the Euler equations*, Lecture notes in physics, **170**, Springer Verlag (1982), p.507.

Sylvie Fabre,
Centre de Mathématiques Appliquées,
Ecole Polytechnique,
91128 PALAISEAU Cedex - FRANCE

Diphasic Equilibrium and Chemical Engineering

FRANÇOIS JAMES

1 – Introduction

Many processes in Chemical Engineering involve matter exchange between two phases in view of separate or analyze multicomponent mixtures. One can mention chromatography [5], [3], distillation [1], or electrophoresis. It is possible, under several hypothesis, to model these processes by a system of first order conservation laws. Consider a 1-dimensional diphasic medium in which phase 1 is moving with a velocity u, and phase 2 with velocity v. Assume u and v to be constant, $u > 0$ and $v \leq 0$: we deal with a countercurrent process. We shall assume also that the whole process is isothermal. Thus the equations of momentum and energy are useless, and we are left with the conservation of matter. So, let c^1 and c^2 be vector-valued functions of x and t, related to the concentrations in phase 1 and 2 respectively. We have

$$\partial_t(c^1 + c^2) + \partial_x(uc^1 + vc^2) = 0. \tag{1.1}$$

The system is for the moment open:we have n equations for $2n$ unknowns. The closure is obtained by a fundamental assumption: we suppose the process to be **quasi-static**. This means that, at each time, the two phases are at stable thermodynamical equilibrium.

This hypothesis introduces a non linear relation between c^1 and c^2, which we investigate in the next section. As we shall see, the system (1.1) will become a nonlinear system of conservation laws, which is proved to be hyperbolic.

2 – Diphasic equilibrium

We give here a few basic thermodynamical tools we shall use widely in the following. Consider two phases, denoted by $i = 1, 2$, and M chemical species, or components, $1 \leq m \leq M$. We adopt the following convention throughout this paper: a *superscript* will denote a phase, and a *subscript* a chemical species. Namely, c^i_m is the amount of component m in phase i, for $i = 1, 2$ and $1 \leq m \leq M$.

Assume that both phases are at thermodynamical equilibrium. According to Gibbs [2], this means

a) each phase, considered as a simple thermodynamical system, is in equilibrium;

b) the internal energy of the system constituted by the two phases is minimum, with respects to several constraints.

It can be given a precise mathematical meaning to these two assumptions, following the classical formalism of Gibbs. We cannot go into the details of modelling here, and we refer for instance to [3] for such a work in the case of chromatography. Therefore our starting point will be the following

Basic assumption. *There exist two functions $\eta_i : \mathbb{R}^M \to \mathbb{R}^M$, strictly convex, of class C^2, such that the equilibrium state is the unique solution of the constrained minimization problem*

$$\min_{c^1+c^2=\text{const.}} \eta_1(c^1) + \eta_2(c^2).$$

Let us just say that the existence of the η_i-s corresponds to assumption a), and that the constrained minimum property to assumption b).

We shall denote in the following by μ_i the gradient of η_i, and by $D\mu_i$ the matrix of its second derivative:

$$\mu_i(c^i) = \eta_i'(c^i), \qquad D\mu_i(c^i) = \eta_i''(c^i).$$

By introducing the Lagrange multipliers corresponding to the constraint $c^1 + c^2 =$ const., one easily check that the equilibrium state is characterized by

$$\mu_1(c^1) = \mu_2(c^2). \tag{2.1}$$

This equality is nothing but the well-known equality of chemical potentials at equilibrium. We intend to study the mathematical properties of this relation, and its consequences on the system (1.1).

Notice first that, since η_2 is strictly convex, μ_2 is monotone on its domain of definition, so that (2.1) can be solved as

$$c^2 = h(c^1). \tag{2.2}$$

Since the functions η_i are twice continuously differentiable, the function h is of class C^1, and we denote by $J(c^1) = h'(c^1)$ its jacobian matrix, which will be called **equilibrium matrix** of the system. But one has a little more.

Theorem 2.1. *The equilibrium matrix is diagonable, and its eigenvalues α_i, $1 \le i \le n$, are positive.*

Proof: The matrix J can be written as the product of two symmetric positive definite matrices: $J(c^1) = \left(D\mu_2(h(c^1))\right)^{-1} D\mu_1(c^1)$. Since η_2 is strictly convex, relation $< u, v >_2 \stackrel{def}{=} D\mu_2(h(c^1))u \cdot v$ for u and v in $\mathbb{R}^n$ defines a scalar product on $\mathbb{R}^n$. It is now easy to prove that J is self-adjoint with respect to this scalar product, and therefore diagonable. Let r_i be an eigenvector of J, α_i the corresponding eigenvalue: we have $D\mu_2 J r_i = \alpha_i D\mu_1 r_i$. Taking the scalar product of this relation with r_i leads to

$$\alpha_i = \frac{D\mu_1 r_i \cdot r_i}{D\mu_2 r_i \cdot r_i},$$

and the positivity immediately follows from the strict convexity of η_1. ∎

We have the following corollary, which allows us to deal with the total amount of matter in the system, namely $w = c^1 + c^2$.

Corollary 2.1. *The mapping $w \to c^1 + h(c^1)$ is a C^1-diffeomorphism from the equilibrium manifold on itself.*

Proof: The result comes from the properties of J. Indeed, one has $dw = (I + J(c^1))dc^1$, I being the identity matrix of $\mathbb{R}^n$. Since $J(c^1)$ is diagonable, with positive eigenvalues, the matrix $I + J(c^1)$ is also diagonable, with eigenvalues greater than 1. It is therefore invertible, and there exists a function g of class C^1 such that $c^1 = g(w)$. The proof is complete. ∎

3 – Conservation equations

Let us go back to system (1.1), to introduce the function h of (2.2):

$$\partial_t[c^1 + h(c^1)] + \partial_x[uc^1 + vh(c^1)] = 0. \tag{3.1}$$

By Corollary 2.1, we can perform the variable change $w = c^1 + h(c^1)$, which consists simply in writing the conservation of total amount. The system (3.1) then becomes

$$\partial_t w + \partial_x \left[ug(w) + vh(g(w))\right] = 0. \tag{3.2}$$

Since $g'(w) = (I + J)^{-1}$, the jacobian $A(w)$ of (3.2) is given by $A(w) = (uI + vJ)(I + J)^{-1}$. One deduces easily that A is diagonable, which insures the hyperbolicity of (3.2). Moreover, the eigenvalues λ_i of A are given by $\lambda_i = \phi(\alpha_i)$, with

$$\phi(\alpha) = \frac{u + v\alpha}{1 + \alpha}.$$

The function ϕ is strictly monotone as soon as $u \neq v$, and, since α takes its values between 0 and $+\infty$, the λ_i-s are uniformly bounded with respect to w. More precisely, we have the

Theorem 3.1. *The system (3.2) is hyperbolic, its eigenvalues are uniformly bounded with respect to w by*

$$\min(u, v) < \lambda_i(w) < \max(u, v) \tag{3.3}$$

for w such that $g(w)$ is in the domain of definition of h.

This result is natural:it means that non linear phenomena of interaction between phases do slow the components with respect to the purely hydrodynamical propagation which is linear (with velocity u or v). In other respects, we do not have any result about strict hyperbolicity (distinct eigenvalues). The system (3.2) is well posed in the following sense:

Theorem 3.2. *The function $\eta(w) \stackrel{\text{def}}{=} \eta_1(g(w)) + \eta_2(h(g(w)))$ is a strictly convex mathematical entropy for system (3.2).*

Proof: Let us first compute the first derivative of η. We have, after (2.1),

$$\begin{aligned}\eta'(w) &= \mu_1(g(w))\left[I + h'(g(w))\right]g'(w) \\ &= \mu_1(g(w)) \qquad\qquad \text{by definition of } g'.\end{aligned}$$

We now want to verify that there exists an entropy flux $q(w)$, i.e. a real function on $\mathbb{R}^n$ such that

$$\eta'(w)A(w) = q'(w).$$

Replace $A(w)$ by its value and write $h'(g(w)) = J$. One obtains

$$\eta'(w)A(w) = \mu_1(g(w))\,(uI + vJ)(I + J)^{-1}.$$

Again apply the definition of $g'(w)$, and use @equil:

$$\eta'(w)A(w) = u\mu_1(g(w))g'(w) + v\mu_2\,(h(g(w)))\,h'(g(w))g'(w).$$

The form of the function q is now obvious:

$$q(w) = u\eta_1(g(w)) + v\eta_2\,((g(w)))\,,$$

and η is indeed an entropy of @ConsTotale.

For the sake of brevity, we skip the proof of the convexity of η, which is obtained by straightforward computation. ∎

We give now, without any proof, two remarks concerning discontinuous solutions of (3.2). First, consider the Riemann problem associated to @ConsTotale. The behaviour of the characteristic fields of A is therefore of some interest. Recall that the i-th field is genuinely non linear (GNL) if $\lambda_i'(w)\cdot r_i(w) \neq 0$, and linerarly degenerate if $\lambda_i'(w)\cdot r_i(w) \equiv 0$. More generally, we are interested in the behaviour of the eigenvalue λ_i along the integral curve of the eigenvector r_i. We have $\lambda_i'(w)\cdot r_i(w) = \phi'(\alpha(w))\alpha'(w)\cdot r(w)$. Since ϕ is strictly monotone, we can state

Lemma 3.1. *The i-th characteristic field of A has the same behaviour as the i-th characteristic field of the equilibrium matrix J.*

Next, consider a piecewise C^1 weak solution propagating with velocity σ. One can prove in a similar way as Theorem 3.3 the following

Theorem 3.3. *The propagation velocity of discontinuities σ satisfies*

$$\min(u, v) < \sigma < \max(u, v).$$

Again, the nonlinear propagation cannot be faster than the hydrodynamical linear propagation: we model retention phenomena.

REFERENCES

[1] *E. Canon – Étude de deux modèles de colonne à distiller*, Thèse de l'Université de Saint-Etienne, 1990.

[2] *J.W. Gibbs – On the Equilibrium of Heterogeneous Substances*, Trans. Connecticut Academy, **III** (1876), 108–248, (1878), 343–524, & Coll. Works, 55–353.

[3] F. James – *Sur la modélisation mathématique des équilibres diphasiques et des colonnes de chromatographie*, Thèse de l'Ecole Polytechnique, 1990.

[4] E. Kvaalen, L. Neel and D. Tondeur – *Directions of Quasi-static Mass and Energy Transfer Between Phases in Multicomponent Open Systems*, Chem. Eng. Sc., **40**(7) (1985), 1191–1204.

[5] P. Valentin and G. Guiochon – *Propagation of Finite Concentration in Gas Chromatography*, Separation Science, **10** (1975), 245–305.

François James,
Centre de Mathématiques Appliquées,
École Polytechnique,
F-91128 PALAISEAU Cedex - FRANCE

Vibrations of a Viscous Compressible Fluid in Bounded and Unbounded Domains

MICHAEL R. LEVITIN

We consider small linearized vibrations of a viscous compressible fluid in a bounded or unbounded domain G with a smooth boundary Γ under different boundary conditions. The system of linear differential equations describing small motions of a viscous compressible fluid in the region G is

$$\frac{\partial \mathbf{v}}{\partial t} = \mu\,\Delta \mathbf{v} + \left(\frac{\mu}{3} + \varsigma\right) \mathbf{grad}\,\mathrm{div}\,\mathbf{v} - \mathbf{grad}\,\rho \ , \tag{1}$$

$$\frac{\partial \rho}{\partial t} = -\,\mathrm{div}\,\mathbf{v} \ , \tag{2}$$

where $\mathbf{v}(\mathbf{x},t) = (v_1, v_2, v_3)$ is a fluid velocity vector, $\rho(\mathbf{x},t)$ is a deviation of a fluid pressure from the equilibrium, $\mathbf{x} = (x_1, x_2, x_3)$ are the Cartesian coordinates in G. Hereinafter we use the dimensionless system of units in which a characteristic curvature of Γ, a fluid density and a speed of sound are equal to 1. By μ and ς we denote the dimensionless viscosity coefficients which must satisfy the thermodynamic assumptions $\mu > 0$ and $\varsigma \geq 0$. Everywhere for simplicity we assume $\varsigma = \beta\mu$, $\beta > 0$. Eq. (1) is the system of three equations of motion, Eq. (2) is the equation of continuity.

We investigate free harmonic vibrations of the fluid depending on time as $\exp(i\omega t)$, where ω is a dimensionless frquency of vibrations. Eliminating the time factor in Eqs. (1), (2), we will rewrite them in the operator form

$$\mathcal{A}^{\mu}\mathbf{F} = \lambda \mathbf{F} \ , \tag{3}$$

where $\mathbf{F} = (\mathbf{v}, \rho)$ is a 4-dimensional vector with the components v_1, v_2, v_3, ρ,

$$\mathcal{A}^{\mu} = \begin{bmatrix} \mu\left[-\Delta - \left(\beta + \frac{1}{3}\right)\mathbf{grad}\,\mathrm{div}\right] & \mathbf{grad} \\ \mathrm{div} & 0 \end{bmatrix} , \tag{4}$$

and λ is the new spectral parameter which will be used below. The parameter λ associates with the frequency ω by the correlation $\lambda = -i\omega$.

We introduce two trace operators,

$$\mathcal{T}_1 \colon \mathbf{F} \to \mathbf{v}|_{\Gamma} \ , \qquad \mathcal{T}_2 \colon \mathbf{F} \to (\rho\,\mathbf{n} + \mu\,\sigma'(\mathbf{v})\,\mathbf{n})|_{\Gamma} \ , \tag{5}$$

where $\mathbf{n}$ is the unit normal vector to Γ towards the infinity and $\mu\sigma'_{jk}(\mathbf{v})$ is the viscous stress tensor of the fluid. We will consider the operators $\mathcal{A}^{\mu}_j$ ($j = 1, 2$) as the L_2–realizations of $\mathcal{A}^{\mu}$ equipped with the boundary conditions $\mathcal{T}_j\mathbf{F} = \mathbf{0}$ (hard or free boundary).

1 – The Inner Problem

Let G be bounded. In this case it is known [1] that the domains $\mathcal{D}(\mathcal{A}_j^\mu) \subseteq (H^1(G))^3 \times L_2(G)$ and the essential spectra of $\mathcal{A}_j^\mu$ consist of two real positive numbers which are proportional to μ^{-1}. Our aim is to consider the discrete spectra.

Multiplying $\mathcal{A}^\mu\mathbf{F}$ by $\mathbf{F}' = (\mathbf{v}', \rho')$ and integrating by parts we obtain the following Green formula for the operator $\mathcal{A}^\mu$ ($\mathbf{F}, \mathbf{F}' \in (H^1(G))^3 \times L_2(G)$):

$$(\mathcal{A}^\mu\mathbf{F}, \mathbf{F}')_{L_2(G)} \equiv \iint (\mathcal{A}^\mu\mathbf{F}) \cdot \overline{\mathbf{F}}' \, dG$$
$$= \mu \, \mathcal{Z}_\beta(\mathbf{v}, \mathbf{v}') - (\rho, \operatorname{div} \mathbf{v}')_{L_2(G)} + (\operatorname{div} \mathbf{v}, \rho')_{L_2(G)} - (\mathcal{T}_2\mathbf{F}, \mathcal{T}_1\mathbf{F}')_{L_2(\Gamma)} \, . \tag{6}$$

Here we have introduced the sesquilinear form,

$$\mathcal{Z}_\beta(\mathbf{v}, \mathbf{v}') = 2\sum_{j=1}^{3} (\mathbf{grad}\, v_j, \mathbf{grad}\, v_j')_{L_2(G)}$$
$$- (\mathbf{curl}\, \mathbf{v}, \mathbf{curl}\, \mathbf{v}')_{L_2(G)} - \left(\frac{2}{3} - \beta\right) (\operatorname{div} \mathbf{v}, \operatorname{div} \mathbf{v}')_{L_2(G)} \, . \tag{7}$$

Let $\lambda \neq 0$ be an eigenvalue of the operator $\mathcal{A}_j^\mu$ ($j = 1$ or 2) and $\mathbf{F}$ be a corresponding eigenfunction. Then, applying Eq. (6) with $\mathbf{F}' = \mathbf{F}$ and eliminating $\rho = \lambda^{-1} \operatorname{div} \mathbf{v}$ (cf. Eqs. (3), (4)), we obtain

$$\lambda^2 \, \|\mathbf{v}\|^2_{L_2(G)} - \lambda \, \mu \, \mathcal{Z}_\beta(\mathbf{v}, \mathbf{v}) + \|\operatorname{div} \mathbf{v}\|^2_{L_2(G)} = 0 \, . \tag{8}$$

We can consider Eq. (8) as the quadratic equation in λ. Then for non real roots of this equation we have

$$\operatorname{Re} \lambda = \frac{\mu \, \mathcal{Z}_\beta(\mathbf{v}, \mathbf{v})}{2 \, \|\mathbf{v}\|^2_{L_2(G)}} \, , \qquad |\lambda| = \frac{\|\operatorname{div} \mathbf{v}\|_{L_2(G)}}{\|\mathbf{v}\|_{L_2(G)}} \, . \tag{9}$$

In order to describe the localization of non real eigenvalues λ we need more information on the functional $\mathcal{Z}_\beta(\mathbf{v}, \mathbf{v})$.

Lemma 1. *For any $\beta > 0$,*

a) $\mathcal{Z}_\beta(\mathbf{v}, \mathbf{v}) \geq \beta \|\operatorname{div} \mathbf{v}\|^2_{L_2(G)} \geq 0$, *for* $\forall \mathbf{v} \in (H^1(G))^3$;

b) $\mathcal{K} \equiv \ker \mathcal{Z}_\beta(\mathbf{v}, \mathbf{v}) = \{\mathbf{a} + \mathbf{b} \wedge x;\ \mathbf{a}, \mathbf{b} \in \mathbb{C}^3\}$; $\dim \mathcal{K} = 6$;

c) $\mathcal{Z}_\beta(\mathbf{v}, \mathbf{v})$ *is coercive on* $(H^1(G))^3$, *i.e.* $\exists\, c_0, c_1 > 0$ *such that*

$$\mathcal{Z}_\beta(\mathbf{v}, \mathbf{v}) \geq c_1 \, \|\mathbf{v}\|^2_{H^1(G)} - c_0 \, \|\mathbf{v}\|^2_{L_2(G)} \, ;$$

d) $\mathcal{Z}_\beta(\mathbf{v}, \mathbf{v})$ *is strongly coercive on* $((L_2(G)^3 \ominus \mathcal{K}) \cap (H^1(G))^3$, *i.e.* $\exists\, \tilde{c}_1 > 0$ *such that*

$$\mathcal{Z}_\beta(\mathbf{v}, \mathbf{v}) \geq \tilde{c}_1 \, \|\mathbf{v}\|^2_{H^1(G)} \, ;$$

e) $\mathcal{Z}_\beta(\mathbf{v}, \mathbf{v})$ *is strongly coercive on* $\{\mathbf{v} \colon \exists \rho \in L_2(G), \mathbf{F} = (\mathbf{v}, \rho) \in \mathcal{D}(\mathcal{A}_j^\mu)\}$.

Further on by $B(z_0, R)$ we will denote the complex circle $\{z \in \mathbb{C} \colon |z - z_0| \leq R\}$ and by $P(\epsilon)$ we will denote the complex angle $\{z \in \mathbb{C} \colon |\arg z| \leq \epsilon\}$.

Theorem 2. *All the eigenvalues of the operators $\mathcal{A}_j^\mu$ ($j = 1, 2$) lie in the right complex half-plane, either on the real non-negative semi-axis or in the sector $S_\mu = B(0, D/\mu) \cap Q_\mu$, where $Q_\mu = \{z \in \mathbb{C}\colon \operatorname{Re} z \geq E\mu\}$. Here the constants D and E are independent of $\mu > 0$. For any fixed $\mu > 0$ and any arbitrary small $\epsilon > 0$ there exist only a finite number of eigenvalues of $\mathcal{A}_j^\mu$ lying in the sector S_μ outside $P(\epsilon)$.*

The proof of Theorem 2 is based on Eqs. (8), (9) and Lemma 1.

The previous Theorem describes the global localization of eigenvalues (and eigenfrequencies) on the complex plane but does not give any methods of their calculation. The problem of special interest is the behavior of eigenfrequencies as the viscosity vanishes.

Let $\mathbf{v}_0$ and ρ_0 be a velocity and a pressure of an inviscid compressible fluid, and $\omega_0 \neq 0$ be a frequency of vibrations. It is well known that the velocity field of an inviscid fluid in the case of vibrations can be chosen purely potential. Let us introduce a fluid displacement potential ψ_0 such that $\mathbf{v}_0 = i\omega_0 \,\mathbf{grad}\, \psi_0$. After eliminating of the time factor $\exp(i\omega_0 t)$ the problem (3), (4), $\mu = 0$ turns into the scalar Helmholtz equation,

$$\Delta \psi_0 + \omega_0^2 \psi_0 = 0 \,. \tag{10}$$

The boundary condition $\mathcal{T}_j \mathbf{F} = \mathbf{0}$ turn into the only boundary condition, $\frac{\partial \psi_0}{\partial n}|_\Gamma = 0$, or $\psi_0|_\Gamma = 0$ for hard and soft walls, respectively. Thereby, in the inviscid case we deal simply with the spectral problem for the Neumann or Dirichlet Laplacian.

Hereinafter we call $\omega_0 \neq 0$ the eigenfrequency of the inviscid problem $\mathcal{A}_j^0$ ($j = 1, 2$) iff the number $-\omega_0^2$ belongs to the spectrum of the Neumann or Dirichlet Laplacian, respectively. The corresponding values $\lambda_0 = -i\omega_0$ are the points of the discrete spectrum of $\mathcal{A}_j^0$.

In order to describe the asymptotic behavior of eigenvalues of the viscous problem we have to consider simultaneously only a finite number of eigenvalues or a bounded region of the complex plane.

Theorem 3. *Let $R > 0$ be a fixed number. Then*

a) *As μ tends to zero, all the eigenvalues λ of the operator $\mathcal{A}_j^\mu$ from $B(0, R)$ lie either in an arbitrary small angle $P(\epsilon)$ near the real positive semi-axis, or in circles $B(\lambda_0, K\mu^{\varkappa})$ with the centers at eigenvalues λ_0 of $\mathcal{A}_j^0$, $|\lambda_0| < R$, where $\varkappa = \frac{1}{2}$ ($= 1$) for the hard (soft) boundary conditions, respectively.*

b) *The total order of the poles of the resolvent $(\mathcal{A}_j^\mu - \lambda I)^{-1}$ in the area enclosed by each contour $B(\lambda_0, K\mu^{\varkappa})$ is equal to the multiplicity of the corresponding inviscid eigenvalue λ_0.*

c) *If $\lambda_0 = -i\omega_0$, $\omega_0 \neq 0$, is a simple eigenvalue of the inviscid problem $\mathcal{A}_j^0$, then there exists the eigenvalue $\lambda = -i\omega$ of the viscous problem $\mathcal{A}_j^\mu$ which admits the asymptotic expansion,*

$$\lambda = \lambda_0 + \sum_{k=k_0}^{\infty} \mu^{\frac{k}{2}} \lambda_k \,, \qquad \mu \to +0 \,. \tag{11}$$

In (11) the summation over natural k starts with $k_0 = 1$ for the hard boundary conditions and with $k_0 = 2$ for the soft boundary conditions. The coefficient λ_1 (resp. λ_2) is expressed explicitly in terms of the inviscid eigenfrequency ω_0 and the corresponding eigenfunction ψ_0. For example, for $j = 1$ (hard walls), and $\omega_0 > 0$

$$\lambda_1 = 2^{-\frac{1}{2}}\, i^{\frac{1}{2}}\, \omega_0^{\frac{3}{2}}\, \mu^{\frac{1}{2}} \left\| \mathbf{grad}\, \psi_0 - \mathbf{n}\, \frac{\partial \psi_0}{\partial n} \right\|^2_{L_2(\Gamma)} \left\| \mathbf{grad}\, \psi_0 \right\|^{-2}_{L_2(G)} .$$

The major idea of the proof is the following. We single out the potential and the solenoidal component in the velocity field. Then the solenoidal component has the boundary layer character and the solution can be constructed by matching asymptotic expansions.

The detailed proofs of Theorems 2, 3 will be published in [2].

2 – The Outer Problem

For simplicity we will consider here the hard boundary condition $\mathcal{T}_1\mathbf{v} = \mathbf{v}|_\Gamma = \mathbf{0}$. We also assum that $\mathbb{R}^3 \backslash G$ is unitary connected.

The structure of the spectrum of the opeartor $\mathcal{A}_1^\mu$ acting in $L_2(G)$ is different from the inner case.

Theorem 4. *The essential spectrum of the outer problem consists of the real positive semi-axis $\{\lambda\colon \operatorname{Im}\lambda = 0, \operatorname{Re}\lambda \geq 0\}$ and the circumference*

$$\left\{\lambda\colon \left(\operatorname{Re}\lambda - \mu^{-1}\left(\frac{4}{3} + \beta\right)^{-1}\right)^2 + (\operatorname{Im}\lambda)^2 = \mu^{-2}\left(\frac{4}{3} + \beta\right)^{-2}\right\}. \tag{12}$$

One of the main problems in unbounded domains is the problem of resonances (scattering frequencies) [3]. In other words, we are looking for $\omega = -i\lambda$, such that a nontrivial solution (not necessarily from $L_2(G)$) of (3), satisfying the boundary conditions and the so-called radiation conditions [6], exists. The following result describes the scattering frequencies of $\mathcal{A}_1^\mu$.

Theorem 5. *All the scattering frequencies of $\mathcal{A}_1^\mu$ lie inside the circle enclosed by (12). When $\mu \to +0$, they tend to the scattering frequencies of the Neumann Laplacian [3].*

Some generalizations of Theorems 2–5 for the problems of vibrations of a viscous compressible fluid in contact with a thin elastic shell can be found in [4]–[6].

BIBLIOGRAPHY

[1] *G. Geymonat & E. Sanchez-Palencia – On the vanishing viscosity limit for acoustic phenomena in a bounded region*, J. Rational Mech. Anal., **75** (1981), 257–268.

[2] *M.R. Levitin – Vibrations of a viscous compressible fluid in bounded domains: spectral properties and asymptotics*, to appear.

[3] *P.D. Lax & R.S. Phillips – Scattering theory*, Academic Press, N.Y., 1967.

[4] *M.R. Levitin – Normal-mode spectrum of a shell filled with a viscous compressible fluid*, Dokl. AN SSSR, **295** (1989), 1355–1358; English transl., in "Sov. Phys. Dokl.", **32** (1987).

[5] *D.G. Vasil'ev, M.R. Levitin,& V.B. Lidskii – Forced vibrations of a thin elastic shell filled with a viscous compressible fluid*, Dokl. AN SSSR, **305** (1989), 329–332; English transl. in "Sov. Phys. Dokl.", **34** (1989).

[6] *D.G. Vasil'ev, M.R. Levitin & V.B. Lidskii – Forcesd vibrations of an elastic thin shell immersed in a viscous compressible fluid*, Funktsional. Anal. i Prilozhen, **25** (1991), no.4; English transl. in "Functional Anal. Appl.", **25** (1991).

Michael R. Levitin,
Institute Problems Mechanics,
Russian Acad. Sciences,
prosp. Vernadskogo, 101,
117526 Moscow – RUSSIA

Shock Wave in Resonant Dispersion Media

YU.I. SKRYNNIKOV

For nonlinear one-dimensional waves propagated in resonant dispersion media, i.e. which contains uniformly distributed identical resonators with eigenfrequency ω_0, the so-called Resonant Dispersion equation (RD-equation) has been obtained [1]:

$$\left(1+\frac{\partial^2}{\partial\tau^2}\right)(\Psi_\tau+\Psi_\eta+2\Psi\Psi_\eta)-\frac{\sigma}{2}\Psi_\eta=0\,, \tag{1}$$

where $\tau=\omega_0 t$, $\eta=\omega_0 x/c$ — dimensionless time and coordinate. We have to choose Ψ in (1) as a wave variable which will satisfy the equation $\Psi_t+c\Psi_x+2\Psi\Psi_x=0$ if there are no any resonators in medium ($\sigma=0$).

The linear terms of the RD-equation is a consequence of the resonant dispersion equation for a plane wave $\propto \exp(-i\omega t+ikx)$:

$$k^2=\frac{\omega^2}{c^2}+\sigma\frac{\omega^2/c^2}{1-\omega^2/\omega_0^2}\,. \tag{2}$$

In (2) we have c — phase velocity in medium without resonators, σ — dimensionless dispersion parameter.

The most typical application of dispersion (2) and, consequently, the RD-equation is the description of waves in a homogeneous liquid containing uniformly distributed identical gas bubbles.

Other examples of the RD-equation are the following:

a) ion-sound waves in plasma;

b) internal waves in constant buoyanoy frequency layer;

c) gas of two atomic molecules;

d) polaritons in ion crystals;

e) waves in nonlinear optical crystals.

As a first step we have found [1] stationary solutions in a form of traveling waves $\Psi=\Psi(\xi)$; $\xi=\eta-V\tau$. Such waves move with velocity V without change their profile. After integrating twice the RD-equation turns into the relation:

$$(\Psi_\xi)^2=-\frac{R(\Psi)}{(2\Psi+\alpha)^2}\,, \tag{3}$$

where R — the 4-th power polynom:

$$R(\Psi)=\Psi^4+\frac{2}{3}(2\alpha_*+\alpha)\Psi^3+(C_1+\alpha_*\alpha)\Psi^2+\alpha C_1\Psi+C_2\,,$$

$\alpha=1-V/c$, $\alpha_*=\alpha-\sigma/2$, C_1 and C_2 — integrand constants.

Integrating (3) yields a number of solutions in a form of both solitary and periodic traveling waves. But here we will concern with the problems of singular solutions of the equation (3) only. As it follows from (3) some of the obtained solutions have infinite derivative at $\xi = \xi_0$, where $\Psi(\xi_0) = -\alpha/2$. The asymptotic behaviour of these solutions in the vicinity of the singularity is

$$\Psi = -\frac{\alpha}{2} + \frac{1}{2}\left(\frac{\alpha^3(\alpha - 4\alpha_*)}{3}\right)^{\frac{1}{4}} |\xi - \xi_0|^{\frac{1}{2}} \operatorname{sgn}(\xi - \xi_0) \ . \tag{4}$$

This singularity is a consequence of a presence of nigher derivative in the nonlinear term of RD-equation.

Of course, such waves exist only in nondissipative media: if any mechanism of energy absorption in the medium is taken into account by the introduction of an odd derivative in the equation (1), rapid energy release will take place due to infinite derivative, causing the wave to die out. But if the absorption is negligible, these waves can propagate without any appreciable change in their profile, although only for a finite time interval. Such behaviour will take place if we consider viscous mechanism of energy absorption.

In such case RD-equation (1) will be transformed into the following one:

$$\left(1 + \frac{\partial^2}{\partial \tau^2}\right)(\Psi_\tau + \Psi_\eta + 2\Psi\Psi_\eta - \epsilon\Psi_{\eta\eta}) - \frac{\sigma}{2}\Psi_\eta = 0 \ , \tag{5}$$

where $\epsilon = \frac{4}{3}\nu\omega_0/C^2$ — dimensionless viscosity (ν — kinematic viscosity). In fact, $\nu \ll 1$ in any physical condition, therefore we can try to find asymptotic solution of the modified RD-equation (5). For simplicity we shall consider only stationary solutions of (5) in a form of traveling waves.

Evidently, the most essential distinction of these asymptotic solutions from the above-mentioned wave profiles that eliminates their singularity we can expect only in a small neighbourhood of ξ_0. Consequently, such waves by analogy with well known Burgers and Burgers-Korteweg-de Vries solutions [2] can be considered as shock waves.

To find the structure of shock waves front (solution of (5) in the vicinity of the singularity point ξ_0) we use a multiple scale method. For this purpose we have to introduce new variables:

$$x_{-1} = \frac{(\xi - \xi_0)}{\epsilon}, \quad x_0 = (\xi - \xi_0), \quad x_1 = \epsilon(\xi - \xi_0), \quad \ldots, \quad x_n = \epsilon^n(\xi - \xi_0) \ ,$$

and to assume that in a small neighbourhood of ξ_0 the solution (inner solution) has a form:

$$\Psi = \Psi^0 + \epsilon\Psi^1 + o(\epsilon) \ ;$$

$$\Psi^i = \Psi^i(x_{-1}, x_0, \ldots, x_n, \ldots) \ , \qquad i = 0, 1, \ldots$$

After all we obtain equation for Ψ^0 and expression for its solution:

$$\Psi^0 = -\frac{\alpha}{2} - \kappa \operatorname{th} \kappa\, x_{-1} \ ,$$

where $\kappa = \kappa(x_0, x_1, \ldots)$ but doesn't depend on x_{-1}. The function κ is found after join of inner (6) and external (4) solutions as $\epsilon \to 0$:

$$\kappa = -\frac{1}{2}\left(\frac{\alpha^9(\alpha - 4\alpha_*)}{3}\right)^{\frac{1}{4}} |x_0|^{\frac{1}{2}} \ .$$

Thus the function

$$\Psi = -\frac{\alpha}{2} + \frac{\beta}{2}|\xi - \xi_0|^{\frac{1}{4}} \operatorname{th} \frac{|\xi - \xi_0|^{\frac{1}{2}}(\xi - \xi_0)}{\epsilon}, \tag{6}$$

$$\beta = \left(\frac{\alpha^9(\alpha - 4\alpha_*)}{3}\right)^{\frac{1}{4}}$$

describes the structure of front of the shock wave propagating in a resonant dispersion media. It is easy to see that the solution (6) and its derivatives of all order have no singularity.

REFERENCES

[1] *S.A. Rybak & Yu.I. Skrynnikov - Nonlinear waves in resonant dispersion media*, Proc. IV Int. Workshop on Nonlin. & Turbul. Processes in Physics, Would Sci., Singapore, **1** (1990), 664-670.

[2] *G.B. Whitham - Linear and nonlinear waves*, New York (1974).

Yu.I. Skrynnikov,
N.N. Andreev Acoustics Institute of the Russian Academy of Sciences
Shvernika Str. 4,
117036 Moscow - RUSSIA

Solitary Vortices – A New Exact Solution of Hydrodynamical Equations

A.T. SKVORTSOV

The analytical description of fluid motion by means of localized vortices is traditional for hydrodynamics, oceanology, plasma physics. Recently it was shown that alongside with the well-known kinds of localized vortices (e.g. Hill's vortex), which are characterized by weak decrease of velocity field (as a power of the inverse distance from vortex center), the vortices with screening (or solitary vortices) may exist. In the latter case the velocity field vanishes exponentially. In this paper a new exact solution of hydrodynamic equation in form of solitary vortices is obtained.

Let us start from the well-known equation [1]

$$\tilde{\Delta}\Psi = -ff' + r^2F' , \quad \tilde{\Delta} = \frac{\partial^2}{\partial z^2} + r\frac{\partial}{\partial r}\frac{1}{r}\frac{\partial}{\partial r} , \tag{1}$$

which discribes the vortex motion in axial geometry (r, z) [1]. Ψ — stream function; $V_r = \frac{1}{r}\frac{\partial\Psi}{\partial z}$; $V_z = \frac{1}{r}\frac{\partial\Psi}{\partial r}$; $V_\phi = f/r$; f, F — arbitrary functions of Ψ.

1) Linear case. Following [2], assume

$$f = K\Psi ; \quad F = \frac{m^2\Psi^2}{2} + d; \quad k, m, d = \text{const.}$$

Substitution of these expressions into (1) yields

$$\tilde{\Delta}\Psi = (m^2r^2 - k^2)\,\Psi . \tag{2}$$

We seek solution of this equation in the form

$$\Psi = Z(z)\,R(r) . \tag{3}$$

Substitution (3) in (2) yields

$$-\frac{Z''}{Z} = \frac{R'' - R'/r}{R} + k - m^2r^2 . \tag{4}$$

For any z and r this equality may be held only when compared values are equal to the constant. Denoting this constant as q^2, on the basis of (4) we obtain the system of two equations [3]:

$$Z'' + q^2Z = 0 , \tag{5}$$

$$R'' - \frac{R'}{r} + (k^2 - q^2 - m^2r^2)R = 0 . \tag{6}$$

The solution of (5) is trivial:

$$Z = A\cos qz + B\sin qz , \quad A, B = \text{const.} \tag{7}$$

For a solution of Eq. (6) let us further introduce a new variable $\xi = mr^2$ and make use of the following substitution

$$R = \xi \exp\left(-\frac{\xi}{2}\right) \omega(\xi) . \tag{8}$$

As a result, Eq. (6) may be represented in hypergeometrical form

$$\xi\omega'' + (2-\xi)\,\omega' + (\kappa - 1)\,\omega = 0 , \qquad \kappa = \frac{(k^2 - q^2)}{4m} .$$

The general solution of the equation is

$$\omega = C_1 F(1-\kappa, 2, \xi) + C_2\, G(1-\kappa, 2, \xi) ,$$

where C_1, C_2 = const; F, G — degenerate hypergeometrical functions of the 1-st and 2-nd kind. In particular, the solution of this equation, being finite at zero and increasing at $\xi \to \infty$ not faster than finite power of ξ, is a degenerate hypergeometrical function [4]:

$$\omega = F(1-\kappa, 2, \xi) \tag{9}$$

at $\kappa - 1 = n$, $n = 0, 1, 2, \ldots$ (at this case F tends to the polynomial [4]). Taking further into account (8) and making use of conventional ideas for degenerate hypergeometrical functions [3] in (9), we finally obtain [3]

$$R = \frac{-\xi}{2\;3 \cdots (n+1)}\, e^{\frac{\xi}{2}} \frac{d^{n+1}}{d\xi^{n+1}} (e^{-\xi} \xi^n) , \qquad \xi = mr^2 . \tag{10}$$

Formulas (3), (7), (10) describe the exact smooth solution of Eq. (1). One can see, that the vorticity distribution, consistent with this solution, is exponentially localised over r and periodically changes along z axis.

2) Nonlinear case. The approach involved can be generalized to the nonlinear case. Let's choose

$$f = K\Psi + \sqrt{\lambda}\Psi\,(\ln\Psi - 1)^{\frac{1}{2}} , \qquad F = \frac{m^2\Psi^2}{2} , \qquad \lambda = \text{const.}$$

Then Eq. (1) may be represented in the form:

$$\tilde{\Delta}\Psi = (m^2 r^2 - k^2)\Psi + \lambda\Psi \ln \Psi . \tag{11}$$

Writing again the solution of (11) in the form (3) and separating variables, we obtain the system of two equations:

$$Z'' + q^2 Z - \lambda Z \ln Z = 0 ,$$

$$R'' - \frac{R'}{2} + (k^2 - q^2 - m^2 r^2)R - \lambda R \ln R = 0 .$$

It may be shown by direct substitution, that at $r, z \to \infty$ the smooth solution of the form

$$Z = \exp(-\alpha z^2), \quad R = \exp(-\beta r^2) , \tag{12}$$

where $\alpha = k^2/2 = q^2/2 > 0$; $\beta_{1,2} = k^2/4 \pm \sqrt{k^2/16 - m^2/4}$ ($\mathrm{Re}\,\beta > 0$); $\lambda = -2k^2$. Thus Eq. (11) discribes a nonlinear solitary vortex with compact structure, i.e. with screening along z and r directions.

REFERENCES

[1] *Batchelor G.K.* - *An Introduction to Fluid Dynamics*, Cambridge Univ. Press (1970).
[2] *Skvortsov A.T.* - *Soviet Technical Physics — Letters*, **14** (1988), 1609.
[3] *Skvortsov A.T.* - *Soviet Technical Physics — Letters*, **17** (1991), 70.
[4] *M. Abramowitz* and *I. Stegun* - *Handook of Mathematical Functions*, Dover Publications (1972).

A.T. Skvortsov,
N.N. Andreev Acoustics Institute of the Russian Academy of Sciences,
Shvernika Str. 4,
117036, Moscow - RUSSIA